Ozone/Chlorine Dioxide Oxidation Products of Organic Materials

Proceedings of a Conference held in
Cincinnati, Ohio, November 17-19, 1976

Sponsored By:
International Ozone Institute, Inc.
U. S. Environmental Protection Agency

Editors:
Rip G. Rice, Ph.D.
Joseph A. Cotruvo, Ph.D.

Copyright 1978 by the
    International Ozone Instiute, Inc.
    14805 Detroit Avenue
    Cleveland, Ohio 44107   (U.S.A.)

Ⓒ under UCC 1978 by the International
Ozone Institute, Inc.

All rights reserved.

Library of Congress Catalog No. 78- 053924

ISBN 0-918650-02-X

Printed in the United States of America
by Syracuse Lithographing Company,
Syracuse, New York

OZONE PRESS INTERNATIONAL
The publishing imprint of the
International Ozone Institute, Inc.
14805 Detroit Avenue
Cleveland, Ohio 44107

TABLE OF CONTENTS

	Page
FOREWORD	v
ACKNOWLEDGEMENTS	vi
INTRODUCTORY COMMENTS.	1
Rip G. Rice and Joseph A. Cotruvo	

OVERVIEW

WATER TREATMENT--THE BACKGROUND FOR PUBLIC HEALTH CONCERN . . . . Boris Osheroff	7

BASIC CHEMISTRY

REACTIONS OF OZONE WITH ORGANIC COMPOUNDS. . . H. Fred Oehlschlaeger	20
OZONE AS A DISINFECTANT OF WATER . . . . Hans L. Falk & James E. Moyer	38
OZONE'S RADICAL AND IONIC MECHANISMS OF REACTION WITH ORGANIC COMPOUNDS IN WATER . . . . Allison Maggiolo	59
BROMOFORM PRODUCTION BY OXIDATIVE BIOCIDES IN MARINE WATERS. . . . . G. R. Helz, R.Y. Hsu & R. M. Block	68
OZONATION OF SEAWATER. . . . . Robert S. Ingols	77

TOXICITY

BIOCHEMICAL ASPECTS OF THE TOXICITY INVOLVED WITH OZONE ORGANIC OXIDATION PRODUCTS IN WATER . . . . Ph. Hartemann, J.C. Block, M. Maugras and J.M. Foliguet	82

EFFECT OF OZONE ON HOSPITAL WASTEWATER
    CYTOTOXICITY.   97
    Riley Kinman, Janet Rickabaugh, Victor
     Elia, Kevin McGinnis, Terrence Cody,
     Scott Clark & Robert Christian

OZONE METHODS AND OZONE CHEMISTRY OF
    SELECTED ORGANICS IN WATER 1. Basic
    Chemistry   115
    Ronald J. Spanggord and Vernon J.
     McClurg

OZONE METHODS AND OZONE CHEMISTRY OF
    SELECTED ORGANICS IN WATER. 2. Mutagenic
    Assays.   126
    Vincent F. Simmon, Sharon L. Eckford
     and Ann F. Griffin

FRENCH METHODS FOR EVALUATING TOTAL MICRO-
    POLLUTANT LOAD IN WATER, AND FOR
    DETERMINING ITS TOXICITY.   134
    Michel Rapinat

GENERAL PAPERS

IDENTIFICATION OF END PRODUCTS RESULTING
    FROM OZONATION OF COMPOUNDS COMMONLY
    FOUND IN WATER   153
    P.P.K. Kuo, E.S.K. Chian and B. J. Chang

COMMENTS ON FIRST DAY SESSION   167
    Harvey M. Rosen

ORGANIC MATERIALS PRODUCED UPON OZONIZATION
    OF WATER   169
    Y. Richard and L. Brener

THE DEGRADATION OF HUMIC SUBSTANCES IN WATER
    BY VARIOUS OXIDATION AGENTS (OZONE,
    CHLORINE, CHLORINE DIOXIDE)   189
    J. Mallevialle, Y. Laval, M. Lefebvre,
     C. Rousseau

THE OXIDATION OF HALOFORMS AND HALOFORM
    PRECURSORS UTILIZING OZONE   200
    Stephen A. Hubbs

OZONATION OF HAZARDOUS AND TOXIC ORGANIC
    COMPOUNDS IN AQUEOUS SOLUTION   210
    Kozo Ishizaki, Richard A. Dobbs &
    Jesse M. Cohen

REACTIONS OF OZONE WITH ORGANIC COMPOUNDS IN
    DILUTE AQUEOUS SOLUTION: IDENTIFICATION
    OF THEIR OXIDATION PRODUCTS   227
    Ernst Gilbert

OXIDATION OF STYRENE WITH OZONE IN AQUEOUS
    SOLUTION.   243
    Floyd H. Yocum

AN ENGINEERING APPROACH TO WATER TREATMENT
    PROCESS SELECTION WITH SPECIAL EMPHASIS
    ON HALOGENATED ORGANICS   264
    M. Schwartz & E.A. Lancaster

OZONIZATION PRODUCTS FROM CAFFEINE IN AQUEOUS
    SOLUTION   284
    Robert H. Shapiro, K.J. Kolonko, P.M.
    Greenstein, R.M. Barkley, and R.E.
    Sievers

REACTION OF ORGANICS NONSORBABLE BY ACTIVATED
    CARBON WITH OZONE   291
    W. A. Guirguis, Y.A. Hanna, R. Prober,
    T. Meister and R.K. Srivastava

OZONE/ULTRAVIOLET

OZONE/UV OXIDATION OF PESTICIDES IN AQUEOUS
    SOLUTION   302
    H. William Prengle, Jr. and Charles
    E. Mauk

BY-PRODUCTS OF ORGANIC COMPOUNDS IN THE
    PRESENCE OF OZONE AND ULTRAVIOLET LIGHT:
    PRELIMINARY RESULTS   321
    William H. Glaze, Richard Rawley and
    Simon Lin

OTHER OXIDANTS (ClO_2)

CHLORINE DIOXIDE: CHEMICAL AND PHYSICAL
    PROPERTIES   332
    David Rosenblatt

USE OF CHLORINE DIOXIDE IN WATER AND WASTE-
    WATER TREATMENT 344
    Sidney Sussman & James E. Rauh

CHLORINE DIOXIDE. AN OVERVIEW OF ITS
    PREPARATION, PROPERTIES AND USES 356
    Ralph Gall

PRODUCTS OF CHLORINE DIOXIDE TREATMENT OF
    ORGANIC MATERIALS IN WATER 383
    Alan A. Stevens, Dennis R. Seeger &
    Clois J. Slocum

OTHER OXIDANTS (BrCl)

COMPETITIVE OXIDATION AND HALOGENATION
    REACTIONS IN THE DISINFECTION OF
    WASTEWATER 400
    Jack F. Mills

OTHER OXIDANTS (Ferrates)

IRON (VI) FERRATE AS A GENERAL OXIDANT FOR
    WATER AND WASTEWATER TREATMENT 410
    Thomas D. Waite and Marsha Gilbert

DRINKING WATER TREATMENT

USE OF OZONE AND CHLORINE IN WATER-
    WORKS IN THE FEDERAL REPUBLIC OF
    GERMANY 426
    Wolfgang Kühn, H. Sontheimer & R. Kurz

COMPARISON OF PRACTICAL ALTERNATIVE TREAT-
    MENT SCHEMES FOR REDUCTION OF TRIHALO-
    METHANES IN DRINKING WATER 442
    James W. Symons, O. Thomas Love, Jr.,
    and J. Keith Carswell

ROUND TABLE WORKSHOP DISCUSSION 455

FOREWORD

In the early 1970s, interest in the quality of drinking water was accelerated as a result of the identification of a large number of organic compounds in the drinking water of New Orleans, Louisiana. This discovery prompted passage in the United States of the Safe Drinking Water Act in late 1974 and increased studies of the use of alternative disinfecting materials for drinking water treatment. Today, more than 400 individual organic compounds, many of them halogenated, have been identified in drinking water supplies of the United States.

As the use of disinfectants other than chlorine is considered, however, the question arises as to the nature and toxicities of the organic oxidation products formed when these alternative materials are introduced into water and wastewater treatment processes.

The purpose of this workshop was to discuss the current status of knowledge of oxidation products of organic materials normally encountered in water supplies and wastewaters when treated with the oxidants ozone, chlorine dioxide, ozone/ultraviolet light, bromine chloride and ferrates. In addition to discussing the identities of organic oxidation products obtained when using these oxidants, we were also interested in discussing the state of current knowledge as to the toxicities of these materials.

Some 33 papers were presented over 2.5 days, then a 2-hour General Discussion was held during which summations were made of the information presented. These Proceedings represent the record of the papers presented, the discussions, plus the General Discussion.

Rip G. Rice, Ph.D.　　　　　Joseph A. Cotruvo, Ph.D.
Ashton, Maryland　　　　　　Washington, D.C.

ACKNOWLEDGEMENTS

This Workshop was sponsored by the International Ozone Institute and co-sponsored by the U.S. Environmental Protection Agency, Environmental Research Information Center, and Office of Water Supply. The editors are grateful to the Environmental Protection Agency for this support.

The editors also are grateful for the typing provided by Mrs. Gertrude E. Mercer, and to Mr. Richard S. Croy, Executive Director of the International Ozone Institute, for arranging publication details.

INTRODUCTORY COMMENTS

Rice: I have a few comments to make regarding what our targets are and what we are going to try to do at this meeting. We are here to talk mainly about ozone and its use in water and wastewater treatment, but also to discuss the use of other powerful oxidizing agents. The program says "ozone/chlorine dioxide," and although most of our 30 technical presentations deal with ozone, we do have four papers dealing with chlorine dioxide.

We also have three papers dealing with the use of combinations of ozone and ultraviolet radiation. We have another paper on the use of bromine chloride and other halogen containing agents as oxidizing agents. We also have a paper dealing with the chemistry of ferrates, which is a "new" class of oxidizing agents that is being looked at for potential application in water and wastewater treatment. Thus, 30% of our technical presentations deal with oxidants other than simply ozone.

In Montreal, Canada in May of 1975 the IOI held the Second International Symposium on Ozone Technology. The City of Montreal is building a new 600 mgd drinking water treatment plant which will use ozone. When it is fully built, about 1980, the amount of ozone that can be generated in Montreal will be 15,000 lbs/day. This will be the largest water supply system using ozone in the world.

- 1 -

But even larger ozone generation facilities are located right here in Cincinnati. Cincinnati is the home of Emery Industries, which has been using ozone for many years to synthesize specialty organics by oxidation of organic compounds such as tall oil. I understand that Emery Industries has several ozone generating plants in Cincinnati, each capable of generating the amount of ozone that Montreal will be generating in 1980. So, we are meeting in the heart of the ozone country, if you will.

Cincinnati is also the home of the U.S. Environmental Protection Agency, and its Technology Transfer Program, Office of Water Supply Technical Support Division Laboratory, the Municipal Environmental Research Laboratory and the Industrial Environmental Research Laboratory, all located here, have programmatic interest in ozone and oxidation technologies.

The objective of our workshop is to talk about the known organic chemistry and the toxicology of oxidized organic compounds. This meeting is not a Forum nor is it a Symposium, in the sense of our authors coming to the meeting with prepared formal papers using data which they developed several months ago. This is a <u>workshop</u> and we hope it will be a workshop in every sense of the word as I define it. We are here to learn the latest results of what people are doing in these two major areas. We want to know who is doing what work, on what organic compounds, with what oxidizing agents, under what conditions, and what compounds they are looking for by what analytical techniques. Finally, what is the known toxicology of these oxidized organics?

As you are aware, in the late 1960's we began to see data published about the toxicity to aquatic life of chlorinated organic compounds in sewage effluents. At that point, and recognizing the number of those chlorinated organic compounds present, the concept of an "alternate disinfectant to chlorine" was born. And the term "alternate disinfectant" in the context of sewage disinfection is a good one, in my personal opinion. Take the halogen out, put something else in that will do the same disinfection job, and you have an "alternate disinfectant."

In November 1974, New Orleans drinking water showed the presence of halogenated organics. Since then, of course, EPA has analyzed the water supplies of something like 130 cities in the USA. All except one use chlorine; Strasburg, Pennsylvania, uses ozone alone. EPA has found halogenated hydrocarbons in all plants tested, including a trace amount of chloroform at Strasburg, Pennsylvania. In my opinion, the concept of an "alternate disinfectant" in drinking water treatment processes is a misnomer, and I'd like to present this thought for your argumentation, discussion, or at least thought.

When we initially started using chlorine for disinfecting drinking water in the early 1900's, the surface water supplies were very low in synthetic organics content. The functions of a disinfectant in water are two-fold: first to kill the microorganisms (disinfect) and second, to provide a residual so that when the water reaches the tap, the home owner is confident that the water supply has not been bacteriologically recontaminated.

Over the years we have seen an increase in the amount of organics in surface water supplies; some of these are natural organics, some are man-made. I don't know how you would want to classify organics from sewage treatment plants, natural, man-made, or both? But the point is that today, many of our surface water supplies contain large amounts of organic compounds.

If we look carefully at the known chemistries of chlorine, which we add ostensibly as a disinfectant and residual-former, we also realize that we are __first__ forming oxidized organic compounds which contain __no__ chlorine, and __secondly__, forming chlorine-substituted organic compounds. __Thirdly__, any bromide present will be oxidized to bromine, which can oxidize organics and/or form brominated organic compounds. (It has been fairly well shown by EPA and others that the major source of brominated organics in drinking water is from bromide oxidation followed by its reaction with organics).

Only __after__ these first three chlorine demands are satisfied will chlorine perform its __fourth__ function, that of disinfection, and then its __fifth__, providing a residual.

In my mind, the term "alternate disinfectant to chlorine" in this kind of a situation is a misnomer, and the real problem becomes one of removing the organics from the original water. If we merely eliminate chlorine and substitute some other disinfectant that is also a powerful oxidizing agent, such as ozone, chlorine dioxide, bromine chloride, ferrates, hydrogen peroxide, permanganate, etc., all of these are going to oxidize those same organic compounds to other organic compounds. Some or all may also oxidize bromide ion to bromine and thus produce the same brominated organics. The non-halogenated oxidizing agents will not produce chlorinated hydrocarbons. They will disinfect and they may or may not provide a residual.

So, by way of an issue-producing introduction, my opinion is that in sewage disinfection we are talking alternatives to chlorine. But in drinking water treatment we should be talking about significantly lowering the organics concentration, or getting them out entirely, by whatever process or combination of processes is appropriate for the specific water, location, etc. Then we should add our disinfectant and/or residual to the treated water, now free of organic contaminants. We should not be talking in terms of alternate disinfectants, because that terminology suggests that current water treatment problems can be solved simply by changing the disinfectant.

To summarize, we are here to discuss what is known about organic oxidation products in water and wastewater. We are here to find out who is doing what research, where it is being done, and by what analytical and toxicological procedures are we looking at these products? Our ultimate objective is to ascertain what are the current information gaps. Where do we now need to put research effort to fill in the total technological picture?

I will now turn the meeting over to Dr. Joseph A. Cotruvo, Director of the Criteria and Standards Division, Office of Water Supply of the U.S. Environmental Protection Agency, Washington, D.C., who will be presenting the remarks of Victor Kimm who is unable to be with us today.

Cotruvo: Thank you Rip. I had expected only to be introducing Victor Kimm, who is the Deputy Assistant Administrator for Water Supply with EPA, but I find myself in the position of standing in for him, so I will be very brief. Mr. Kimm will be here on Friday, and will address us then.

We are really very pleased and very proud to be associated with the organization that has developed this meeting, this symposium, or workshop, or whatever Rip feels it ought to be called. We are also very pleased that such a large number of people and such a wide range of interests are represented here.

The main reason for this kind of gathering, of course, is to develop communications and information flow between the people who are intimately involved in the treatment processes we are dealing with, be it from the points of view of regulatory activity, or of developing the technical information, or of applying technology at the water treatment plant, or of providing the technology from the industrial side.

I can also tell you that this Workshop really could not have come at a more opportune moment, as far as we at EPA are concerned for our standard setting operation. There is really no issue that is nearer to our hearts as regulators in the EPA Office of Water Supply than the issue of the best methods of treating water to make it palatable and safe as far as consumers are concerned.

As Rip has already mentioned, the issue really became acute several years ago with the identification of a number of chlorinated compounds in water and a number of other industrial chemicals too. That raises many questions, and I think it is obvious that the questions that are raised concerning chlorine and its effect on drinking water quality and ultimate human consumption and the chemistry and toxicology of the substances, are the identical questions that have to be asked about every other possible substitute, or alternative, regardless of whether it is there primarily for disinfection purposes or for chemical oxidation.

The necessity, obviously, is to produce the best possible product as far as the consumers are concerned, that will provide good levels of health protection, certainly primarily from the aspect of microbiological contamination, but also from the new factor of treatment-induced chemical contamination of the water. And I know that this gathering of people, representing some of the best minds around who have been looking at this question, identifying those materials that are in the water, determining where they come from and their toxicological effects, is going to be of great help to us at EPA in our deliberations within the next few weeks, in weeding out the alternatives that we have to deal with, the risks that we have to consider, and the ultimate benefits that we hope to achieve by assuring the quality of drinking water.

Again, I am really happy to see you all here. I think this is going to be a very fruitful meeting, and I welcome you all.

WATER TREATMENT
THE BACKGROUND FOR PUBLIC HEALTH CONCERN

Boris J. Osheroff
Principal Environmental Officer

Public Health Service
U.S. Department of Health, Education & Welfare
5600 Fishers Lane, 17A-46
Rockville, Md. 20852 U.S.A.

All animals except for man instinctively drink water upstream from where they bathe and excrete. Man, on the other hand, has been forced by modern civilization to utilize water which has been excreted by other men, has been discharged from industrial plants, contains a myriad of chemicals and organic waste material and carries with it the organic and inorganic run-off including highly toxic poisons from thousands of square miles of farm land. Our public health concerns are simple - How do we render and assure the consumer of this water of the continuing availability of a safe product? You are all here because of your expertise in this field or related fields - I don't think any of you can supply the factual data necessary for this reassurance.

Water supply and usage is extremely dynamic, constantly shifting in response to demographic and industrial demand. Increasing population concentrations and new industry both compete for existing supplies and require complex and sophisticated manipulation before it can be reused.

The water used for household purposes, for the most part, can be recycled after disinfection and simple separation of solids. Some household chemicals do give us problems - such as the infamous non-degradable detergents. On the other hand, industrial discharge, run-off from city streets carrying suspended fallout from auto traffic, run-off from suburban lawns containing pesticides, herbicides and fertilizers, run-off from asbestos shingles and asbestos linings add a new dimension to the needs for new improved methods of water treatment. Compound this with contamination of water supplies and reservoirs by fallout from clouds emitted by industrial plants and potential hazards of natural contaminants (some of which are very persistent) and the task becomes much more difficult -- and much more essential to the protection of health and safety. Is technology to accomplish this task available? If it is, let us not relax -- the implications to the safety of water supplies from new energy technologies such as nuclear power plants and run-off of leachants from processed oil shale will again confound the sanitary engineer, the chemist and the public health man.

Present technology, as I stated previously, is adequate for assuring that public drinking water supplies are not a vector for infectious diseases. In 1974, 28 waterborne disease outbreaks involving 8,413 cases were reported to CDC. Roughly 5,000 of these cases were involved in 4 outbreaks -- in all of the outbreaks which involved more than 20 persons, the cause of the system deficiency was untreated or inadequately treated water -- i.e. untreated surface water, untreated ground water or other treatment deficiencies. These accounted for 99% of the total cases of waterborne disease in 1974.

We do not know what percentage of food-borne disease may be attributed to improperly or inadequately processed water used in food preparation. The most outstanding case I can recall is the typhoid epidemic in Great Britain about 10 years ago in which raw river water was used to cool tins of corned beef after processing and canning.

Relating levels of chemical contaminants in water to water intake and thus body burden does not work. Biomagnification in marine organisms is possible at up to 5 orders of magnitude. Uptake by plants from irrigation water and stock watering may mean entrance into the food chain. Body burdens must be calculated on the holistic exposure of man -- his food, his air, purposeful exposure through drugs and even absorption through the skin. Synergism and additive effects of these compounds must be determined in the laboratory, confirmed by retrospective epidemiology, if possible, and constant surveillance and monitoring maintained to assure confirmation of laboratory established safety levels.

Recently available data of this type indicate that the water treatment methodology used today may need re-evaluation -- the benefit/risk of treating water as we are needs to be re-established. Based on the newly identified potential formation of reaction products in the treatment process itself, there may be a significant risk to public health and safety.

Recent survey data indicate that the risk in water treatment may be high -- while not exceeding the risk of untreated water -- and that new methods of treatment must be found.

The technology of sewage and water treatment has not kept apace of the need for methodology to control infectious disease problems while not creating new health risks.

As you are all aware, we are now faced with the task of developing new sources of energy. This includes among others the tremendous logistical job of extraction of oil from oil shale, the opening up of new strip coal mines and oil wells in the western plains and the transportation of these fuels, or their conversion to energy on site. These projects require use of huge quantities of the scarce resource - water. New contaminants will enter watersheds. In order to make these projects feasible both economically and environmentally, the water should be reprocessed and available as potable water. A multitude of proposals has been made for improvement of water quality through new treatment plants and dams. Assessments of the impact of these projects on quality and availability are being made.

The concern of the Public Health Service is quite simple -- we must have available to us qualitative and quantitative profiles of the finished water product so that we may identify and assess the health risk, if any.

Chlorinated hydrocarbons, epoxides, peroxides as well as viruses have been found in finished water. The benefits of treating the water are well known -- the virtual disappearance of typhoid and other infections, gastrointestinal diseases of bacterial origin has been a remarkable step forward in public health. Recent reports that we have, in chlorinating water, created a new family of significant health hazards have yet to be confirmed. It is cogent and timely for all of us to look at our sewage and water treatment to review and explore alternatives to chlorination which will be at least equally effective as a disinfectant, as easy to use and will leave us with an indicator of safety, such as the residual chlorine. The process should additionally destroy viruses and should not generate chemical compounds which may result in disease decades after the ingestion of this water.

The Public Health Service Act provides, among other things, for the establishment of regulations to insure the safety of potable water used and served aboard interstate common carriers. The U.S. Department of Agriculture, under the authority of the U.S. Meat and Poultry Act, approves and establishes good manufacturing practice for the use of sanitizing and processing chemicals in reuse water in Federally inspected plants. A number of substances used in the processing of sewage and wastewater have been classified as economic poisons. The responsibility for insuring that these substances are used in a manner that will result in minimum impact on human health is placed with the Environmental Protection Agency.

Water is a food within the meaning of the Food, Drug, and Cosmetic (FD&C) Act and, as such, any functional ingredient added to water, with certain exceptions, must comply with the appropriate food additive provisions of the Act.

The Commissioner of Food and Drugs is delegated the authority for development and administration of regulations to assure that no unsafe foods are permitted to be marketed in this country. It is acknowledged that while adequate authority now exists to insure a safe food and water supply in the United States, certain economic and environmental factors, related particularly to food processing, have required that new and expanded legislation be written. Such new legislation has resulted in overlapping agency program activity and authority, particularly in the area of water treatment.

There is no doubt of the necessity of maintaining a safe water supply, municipal as well as potable water. In the past, the FDA has issued opinion letters, established regulations, and has maintained a generally recognized as safe list of substances permitted in contact with water. Where overlapping authority existed, the agencies with such authority have issued guidelines on their own responsiblility in line with FDA regulations and opinions. Recent events, such as the National Cancer Institute study on chloroform, require that a more positive effort be undertaken to identify the areas of overlappping responsibility. This effort has resulted in the establishment of an FDA-EPA combined task force on water use and safety. The results will be a better coordinated effort in responding to safety requirements and in taking appropriate action.

Of particular concern to FDA, with respect to water safety, is the expanding use of germicidal agents in reuse water, particularly in modern food industries, for purposes of food processing and general sanitation. FDA is also aware of the expanding number of substances that are being proposed and used for disinfecting bottled and recycled water, and in some instances claims that these substances are essential to prevent or reduce contamination by harmful microorganisms that gain entry to processing water from the environment. Ozone, chlorine, and chlorine dioxide are recognized as three of the most important germicidal agents used in water treatment. While ozone is reported to be superior to chlorine in disinfection capability, chlorine is reported to have gained dominance due to its efficacy, economy, and the practice of maintaining a residual germicide throughout the distribution system. Chlorine dioxide, in addition to its germicidal properties, is reported to be especially effective against odor-causing substances in water.

Petitions requesting affirmation of GRAS status for uses of chlorine, chemically available sources of chlorine, and ozone have recently been filed with the FDA. These petitions are requesting concurrence that the increasingly wide variations of exposure of food to these substances are safe. The paucity of safety data to justify the request for some uses and the lack of support data in the published literature makes it advisable that close scrutiny be given to all current and future uses of these substances. No changes in currently regulated uses of these substances are anticipated. However, FDA is taking the following actions:

A notice was published in the <u>Federal Register</u> on July 7, 1976, in which the Food and Drug Administration requested the submission of data, information, and views on the safety and use of chlorine in food processing. These data and information will be used to determine which conditions of use of chlorine are generally recognized as safe (GRAS). All data submitted in response to this notice will be available for public inspection in the office of the Hearing Clerk.

The FDA is currently evaluating a petition requesting use of ozone as a chemical sterilant for bottled water. One problem for this proposed use could relate to the use of plastic bottles for the bottling of ozone treated water. Currently, in § 128d.7(d) (4) (21 CFR 128.7(d) (4) the FDA permits 0.1 part per million ozone/water solution in an enclosed system for at least five minutes in sanitizing operations to effect sanitizing of contact surfaces used in processing and bottling of bottled drinking water.

It is expected that until additional safety data is available, such as the data being solicited for chlorine, the permitted uses of ozone will be limited to potable water.

Aquatic pollutants have been a matter of concern to biomedical scientists, public health authorities, and regulatory and legislative groups. Contaminants in waterways from which municipal water supplies are derived are also of concern, nationally and internationally, because of destruction of marine life (fish and shellfish) with its socioeconomic impact, and the observations that cancer develops in marine animals from such pollution. Marine life neoplasia may have some association to epidemiological surveillance in human cancer incidence under specified environmental conditions. Raw water contaminants which are not completely removed in water treatment plants or are magnified in types of molecular species and amounts through processing (i.e., chlorination) become the biorefractories identified and quantified in municipal drinking water. Some of these biorefractories are recognized or suspect carcinogens.

Indirect and direct approaches are used to record the presence of carcinogenic contaminants in water. The direct approach is to identify and monitor their presence in water while an indirect approach involves acquisition of data and evidence that such chemical contaminants are carcinogenic in experimental animals when administered in drinking water or establishment of their carcinogenic activity in marine test animals by observations on tumorigenicity. Some of these identified biorefractories in water have been tested for their carcinogenic activity through bioassay in rodents at the National Cancer Institute. As early as 1954, Hueper and coworkers (1) and Dunham and coworkers (2) in 1967 recognized the importance of evaluating the carcinogenic activity of chemicals in various raw water samples from certain rivers in the U.S.A. At that time technologies for recovery and isolation of carcinogenic contaminants in extracts and eluates were not too well developed and concentrates tested were not truly reflective of carcinogenic potency; consequently, the assay results were somewhat inconclusive.

In late 1972, renewed interest in this area of environmental cancer developed at the National Cancer Institute resulting in the establishment of close cooperation and collaboration with the Water Supply Health Effects Laboratory of the Environmental Protection Agency at Cincinnati, Ohio, and the Washington Program Staff of the Environmental Protection Agency. This collaborative effort led to a review of various reports and arbitrary classification of organic biorefractories in drinking water and organic contaminants in raw water as either recognized, suspect or potential carcinogens. The Environmental Protection Agency periodically distributes lists of organic biorefractories in drinking water supplies of various municipalities. The initial list of 66 chemicals so identified soon grew to 162 and as of April 1, 1976 numbered 299 organic biorefractories. Today, I understand, the number now is over 400. The National Cancer Institute is continuously evaluating these listings, sorting out those chemicals that are considered recognized or suspect carcinogens.

At the National Cancer Institute we have scanned this list along with the Environmental Protection Agency and recommended for carcinogenic bioassay those chemicals on which we do not have sufficient data; using as criteria those chemicals that occur in the greatest amounts in drinking water and with a significant frequency of occurrence. An example of such a nominated chemical would be bis-(2-chloroisopropyl) ether, a contaminant that frequently appears in water supplies in the U.S.A. and Europe. Some ethers are recognized as carcinogens and thus, on the basis of structural similarity, this ether was recommended for test and a bioassay is in progress. Some studies on this chemical are also being undertaken by the Health Effects Laboratory of the Environmental Protection Agency in Cincinnati, Ohio.

Previous reference was made to the occurrence of neoplasms in finfish and shellfish and that marine animals may reflect the extent of contamination of water; in the extreme cases by lethalities, to a lesser degree by effects on reproduction or propagation of marine animals. By observations on tumorigenicity marine animals may provide presumptive evidence of contamination and have possible relevance of such epizootics to geographic distribution of cancers in human populations. The National Cancer Institute maintains a contractual project with the Smithsonian Institution in Washington for studies on tumors in invertebrates and a tumor registry on these species. This registry should prove out to be a useful resource in these broad areas of environmental cancer.

Through close collaboration with the Environmental Protection Agency in the Water Quality Program, we are currently assisting this agency by our work and consultation on the Subcommittee on Organic Contaminants of the National Drinking Water Advisory Committee, Assembly of Life Sciences of the National Academy of Sciences - National Research Council. At three meeting thus far, we have helped in the complication of dossiers of some seventy chemicals on which toxicological appraisals and carcinogenicity assessments must be made, leading ultimately to the development of risk factors and "acceptable" levels for these chemical contaminants in drinking water. The Environmental Protection Agency has made an evaluation on the carcinogenicity of chloroform and a risk assessment, which will be reviewed by the National Cancer Institute. Similarly, evaluations will be made of benzene and other carcinogenic contaminants. For the noncarcinogenic contaminants, whenever possible, Acceptable Daily Intakes will be developed.

On February 21, 1975 we participated, along with representatives of the Environmental Protection Agency and the Environmental Defense Fund, in hearings before the Committee on Health and Welfare of the Louisiana House of Representatives in New Orleans. The National Cancer Institute testimony, in the form of prepared statements and a fact sheet (chemicals in drinking water), attempted to put all issues into proper perspective and to delineate our abilities or inabilities, from current scientific methodologies, to either define, measure or solve this problem on alleged causal relationships of water contamination and cancer excess. In essence, it was stressed that further work would be required to define better the relationship in question and thus a large case-controlled retrospective study would be advocated. (Copies of statements are available from my office on request).

Epidemiology

The National Cancer Institute, in collaborative effort with the Environmental Protection Agency, is planning activities in the following areas:

(a) Analyses of the results of several EPA water analysis surveys in relationship to rates of malignancy at various sites. The analyses are based on the 80 water system surveys for organics done one year ago, the current survey of 112 water systems for organics and the on-going analysis of water systems used to supply interstate carriers.

(b) Additionally, field studies will be conducted over the next year which will include evaluation of the sources of drinking water and analyses of the content of these water sources in areas experiencing unusual cancer mortality.

Collaborative Conference - 1976

The culmination of this interest in aquatic pollutants and municipal water biorefractores over the past few years will be reflected also in a collaborative effort between the Environmental Protection Agency and the National Cancer Institute, with some support from the Energy Research and Development Administration and the National Institute of Environmental Health Sciences. A conference by the New York Academy of Sciences entitled "Aquatic Pollutants - Biological Effects with Emphasis on Neoplasia" was held on September 27, 28 and 29, 1976 in New York City. This conference had as its primary objective an orientation on the dimensions of the water contamination problem, the effects on eco-systems, and the assessment of potential public health risk, if any, from exposure to aquatic pollutants.

Ozone as a Disinfectant of Water

Chlorine, first used to disinfect water supplies in America in 1908, has grown so that the majority of water treatment plants in the United States now use chlorine to disinfect their drinking water. Recent research by the Environmental Protection Agency has resulted in a list of 187 organic compounds that have been positively identified in drinking water in trace amounts. This list contains chlorinated organic compounds, some of which are most likely the result of water disinfection with chlorine. It has been pointed out that these chlorinated compounds are more resistant to biodegradation than their respective unchlorinated analogs. A few of these chlorinated organic compounds have been identified by EPA as carcinogens.

For these reasons, we must consider the possibilities of using other disinfectants, such as UV or gamma radiation, or an oxidizing agent such as ozone.

The advantages of ozone over chlorine as a disinfectant of water are well documented. It has been shown to oxidize taste-and odor-causing compounds, to add no smell or taste of its own, and has been shown to improve the appearance of water. Chlorine, on the other hand, may contribute its own smell to the treated water and has been known to react with certain malodorous compounds, causing them to smell even worse, as with the phenols.

Organic compounds which result from ozonation are more biodegradable than chlorinated organic compounds. Ozone itself is very unstable and has a reported half-life of twenty minutes in water.

It has been reported (3), from the Saint-Maur water treatment plant in Paris, that ozonization has been able to effectively eliminate E. Coli from treated water. Upon switching from chlorine to ozone in 1955, a further reduction in the Clostridium perfringens content of the water was evidenced. Statistics from a Philadelphia plant further support the effectiveness of ozone in destroying coliform organisms and bacteria. Ozone has proven capable of destroying Bacillus cereus and Bacillus megaterium, and their spores. Ozonization has also proven able to inactivate coliphage T2 and polio virus. Some consider ozonization to be the most effective treatment for viruses in regards to thoroughness and economy.

Any ozone released to the atmosphere may react with pollutants already in the air, thus making a contribution to the reactions which occur in air polluted areas and has been shown to destroy nearby plants and shrubbery.

In water ozonization plants, therefore, efforts must be, and are made to capture and recycle the gaseous, ozone-containing effluent, or to destroy it.

At concentrations in air above 0.1 ppm, ozone has a distinct odor and on continued exposure at this level people have experienced headaches, dryness of throat and irritation of respiratory passages and eyes.

Data have been assembled which document high toxicity of ozone in several rodent species. Male Swiss 25 g. mice exposed to ozone at levels from 0.62 to 4.25 ppm, for four hours, showed significant decreases in pulmonary bactericidal activity towards Staphylococcus aureau at levels <1.10 ppm ozone, as compared to control mice. In a similar study using the same type mice, bacteria and time exposure, with ozone levels from 0.38 $\pm$ 0.05 to 1.59 $\pm$ 0.13 ppm, the same results were obtained, except that a significant difference was found at levels of ozone <0.41 ppm. Hueter and Fritzhand add that , "Ozone has been shown to....weaken lung structure, reduce pulmonary function and disrupt cellular biochemistry."

It has been demonstrated (4) that ozone inhalation reduces lysozyme and phagocytic protective activity in the lungs of experimental animals as well as ability of ozone to cause chromosome aberrations in Chinese hamsters, which suggests a possible effect upon humans. The break frequency which resulted agreed well with the value expected from calculations based upon in vitro exposure of human cells. Zelac, et al., point out that:

Presently permitted human ozone exposures would be expected to result in break frequencies that are orders of magnitude greater than those resulting from permitted human radiation exposures if the results of this animal study were directly extended to the human case.

Ozone, however, decomposes very rapidly, so that there is none left as a toxic residue. This can present a problem, however, in that there is a question of proof of disinfection. The existence of even a trace of chlorine residual, regardless of form, drastically reduces or eliminates total coliforms from the distribution system samples. The presence of a chlorine residual as proof of disinfection would minimize surveillance problems and risks for those small but numerous water systems, for example, ski and other resorts, roadside restaurants, bus stops, motorway rest areas, trailer camps, farms, suburban homes, isolated institutions, organized summer camps, and other similar water systems.

While it is clear that there are several reasons why ozone may be more desirable than chlorine as a disinfectant for water treatment plants, before advocating switching to ozone as the major disinfectant, we must consider what chemical compounds may be produced in water upon ozonization. This matter will be more fully addressed later in the program.

In assessing the health risks of the presence of these chemicals in water, one must also consider the body burden of these chemicals as derived from other sources (air, water, food and drugs), and must minimize exposure from all sources. The synergism of these compounds and their additive effects must also be evaluated, biological significance of absorption, retention and the pharmacodynamics of the compound or compounds must be elucidated. We do not want to make an error by substituting one hazard for another, resulting in a negative benefit to public health.

We look forward to such groups as this to help us with the assessment of alternative methods of sewage treatment and water treatment. A safe, effective method will provide us the capability to recycle water, to maintain an ample supply of clean, safe, potable water and may furthermore enable us, judiciously, to remove some constraints on development of energy resources.

LITERATURE CITED

(1) Hueper and Ruchhoft, AMA Archieves of Industrial Hygiene and Occupational Medicine 9:488-495 (1954).

(2) Dunham, Ogara and Taylor, American Journal of Public Health 57(12):2178-2185 (1967).

(3) Guinvarc'h, P., "Three years of ozone sterilization of water in Paris", in Ozone Chemistry & Technology, Am. Chem. Soc., Washington, D.C. (1959), p. 416-429.

(4) Heuter, F.G., and Fritzhand, M., "Oxidants and lung biochemistry, a brief review." Archives of Internal Medicine 128:51 (1971).

DISCUSSION

Julian Josephson, Environmental Science & Technology: Did I understand you correctly to say that there is now a list of 400 substances that are suspected carcinogens?

Osheroff: I'm glad you asked that question because it's really easy for me to answer. Joe Cotruvo just told me about this before I got up here, so I can refer the question to Joe.

Cotruvo: There is a listing compiled by the Cincinnati Health Effects Research Laboratory and the Municipal Environmental Research Laboratory. It's a running list, it keeps growing almost daily, and it's up to about 400 total compounds that have been identified in water throughout the United States. Only a very small number of those 400 compounds have received any kind of substantial toxicological evaluation. A few of them, thus far, have been implicated as animal carcinogens, based on certain feeding studies that have been done, and a couple have been implicated as human carcinogens from other kinds of studies that have been conducted. But the majority of them have not received any kind of comprehensive evaluation as yet.

REACTIONS OF OZONE WITH ORGANIC COMPOUNDS

H. F. Oehlschlaeger

Emery Industries, Inc.
4900 Este Avenue
Cincinnati, Ohio 45232
U.S.A.

Many of the papers to be presented at this meeting will deal with the problems associated with the use of ozone in water purification. Most of the reactions of ozone in these cases take place with organic compounds. Thus, a brief review of the chemistry of the reactions of ozone with organic compounds seems to be in order.

Since most of the work on ozonation has been done in non-aqueous solvents, these reactions will be emphasized, although several examples of ozonation carried out in aqueous media will be presented. However, it is felt that many ozonation reactions carried out in non-aqueous polar solvents will also proceed in aqueous media.

Ozone is an allotropic form of oxygen containing three atoms of oxygen per molecule. It is an extremely powerful oxidizing agent which reacts with a great number of organic compounds.

Ozone generally is regarded as a resonance hydrid of the four contributing structures, as shown in Figure 1.

Ozone usually reacts as a 1,3-dipole, as shown in structures III and IV, or as an electrophilic reagent through the electron deficient terminal oxygen atoms represented by structures III and IV (1).

Reactions with Olefins

One of the most important reactions of ozone with organic compounds, and one which certainly has received a great deal of research attention, particularly in the last two decades, is the reaction of ozone with the olefinic double bond. The reaction of ozone with most olefinic materials is rapid and essentially quantitative. Subsequent treatment of the intermediate products of ozonation can lead to alcohols, aldehydes, ketones, acids and esters, and amines. However, because this group is primarily concerned with the removal of organic compounds via oxidative cleavage with ozone, the various methods of converting intermediate ozonides via hydrogenation and/or other processes into alcohols, aldehydes or amines will not be considered in this paper.

Any discussion of the reaction of organic compounds with ozone would not be complete without at least a brief look at the mechanism proposed to explain the reaction between ozone and olefins by Dr. Rudolph Criegee in the 1950's (2).

Criegee postulated that the initial reaction of ozone and an olefin produced a normally transitory material, the so-called primary ozonide, which rapidly decomposed to yield a reactive zwitterion and a carbonyl-containing fragment. Criegee did not concern himself greatly with the structure of the primary ozonide, but recent work by Bailey and co-workers has indicated the primary ozonide to exist usually as the 1,2,3-trioxolane structure, shown in Figure 1 (3).

The zwitterion formed by decomposition of the primary ozonide is the key to the Criegee mechanism. It is generally depicted today as a carbonyl oxide structure (Figure 1). It is extremely reactive and can undergo a variety of subsequent reactions depending mainly upon the nature of the solvent used for the ozonolysis reaction. If the solvent is an alcohol, an organic acid or water, alkoxy hydroxides, acyloxy hydroperoxides and hydroxy hydroperoxides, respectively, are the main products. These solvents are referred to as participating solvents (Figure 1). The hydroxy hydroperoxide structure formed when the reaction is carried out in aqueous media usually is rather easily decomposed to an aldehyde and ketone or to carboxylic acid; see Figure 2 (4).

If the ozonolysis solvent is a non-participating solvent, recombination of the zwitterion with the carbonyl fragment yields the normal ozonide, the 1,2,4-trioxolane structure shown in Figure 2. Some dimer or linear polymeric peroxides may also be formed under these conditions. These peroxides are considered undesirable because they are difficult to process further into useful products (2).

Fragmentation of a primary ozonide from an unsymetrical olefin can occur in two ways so that two different zwitterions and two different carbonyl fragments can be formed. The dotted lines in Figure 3 indicate the two methods of cleavage which are possible.

Recombination of these fragments into ozonides, which incidently can also exist in cis and trans forms, increases the complexity of the mixture resulting from ozonolysis (Figure 3). For example, the ozonolysis of methyl oleate in a non-participating solvent, petroleum ether, has been shown to yield the three cis-trans pairs of ozonides predicted by the updated Criegee mechanism; a total of six ozonide structures (5).

The Criegee mechanism has stood the test of time reasonably well, but a number of modifications which are beyond the scope of this paper have been proposed to explain some of the results which were not covered by the original Criegee proposal.

Let us examine briefly now the important use of ozonation in synthesis. Ozone has been found to be an extremely useful oxidant in both laboratory and industrial scale synthesis.

Ozone has proven particularly useful in the preparation of aliphatic acids from the corresponding olefins. Perhaps the most important reaction utilizing ozone from a commercial standpoint is the one which my company, Emery Industries, has been involved with for a number of years, that is the ozonolysis of oleic acid. This reaction has been carried out on a multi-million pound/year scale for a number of years. In this process, oleic acid is dissolved in pelargonic acid, which is also one of the products, and the solution is reacted with ozone to yield a complex mixture of ozonides and acyloxy hydroperoxides of various structures. For simplicity, only the normal ozonide is shown in Figure 4. These compounds are oxidized via oxygen to yield the desired products, pelargonic and azelaic acids (6).

In similar fashion Emery ozonizes erucic acid, the 22 carbon monounsaturated acid with the double bond in the 13 position to yield brassylic acid, the 13 carbon dibasic acid. Note that the double bond in erucic acid is 9 carbons from the non-carboxylic acid end of the molecule, so again pelargonic acid is the co-product (6).

Considerable additional work has been carried out on oxidative ozonization, that is ozonolysis followed by oxidation of the ozonolysis mixture, usually to yield carboxylic acids. The following dicarboxylic acids are conveniently prepared by ozonolysis of cyclo-olefins followed by oxidation usually via oxygen or hydrogen peroxide. For example, adipic acid has been prepared from cyclohexene by a number of investigators (7) (8) (9). Ozonation of cyclododecene, which has recently become available from the trimerization of butadiene followed by selective hydrogenation, yields dodecanedioic acid, an important intermediate in the preparation of 6,12 nylon (10). 4-vinyl cyclohexene-1 yields butane 1,2,4-tricarboxylic acid (11). See Figure 5.

The ozonation of maleic acid in aqueous acid solutions is of interest since it only yields one mole of glyoxylic acid where two are to be expected (12). Glyoxylic acid is an important intermediate in the synthesis of vanillin. The reaction is illustrated by the equation in Figure 6.

Bernatek et al. postulate a rearrangement of the intermediate zwitterion to formic acid and carbon dioxide (13).

Bicyclic olefins have also yielded some interesting polycarboxylic acids when subjected to oxidative ozonolysis (Figure 6). Dicyclopentadiene, an inexpensive raw material, yields the tetracarboxylic acid 2,3,5-tricarboxy cyclopentane acetic acid (14). Tetrahydrophthalic anhydride, when ozonized in a participating solvent, yields 1,2,3,4-butane tetracarboxylic acid in excellent yields (15). Alpha-pinene produces the ketonic acid, pinonic acid which can be further oxidized to pinic acid (16) (17).

Acetylenic compounds undergo ozonolysis at the triple bond at a rate much slower than that for olefins, and the final product is usually two moles of carboxylic acids. Very little work has been done on this reaction (18).

The carbon-nitrogen double bonds of compounds such as oximes, imimes, hydrazones and related compounds are rapidly cleaved by ozone to yield a variety of products, including carboxylic acids and ketones. For leading references and additional details see Reference 4.

Aromatic Compounds

Most aromatic compounds are reactive towards ozone although the reaction is considerably slower than the reaction of ozone with olefins. The order of decreasing reactivity towards ozone (19) is as follows:

olefins > phenanthrene > anthracene > naphthalene > benzene

Substituents which withdraw electrons from the ring, such as the nitro group, halogens, sulfonic acid and carbonyl groups, deactivate the ring towards ozone slowing down the rate of ozone attack; whereas electron-releasing substituents, such as the alkyl group, the methoxy group, and the hydroxy group, activate the ring towards ozone. For example, the rate of ozone attack on methyl substituted benzenes increases as the number of methyl substitutents increases, e.g., toluene is more rapidly attacked than benzene, xylene more rapidly than toluene, etc. (19). Phenol, of course, is attacked very readily.

Benzene reacts with three moles of ozone with destruction of the ring to yield the expected products glyoxal, glyoxylic acid and oxalic acid, plus the fragmentation products formic acid and carbon dioxide (Figure 7). (See numerous papers by Wilbaut and co-workers in Reference 19.) Alkylbenzenes yield similarly substituted products.

Phenols have been studied extensively because of the recent interest in the use of ozone to purify wastewater. In aqueous media phenol itself is oxidized by ozone to the same mixture of products which resulted from the ozonolysis of benzene, including glyoxal, glyoxylic acid, oxalic acid, formic acid and carbon dioxide (20).

The route to these decomposition products is very complex and apparently involves muconic acid, catechol and ortho-quinone as intermediates. A simplified version of the main reactions is shown in Figure 8. The right hand path involves normal 1,3-dipolar cyclo-addition followed by reaction with water to the hydroxy hydroperoxide, which eventually is converted to muconic acid. Simultaneously catechol is formed, as shown on the left hand path, via an electrophilic attack on phenol. Catechol is then rapidly converted via additional reaction with ozone to also yield muconic acid. The two double bonds of muconic acid are then quickly attacked by ozone to yield the glyoxalic type products already mentioned. o-Quinone from the oxidation of catechol also is apparently an intermediate in the conversion to muconic acid, but this is probably a minor reaction since it has been shown that catechol is rapidly converted to muconic acid via ozonation. For more detailed discussion and references, see Reference 4.

Polycyclic aromatic compounds

The ozonolysis of many polycyclic aromatic hydrocarbons has been studied. We will examine only a few. With these compounds several types of ozone attack can apparently occur. With reactive double bonds 1,3-dipolar cyclo-addition leading to normal ozonolysis products is the most prevalent reaction, but in some cases there is an electrophilic attack by ozone at highly reactive carbon atoms to give quinones.

The 9,10 bond of phenanthrene and the 1,2 and 3,4 bonds of napthalene have considerable double bond character, and ozonolysis is the major reaction with these compounds.

With phenanthrene in a participating solvent such as methanol, an internal cyclic peroxide compound is formed from the zwitterion which can then be decomposed under alkaline conditions to the diphenaldehydic acid or oxidized to diphenic acid (Figure 9) (21).

In aqueous solution the dihydroxy analog appears to be formed which is easily decomposed to diphenic dialdehyde by heating (22). Additional exposure to ozone would probably yield diphenic acid.

Napthalene reacts similarly, being attacked at the 1,2 and 3,4 bonds to yield phthalaldehyde, phthalaldehydic acid and phthalic acid (23). In non-participating solvents a normal diozonide is formed as the main intermediate product. In a participating solvent, such as methanol, another intramolecular hydroxy cyclic peroxide appears to be the intermediate product. Either intermediate may be thermally decomposed to a mixture of phthalaldehydic acid and phthalic acid with eventual conversion to phthalic acid via further oxidation (Figure 10).

Again in aqueous media, ozonization produces phthalaldehyde by hydrolytic decomposition, with hydrogen peroxide as by-product (24). Again, additional exposure to ozone/oxygen would undoubtedly yield largely phthalic acid.

Ozonization of anthracene in a participating solvent, such as acetic acid-water, yields anthraquinone as the main product in good yields, probably by electrophilic attack by ozone at the 9,10 position; three moles of ozone are utilized and three moles of oxygen are produced (Figure 10). A co-product obtained in low yields is phthalic acid, apparently formed by normal ozonolysis at the 1,2 and 3,4 bonds, followed by ozone attack on the center ring (25).

Aromatic-Aliphatic Compounds

Ozonization of aromatic unsaturated aliphatic condensed ring systems occurs exclusively in the more reactive aliphatic ring. For example, the ozonolysis of indene in a participating solvent (ethanol) proceeds through the now-familiar cyclic peroxide which can be further oxidized to homophthalic acid or reduced to the corresponding homophthalaldehyde (26) (See Figure 11).

Heterocyclic Compounds

Most aromatic heterocyclic compounds are quite readily attacked by ozone, except for pyridine with its very stable ring system which is only very slowly attacked.

Quinoline for example, when subjected to ozonolysis in the presence of a mineral acid in water or aqueous acetic acid solution, yields nicotinic acid in good yields (Figure 11). In the patent which describes the reaction, the mineral acid is stated to be essential to stabilize the heterocyclic ring. The reaction with ozone then takes place at the carbocylic ring. Quinolinic acid is an intermediate and can be isolated in good yields prior to decarboxylation to nicotinic acid. Nitric acid, which is used to solubilize the compound and to prevent attack on the nitrogen-containing ring, also serves as the oxidant for the intermediate peroxy compounds. These intermediates are not shown, but are probably hydroxy hydroperoxide or cyclic peroxides, similar to some of the structures already discussed.

In most cases of aromatic condensed heterocyclic compounds, however, the heterocyclic ring is more readily attacked, giving a variety of compounds, usually aromatic aldehydic and acidic derivatives. For example, ozonation of benzofuran yields salicylic acid, salicylaldehyde and catechol (28). Benzothiophene is converted to diphenyl disulfide derivatives containing hydroxy, carboxy, and aldehydic groups (28). Further ozonization would undoubtedly result largely in acidic products.

Simple heterocyclics, such as pyroles and furans are readily attacked with destruction of the ring to yield a wide variety of oxidative products including formic acid, acetic acid, and in the case of pyroles, amines or ammonia. For references to numerous papers by Wilbaut and co-workers, see Reference 2, p. 971.

Nucleophiles

Most organic substances containing a nucleophilic atom in their structures are usually quite rapidly oxidized by ozone with the release of one mole of oxygen.

Sulfides are converted to sulfoxides and eventually to sulfones (24); tertiary amines, phosphines and arsines are converted to their corresponding oxides (29)(30). (Figure 12).

Phosphites are smoothly converted to phosphates through an adduct which in some cases can be characterized (31). These reactions can be explained by postulating an electrophilic attack by a terminal oxygen of the ozone molecule to form a new bond with the S, N, or P atom followed by liberation of the 2nd and 3rd atom of oxygen as molecular oxygen (29).

The ozonation of primary and secondary amines proceeds rapidly, and the reaction is extremely complex, yielding nitro compounds, nitroxides, and a variety of oxidative products resulting from carbon chain attack. For example, ozonation of isopropyl amine in a non-participating solvent, pentane, yields acetone, 2-nitropropane and isopropyl ammonium nitrate among the products (32). Ozonation of amines in aqueous media has not been studied extensively, but would probably result in attack of the side chains. In aqueous acidic media, amines are unreactive towards ozone because of salt formation which causes unavailability of the nitrogen electron pair. This prevents electrophilic attack by the ozone molecule.

Ozone Initiated Oxidations

The last reaction of ozone with organic compounds to be considered includes reactions in which ozone acts primarily as the initiator and oxygen enters into the reaction. These include the reaction of aldehydes, ketones, alcohols, ethers, saturated hydrocarbons and others. Aldehydes are converted to acids and peroxides, ketones to carboxylic acids, hydrocarbons, alcohols and ethers yield mainly aldehydes, acids, peroxides, and carbon dioxide; (for additional detail see Reference 4). These reactions in general proceed very slowly at the usual ozonation temperatures and not at all in the presence of a reactive group such as the double bond. The mechanisms are somewhat in doubt but probably involve ozone attack yielding a free radical from the carbon chain which, in turn, initiates an auto-oxidation process by molecular oxygen.

In conclusion, many of the reactions presented in this paper are based on classical methods of synthesis where exposure to excess ozone is carefully avoided so that intermediate products can be isolated in good yields. However, in the purification of water, over-ozonation conditions are frequently resorted to, and in fact, many times are desirable. Thus, intermediate products can be expected to be converted to higher oxidative states, or to be eventually destroyed.

Literature Cited

(1) R. Trambarulo, S. N. Ghosh, C. A. Burrus Jr., and W. Gordy. J. Chem. Phys. 21:851 (1953).

(2) P. S. Bailey. Chem. Revs. 58:925-1010 (1958).

(3) P. S. Bailey, J. A. Thompson, and B. A. Shoulders. J. Am. Chem. Soc. 88:4098 (1966).

(4) P. S. Bailey, "Ozone in Water and Wastewater Treatment", F. L. Evans III, Ed. Ann Arbor Science Publishers, Ann Arbor, Michigan (1972), Chap. 3.

(5) G. Riezebis, J. C. Grimmelikhuysen, and D. A. Van Dorp. Rev. Trav. Chim. 82:1234 (1963).

(6) C. G. Goebel, A. C. Brown, H. F. Oehlschlaeger, and R. P. Rolfes. U. S. Patent 2,813,113 to Emery Industries, Inc. (1957).

(7) A. L. Henne, and P. Hill. J. Am. Chem. Soc. 65:752 (1943).

(8) H. Wilms. Ann Chem. 567:96 (1950).

(9) P. S. Bailey. J. Org. Chem. 22, 1548 (1957). P. S. Bailey. Ind. Eng. Chem. 50, 993 (1958).

(10) A. Maggiolo. U.S. Patent 3,280,183 to Wallace and Tiernan, Inc. (1966).

(11) R. H. Perry and H. Kail. British Patent 945,428 (1963).

(12) W. T. Black and G. A. Cook. Ind. Eng. Chem. Prod. Res. Dev. 5:350 (1966).

(13) E. Bernatek, P. Groenning, J. Ledaal. Acta Chem. Scand. 18:(8) 1966-74 (1964). (In English).

(14) A. Maggiolo and A. L. Tumolo. U.S. Patent 3,023,233 to W. R. Grace & Co. (1962).

(15) J. E. Franz, W. S. Knowles, and C. Osuch. J. Org. Chem. 30:4328 (1965).

(16) G. S. Fisher and J. S. Stinson. Ind. Eng. Chem. 47:1569 (1955).

(17) F. Holloway, H. J. Anderson, and W. Rodin. Ind. Eng. Chem. $\underline{47}$, 2111 (1955).

(18) F. Bohlmann and H. Sinn. Chem. Ber. 88:1869 (1955).

(19) P. S. Bailey, "Ozonolysis of Aromatic Compounds". Chem. Rev. $\underline{58}$, 957-965 (1958).

(20) E. Bernatek, and C. Frengen, "Ozonolysis of Phenols". Acta. Chem. Scand. $\underline{15}$, 471 (1961).

(21) P. S. Bailey. J. Am. Chem. Soc. 78:3811 (1956).

(22) M. G. Sturrock, E. L. Cline, and K. R. Robinson. J. Org. Chem. 28:2340 (1963).

(23) P. S. Bailey, S. S. Bath, F. Dobinson, F. J. Garcia-Sharp, and C. D. Johnson. J. Org. Chem. 29:697 (1964).

(24) M. G. Sturrock, B. J. Cravy, and V. A. Wing. Can. J. of Chem. 49:3047 (1971).

(25) P. S. Bailey, P. Kolsaker, B. Sinha, J. B. Ashton, F. Dobinson, and J. E. Batterbee. J. Org. Chem. 29:1400 (1964).

(26) J. L. Warnell and R. L. Shriner. J. Am. Chem. Soc. $\underline{79}$, 3165 (1957).

(27) M. G. Sturrock, E. L. Cline, K. R. Robinson, and K. A. Zercher. U.S. Patent 2, 964, 529 to Koppers Co. Inc. (1960).

(28) A. Von Wacek, H. O. Eppinger, and A. Von Bezard. Ber. 73:521 (1940). See also Reference (2).

(29) A. Maggiolo and S. J. Niegowski. Ozone Chemistry and Technology. Advances in Chemical Series. 21:200-202 (1958).

(30) L. Horner, H. Schaefer, and W. Ludwig. Ber. 91:75 (1958).

(31) Q. E. Thompson. J. Am. Chem. Soc. 83:845 (1961).

(32) P. S. Bailey, T. P. Carter, Jr., and L. M. Southwick. J. Org. Chem. 37:2997 (1972).

DISCUSSION

David Rosenblatt, U. S. Army, Fort Detrick: I would like to know how you feel about the question of tolerating hydrogen peroxide as an end product in the ozonolysis of olefins in aqueous solution? Do you think that this presents a problem that we haven't really faced in ozonolysis treatments?

Oehlschlager: It is undoubtedly present. I am not really sure how much of a problem peroxide is in wastewater. In this case you will have aldehydes and peroxide. I think that eventually peroxides will react with residual aldehyde and some of it will be destroyed.

Rosenblatt: At those concentrations?

Oehlschlager: I'm speaking of the higher concentrations. In the very low concentrations I am not really sure. I think that is a good point for further research, though.

Aaron Rosen, EPA: An important class of water pollutants is tertiary alcohols. Will they react with ozone?

Oehlschlager: I should imagine that they would. I don't know of any papers that I have read covering that, but I would think they would be slowly attacked.

FIGURE 1. RESONANCE STRUCTURES OZONE MOLECULE

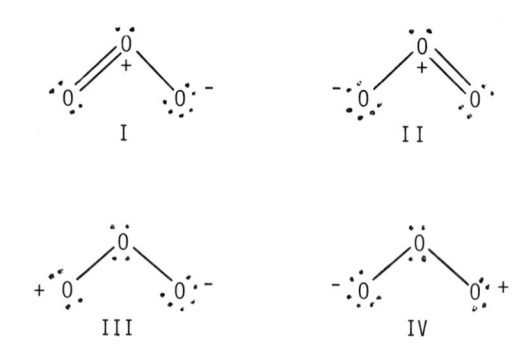

CRIEGEE OZONOLYSIS MECHANISM
INITIAL PHASES

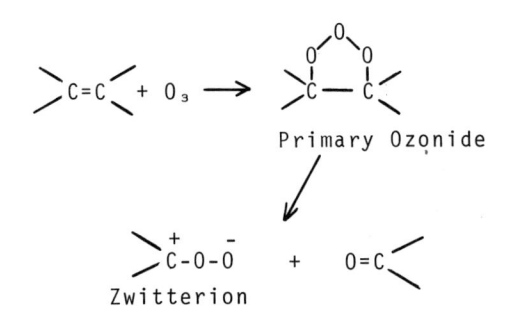

Primary Ozonide

Zwitterion

PART OF FIGURE 1.

REACTION OF ZWITTERION WITH PARTICIPATING SOLVENTS

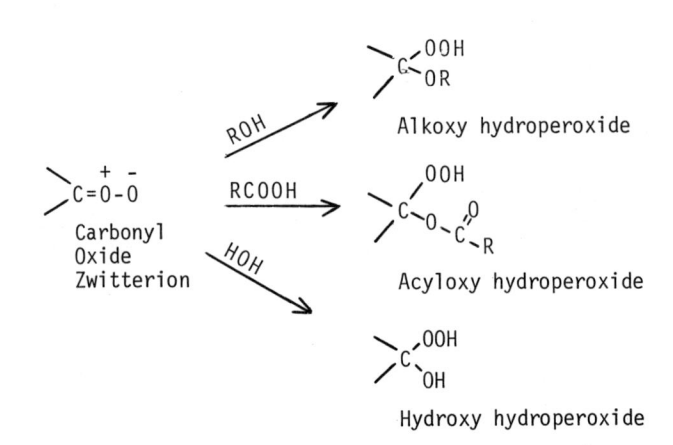

Carbonyl Oxide Zwitterion

ROH → Alkoxy hydroperoxide

RCOOH → Acyloxy hydroperoxide

HOH → Hydroxy hydroperoxide

FIGURE 2.　ZWITTERIONS AND CARBONYL FRAGMENTS

OZONOLYSIS IN AQUEOUS MEDIA

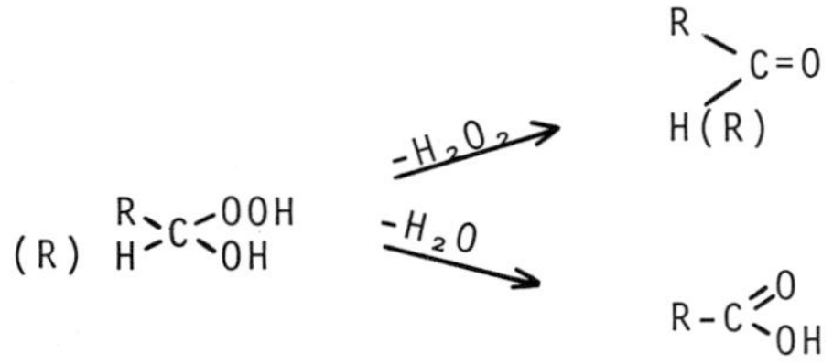

FIGURE 3.　OZONOLYSIS OF UNSYMETRICAL OLEFINS

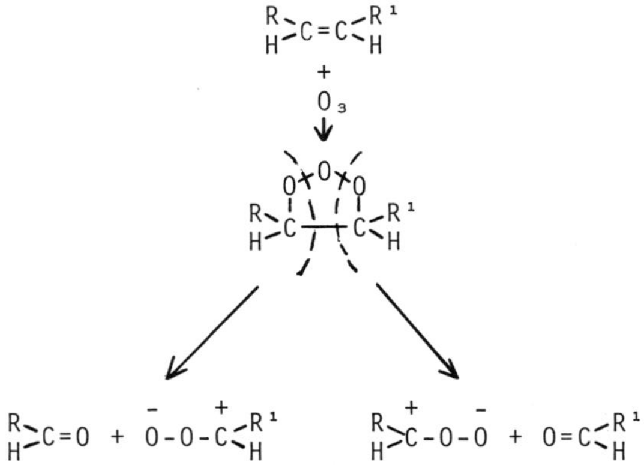

ZWITTERION REACTIONS
NON PARTICIPATING SOLVENT

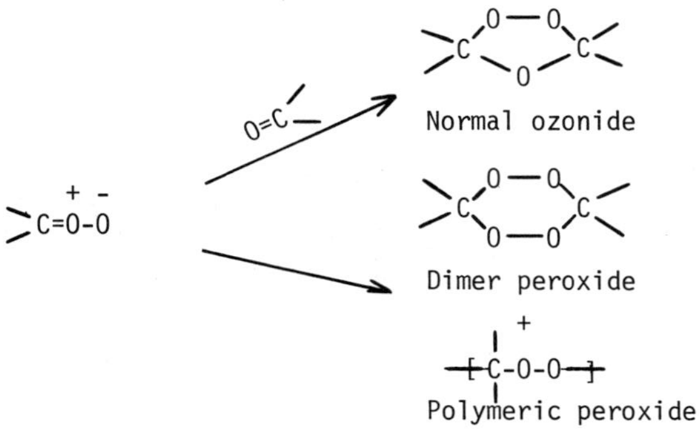

OZONIDES FROM UNSYMETRICAL OLEFINS

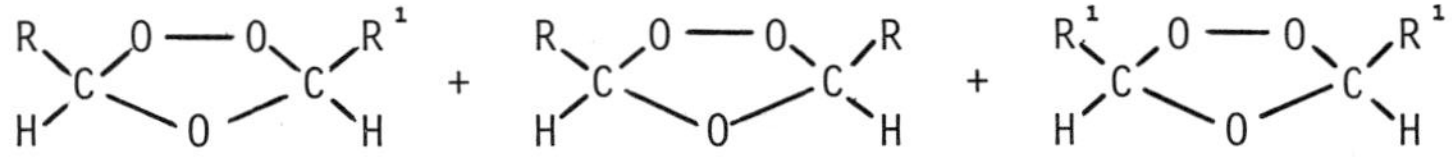

Normal ozonide　　　Crossed ozonides

Normal ozonide
Dimer peroxide
Polymeric peroxide

-33-

FIGURE 4.   OZONOLYSIS OF OLEIC ACID

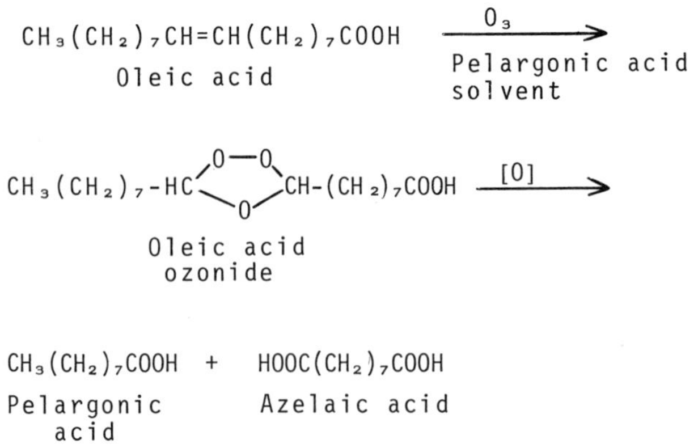

$$CH_3(CH_2)_7CH=CH(CH_2)_7COOH \xrightarrow[\text{solvent}]{O_3} \text{Pelargonic acid}$$

Oleic acid

$$CH_3(CH_2)_7-HC\underset{O}{\overset{O-O}{\diagup\diagdown}}CH-(CH_2)_7COOH \xrightarrow{[O]}$$

Oleic acid ozonide

$$CH_3(CH_2)_7COOH \;+\; HOOC(CH_2)_7COOH$$

Pelargonic acid   Azelaic acid

OZONOLYSIS OF ERUCIC ACID

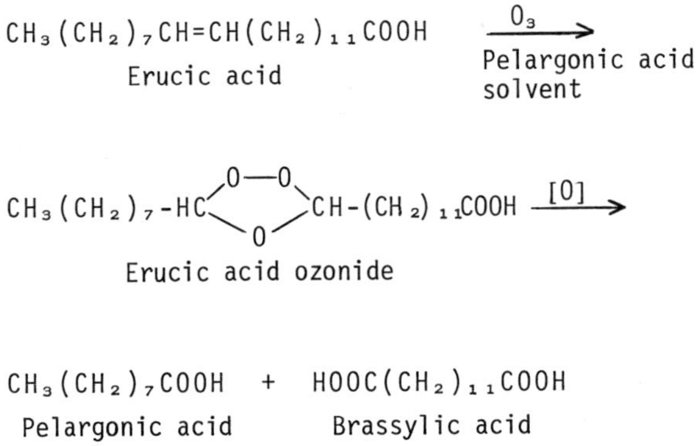

$$CH_3(CH_2)_7CH=CH(CH_2)_{11}COOH \xrightarrow[\text{solvent}]{O_3} \text{Pelargonic acid}$$

Erucic acid

$$CH_3(CH_2)_7-HC\underset{O}{\overset{O-O}{\diagup\diagdown}}CH-(CH_2)_{11}COOH \xrightarrow{[O]}$$

Erucic acid ozonide

$$CH_3(CH_2)_7COOH \;+\; HOOC(CH_2)_{11}COOH$$

Pelargonic acid   Brassylic acid

FIGURE 5.   OZONOLYSIS OF CYCLIC OLEFINS

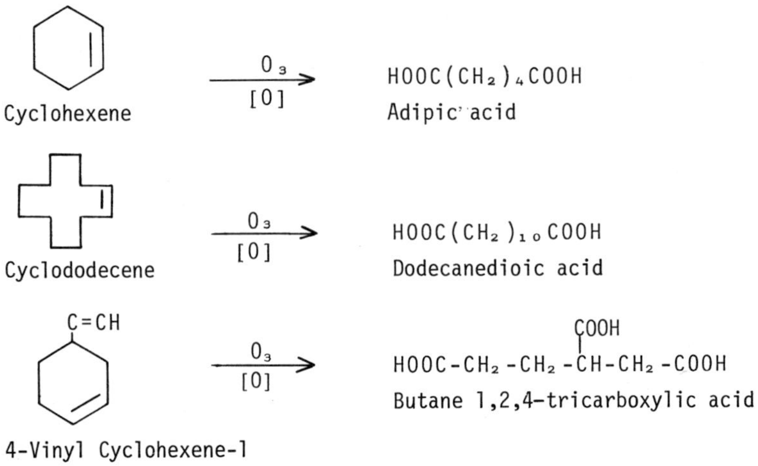

Cyclohexene $\xrightarrow[\text{[O]}]{O_3}$ HOOC(CH_2)_4COOH   Adipic acid

Cyclododecene $\xrightarrow[\text{[O]}]{O_3}$ HOOC(CH_2)_{10}COOH   Dodecanedioic acid

4-Vinyl Cyclohexene-1 $\xrightarrow[\text{[O]}]{O_3}$ HOOC-CH_2-CH_2-CH(COOH)-CH_2-COOH   Butane 1,2,4-tricarboxylic acid

-34-

FIGURE 6. OZONOLYSIS OF MALEIC ACID

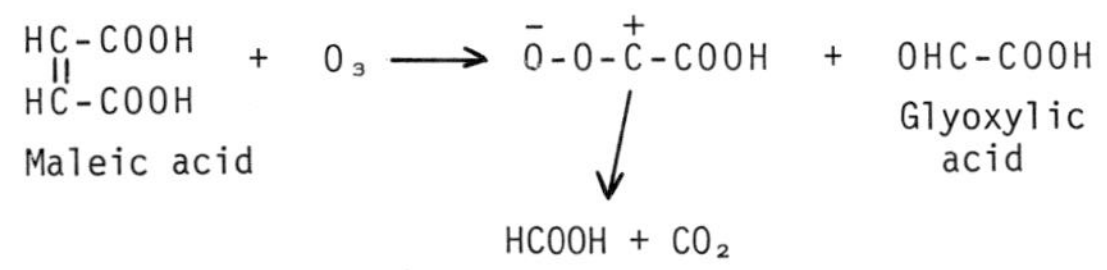

HC-COOH
|| + O₃ ⟶ Ō-O-Ċ⁺-COOH + OHC-COOH
HC-COOH   Glyoxylic
Maleic acid   acid
                        ↓
                   HCOOH + CO₂

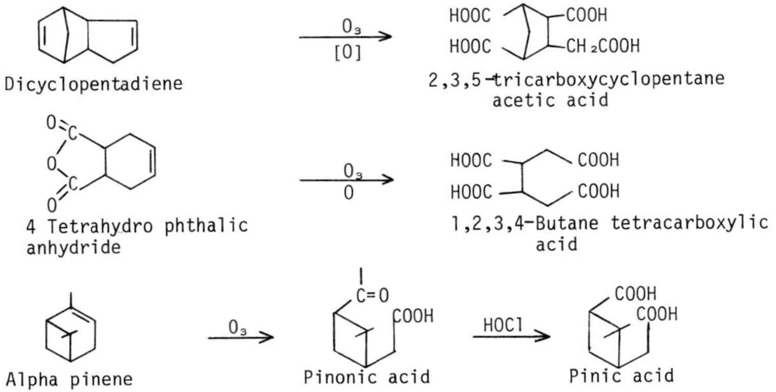

FIGURE 7. OZONOLYSIS OF BENZENE

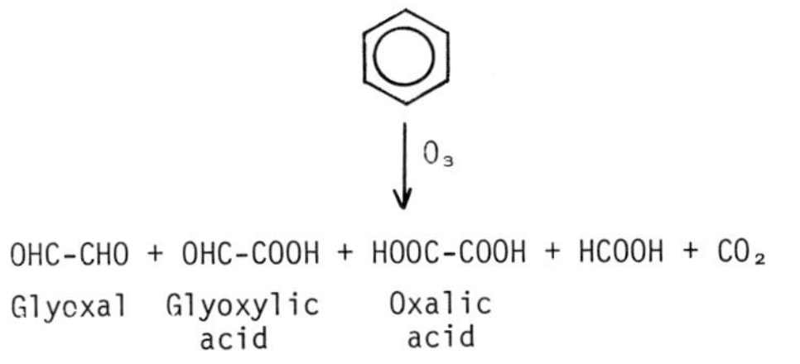

⟶ OHC-CHO + OHC-COOH + HOOC-COOH + HCOOH + CO₂
   Glyoxal Glyoxylic   Oxalic
             acid   acid

Figure 8. OZONOLYSIS OF PHENOL IN WATER

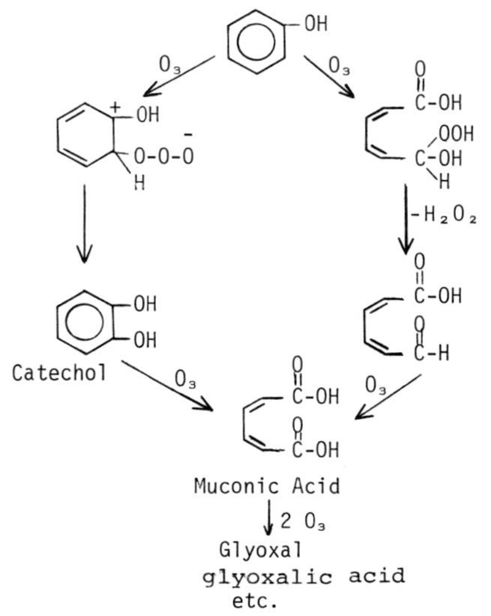

Catechol

Muconic Acid
Glyoxal
glyoxalic acid
etc.

OZONOLYSIS OF BICYCLIC OLEFINS

Dicyclopentadiene →[O₃, [O]] 2,3,5-tricarboxycyclopentane acetic acid

4 Tetrahydro phthalic anhydride →[O₃/O] 1,2,3,4-Butane tetracarboxylic acid

Alpha pinene →[O₃] Pinonic acid →[HOCl] Pinic acid

-35-

FIGURE 9. OZONOLYSIS OF PHENANTHRENE

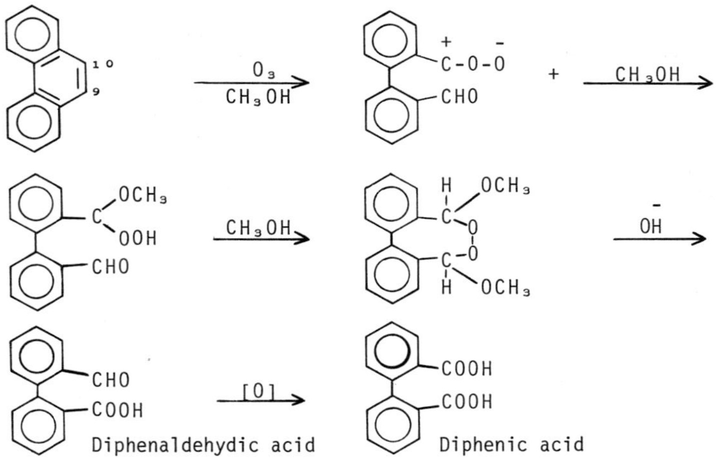

FIGURE 10. OZONOLYSIS OF NAPHTHALENE

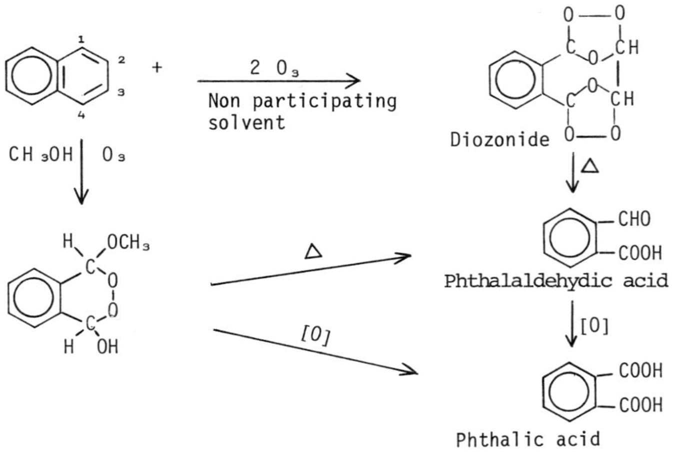

OZONOLYSIS OF PHENANTHRENE IN AQUEOUS SOLUTION

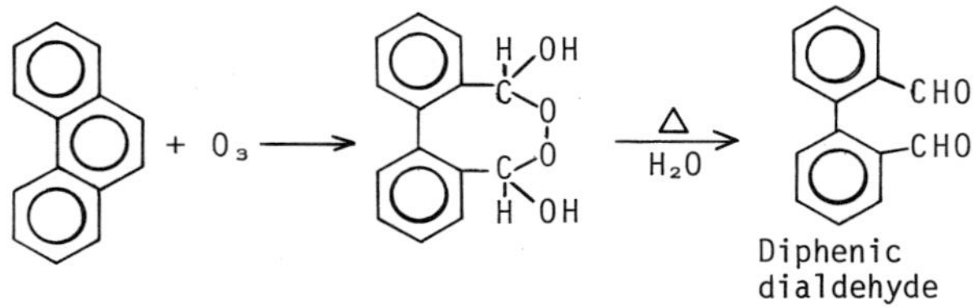

OZONATION OF ANTHRACENE

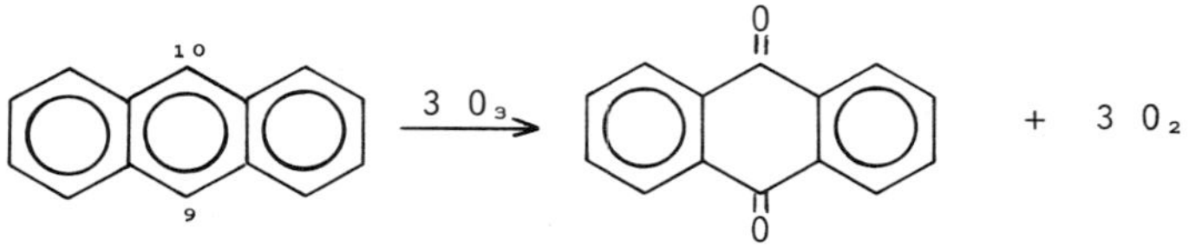

-36-

FIGURE 11.   OZONOLYSIS OF INDENE

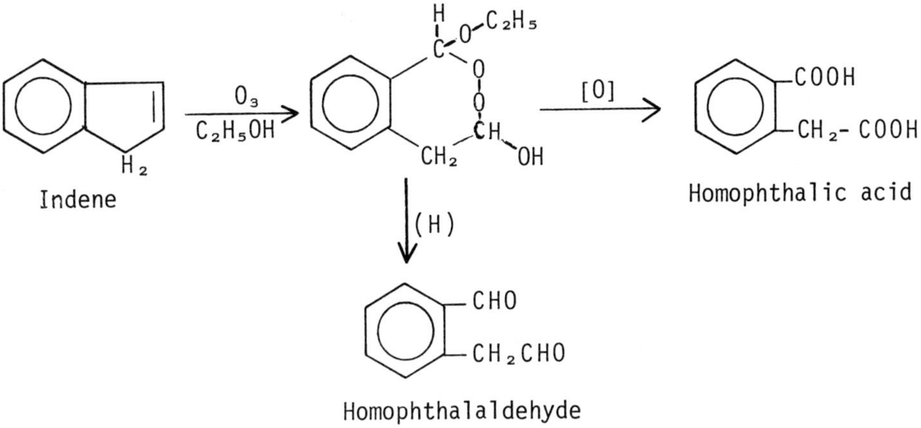

Indene → (O$_3$, C$_2$H$_5$OH) → [ethoxy hydroperoxide intermediate] → [O] → Homophthalic acid (benzene with COOH and CH$_2$-COOH)

↓ (H)

Homophthalaldehyde (benzene with CHO and CH$_2$CHO)

OZONOLYSIS OF QUINOLINE

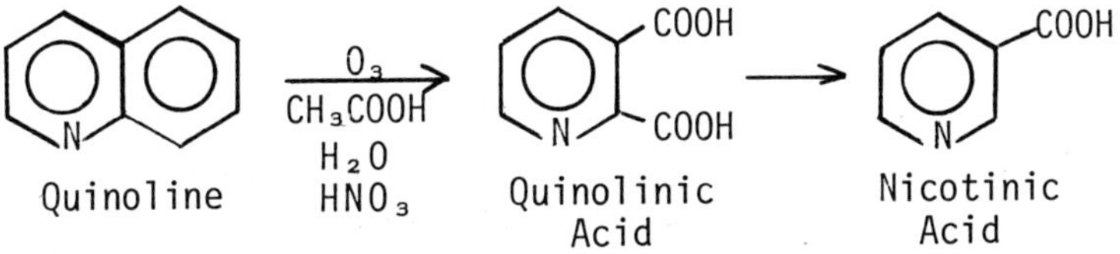

Quinoline → (O$_3$, CH$_3$COOH, H$_2$O, HNO$_3$) → Quinolinic Acid (pyridine-2,3-dicarboxylic acid) → Nicotinic Acid (pyridine-3-carboxylic acid)

FIGURE 12.   OZONATION OF NUCLEOPHILES

$$R_2S + O_3 \longrightarrow R_2SO + O_2 \xrightarrow{O_3} R_2SO_2 + O_2$$

$$R_3N + O_3 \longrightarrow R_3N \rightarrow O + O_2$$

$$R_3P + O_3 \longrightarrow R_3P \rightarrow O + O_2$$

$$R_3As + O_3 \longrightarrow R_3As \rightarrow O + O_2$$

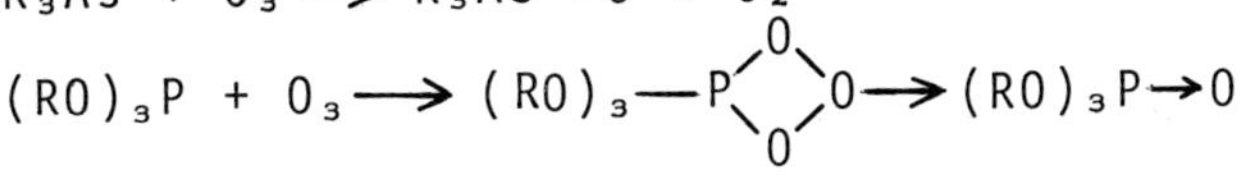

$$(RO)_3P + O_3 \longrightarrow (RO)_3 - P\begin{smallmatrix}O-O\\O\end{smallmatrix} \longrightarrow (RO)_3P \rightarrow O$$

OZONE AS A DISINFECTANT OF WATER*

Hans L. Falk and James E. Moyer
Department of Health, Education & Welfare
Public Health Service
National Institute of Environmental Health Sciences
Research Triangle Park, North Carolina 27709

I. Introduction

Chlorine was first used to disinfect water supplies in America in 1908 (28), and has grown so that the majority of water treatment plants in the United States now use chlorine to disinfect their drinking water (27). Recent research by the Environmental Protection Agency has resulted in a list of 187 organic compounds (18) that have been positively identified in drinking water in trace amounts. This list contains chlorinated organic compounds, some of which are most likely the result of water disinfection with chlorine (40). It has been pointed out that these chlorinated compounds are more resistant to biodegradation than their respective unchlorinated analogs (40). A few of these chlorinated organic compounds have been identified by EPA as carcinogens (19).

---

*Paper presented by Dr. Joseph A. Cotruvo, U. S. Environmental Protection Agency, Washington, D. C.

For these reasons, we must consider the possibilities of using other disinfectants, such as ultraviolet or gamma radiation (13), or an oxidizing agent such as ozone. The following is a consideration of ozone as a water disinfectant.

II. Advantages of Ozone Over Chlorine as a Disinfectant of Water

A. Taste and odor oxidizing ability

Ozone has been shown to oxidize taste- and odor-causing compounds (28,32), adds no smell or taste of its own and has been shown to improve the appearance of water (24,32). Chlorine, however, may contribute its own smell to the treated water and has been known to react with certain malodorous compounds, causing them to smell even worse (50), as with the phenols.

B. Ability to form biodegradable products

Organic compounds which result from ozonation are more biodegradable than chlorinated organic compounds (40). Ozone itself is very unstable and has a reported half-life of twenty minutes in water (37).

C. Ability as oxidant of bacteria and virus

It has been reported, from the Saint-Maur water treatment plant in Paris, that ozonization has been able to effectively eliminate E. Coli from treated water (24). Upon switching from chlorine to ozone in 1955, a further reduction in the Clostridium perfringens content of the water was evidenced (24). Statistics from a Philadelphia plant further support the effectiveness of ozone in destroying coliform organisms and bacteria (7). Ozone has proven capable of destroying Bacillus cereus and Bacillus megaterium, and their spores (28). Ozonization has also proven able to inactivate coliphage T_2 and polio virus I (28). Some consider ozonization to be the most effective treatment for viruses in regards to thoroughness and economy (32,37).

III. Disadvantages of Ozone as a Water Disinfectant

　　A.　Escape to air possibilities

　　　　1.　Smog contribution

　　Any ozone released to the atmosphere may react with pollutants already in the air, thus making a contribution to the reactions which occur in air polluted areas (20,30). Ozone released into the ambient air has been shown to destroy nearby plants and shrubbery (7).

　　In water ozonization plants, therefore, efforts must be, and are made to capture and recycle the gaseous, ozone-containing effluent (7,24). This is done to keep ozone from escaping to the environment, where it may exhibit its undesirable effects (30,46), but also for cost-effectiveness. Since energy is expended in producing ozone in the first place, many modern ozonation plants recycle ozone-rich contactor off-gases so as to utilize ozone to the maximum extent possible.

　　　　2.　Mutant and irritant; toxicity

　　At concentrations in air above 0.1 ppm, ozone has a distinct odor. At continued exposure to ozone at this level humans have experienced headaches, dryness of throat, irritation of respiratory passages and eyes (46).

　　Data have been assembled which erase all doubt regarding the high toxicity of ozone in several rodent species (12). Male Swiss 25 g. mice exposed to ozone at levels from 0.62 to 4.25 ppm, for four hours, showed significant decreases in pulmonary bactericidal activity towards Staphylococcus aureus at levels ⩾1.10 ppm ozone, as compared to control mice (23). In a similar study using the same type mice, bacteria and time exposure, with ozone levels from 0.38 ± 0.05 to 1.59 ± 0.13 ppm, the same results were obtained, except that a significant difference was found at levels of ozone ⩾0.41 ppm (22). Hueter and Fritzhand add that, "Ozone has been shown to...weaken lung structure, reduce pulmonary function, and disrupt cellular biochemistry" (26).

It has been demonstrated that ozone inhalation reduces lysozyme and phagocytic protective activity in the lungs of experimental animals (26). Adding to this the documented ability of ozone to cause chromosome aberrations in Chinese hamsters, suggests a possible effect upon humans (52). The resulting break frequency agreed well with the value expected from calculations based upon in vitro exposure of human cells (52). Zelac, et al., point out that:

> "Presently permitted human ozone exposures would be expected to result in break frequencies that are orders of magnitude greater than those resulting from permitted human radiation exposures if the results of this animal study were directly extended to the human case" (52).

B. Non-persistence in water

    1. Need for chlorination following ozonization of water

Ozone, however, decomposes very rapidly, so that there is none left as a toxic residue (37). This can present a problem, however, in that there is no disinfectant left in the water to destroy bio-contaminants which may enter the water between treatment and consumption. The injection of small amounts of chlorine subsequent to ozonization, to serve as a residual disinfectant, is one way to deal with this problem (37). Others argue that the risks of subsequent water contamination are insignificant (24).

It is clear that there are several reasons why ozone may be more desirable than chlorine as a disinfectant for water treatment plants. Before we use this as a basis for switching to ozone as the major disinfectant, however, let us consider what chemical compounds may be produced in water upon ozonization.

IV. Chemical Reactions of Ozone with Organic Compounds Present in Water

All of the substances considered in this section are taken from a list of organic compounds that have been found in water, which was compiled by EPA (18).

A. Alkanes

Ozonolysis of alkanes yields a variety of alcohols, ketones, aldehydes, peracids and acids, along with unaffected alkanes. This is illustrated by the ozonization of methane, which yields formic acid, methanol, CO_2, CO and water (15,45). Ozonolysis of tetradecane yields alcohols, ketones and peracids, and acids with a median chain length of C-10 (39).

B. Olefins: explanation of mechanism

Ozone is most reactive with double bonds of various olefins. It is generally agreed that ozone forms an obtuse angle in its molecular structure and is a hybrid of four structures (Figure 1) (1). The usual method of attack is understood to be through an electrophilic terminal oxygen (2), and it reacts to form a variety of products (figure 2).

C. Alcohols and ketones

Ozonolysis of alcohols is known to produce aldehydes and ketones (41), while ketones are known to yield acids (6). Ozonolysis of acetone has been shown to yield formic and acetic acids (6), while the products of methyl ethyl ketone ozonolysis have been identified as formic, acetic and propanoic acids (6). Cycloheptanone can be expected to form heptanedioic acid (6). [1].

D. Aldehydes

Aldehydes are more reactive than ketones (6) and have been shown to form peracids, acids and other aldehydes. Ozonolysis of crotonaldehyde (2-butenal) [2] forms formic acid, acetaldehyde and glyoxal (ethanedial) (34). The aldehyde groups of benzaldehyde (6, 35) [3], 2-methyl-propanal [4], 3-methylbutanal [5] and acetaldehyde [6] are converted to mixtures of their respective peracids and acids (6).

E. Acids

Carboxylic acids are not highly reactive with ozone, although acetic acid ozonolysis has yielded low levels of peracetic acid (6) [7].

F. Radical formation

It should be kept in mind that the proposed mechanism of reaction of ozone often involves radical formation (6). For example, note that the proposed mechanism for ozone oxidation of an aldehyde involves radical formation (Figure 3), and these radicals react with other aldehyde molecules. Assuming that these aldehydes will not be appearing in nearly so concentrated a form in the water treatment situation as they are under experimental conditions, these radicals may react with other organic molecules in the water. The possible products of ozonolysis reactions are, therefore, obviously many and varied.

G. Sulfur containing Compounds

Dialkyl and diaryl sulfides (XIV) can be expected to react with ozone to form sulfoxides (XV), which react with ozone to form sulfones (XVI)(Figure 4) (33). Benzothiophene (XVII) ozonolysis yields XVIII (50%), XIX (30%) and XX (20%) (Figure 5) (4).

H. Phosphorus containing compounds

Reactions of trialkyl- and triarylphosphides are rapid and quantitative, proceeding as in Figure 6 (8).

I. Nitrogen containing compounds

Amines are usually not very reactive with ozone. Bailey states, however, that:

"Primary and secondary aliphatic amines...are decomposed by ozone, and tertiary amines are converted to the corresponding amine oxides" (5).

J. Benzene and its methylated derivatives

Ozonolyses of benzene and its methylated derivatives have yielded mixtures of short chain acids and aldehydes as primary products. Benzene reactions with ozone have produced a triozonide, which decomposes with water to yield two moles of glyoxal per mole of triozonide (21) [12]. Methylglyoxal comprises 85% of the carbon product from the reaction of mesitylene (1,3,5-trimethylbenzene), with small amounts of formic and acetic acids, CO and CO_2 (38) [13].

1. Glyoxal; formation and effects

We have seen that glyoxal is an ozonolysis product of 2-butenal and benzene, and we shall see that it is also a product of naphthalene ozonolysis. Experiments have shown that glyoxal reacts to denature the native DNA of T_2 phage, primarily by unwinding the guanosine-cytosine rich regions of the DNA, at a temperature below Tm (9).

2. Methylglyoxal; formation and effects

Observations have identified methylglyoxal as the primary ozonolysis product of mesitylene, and it is one of the likely products of some other methylated benzene derivatives. Albert Szent-Gyorgyi discusses what he calls "retine", a cell growth inhibitor, and "promine", the antagonist of retine (49). Both appear to exist in most, if not all cells (49). Since they are both highly reactive and difficult to isolate, their exact structure has not yet been elucidated (49). It has been found, however, that methylglyoxal reacts in a manner similar to retine, as do α-ketoaldehydes in general, so as to inhibit cell growth (16). The inhibition results from the reaction of methylglyoxal with SH groups, which are necessary for cell division to occur (16).

Experiments show marked decreases in cell division of E. coli in concentrations of methylglyoxal from $1 \times 10^{-3}M$ to $1 \times 10^{-4}M$ (16). It must be noted, however, that:

> "Most cells contain a very active enzymic system, consisting of glutathione, glyoxalase I and II, which system readily transforms methylglyoxal into lactic acid and methyglyoxal derivatives into the corresponding α-hydroxy acid" (17).

It has also been stated that "organic residues in drinking water range at about the 10 ppm level. Specific compounds, however...may be in the parts per billion or trillion range." (21).

K. Other benzene derivatives and aromatic compounds

The products of ozonolysis of the various benzene derivatives and the higher aromatic hydrocarbons cover a wide spectrum. The possible products include alcohols, aldehydes, ketones, acids, peracids, ozonides and epoxides.

Isopropyl benzene (XXI) (cumene) reacts to form cumene hydroperoxide (XXII) (Figure 7) (6). It should be noted that cumene peroxide has been reported to have mutagenic properties (29).

Naphthalene (XXIII) reacts readily with the first two moles of ozone to form a diozonide (21) and then decomposes to a mixture of o-phthalaldehyde (XXIV), phthalic acid (XXV) and glyoxal (XXVI) (Figure 8) (3,31). Acenaphthalene (XXVII) has been shown to form XXVIII upon ozonolysis (Figure 9) (8).

The 8-9 double bond of limonene (XXIX) ozonizes in preference to the 1-2 double bond, which breaks down into formaldehyde and a ketone (XXX) (35). The 1-2 double bond then reacts with ozone to form an ozonide, and then decomposes to products which have not been positively identified (Figure 10) (35).

L. Pesticides

It is important that the possibility of epoxide formation be considered, as certain epoxides have been shown to be carcinogenic (29). The reactions with ozone of the pesticides aldrin, heptachlor, heptachlor epoxide, chlordane, dieldrin and DDT were tested experimentally. Aldrin and heptachlor were found to react quantitatively with ozone, while the others were hardly touched (25).

In consideration of the inhibiting effects of chlorine upon the electrophilic attack of ozone (42), one can be assured that the 6-7 double bond of aldrin (XXXII) will undergo ozonolysis in preference to the 2-3 double bond. One of the theoretically possible products would be an epoxide (XXXIII) (Figure 11). We know that aldrin is converted to its epoxide (dieldrin) and heptachlor (XXXIV) is converted to heptachlor epoxide (XXXV). in plant and animal tissues (51). Heptachlor (XXXIV) should have heptachlor epoxide (XXXV) as one of its possible ozonolysis products (Figure 12).

A paper by Bartlett and Stiles (6a) has yielded more information on the possibility of epoxide formation with ozonolysis. The article states that the ozonization of olefin A in dry ethyl acetate at -20° "resulted in 50% of the product appearing as an epoxide." The authors add that:

"The direct production of an epoxide in ozonization also has been observed before in highly hindered compounds. Backer (A) obtained an epoxide ($C_{12}H_{24}O$) from the ozonization of an olefin obtained by partially hydrogenating 2,3-di-t-butylbutadiene."

Olefin B produced a liquid epoxide at an 81% yield after 86 hours at 0° with perbenzoic acid in a benzene solution. It was also found that "ozone in ethyl acetate at 0° gave a mixture" composed primarily of this epoxide.

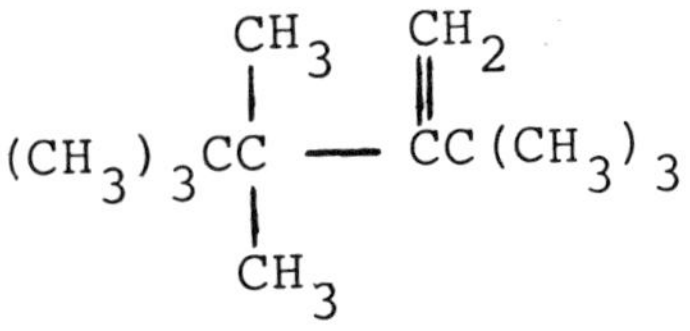

$$(CH_3)_3CC(CH_3)(CH_3) - CC(CH_3)_3 \text{ with } CH_2$$

Olefin A

$$CH_3C(CH_3)(CH_3) - C(CH_3)(CH_3) - C(=CH_2)(CH_3)(CH(CH_3)_2)$$

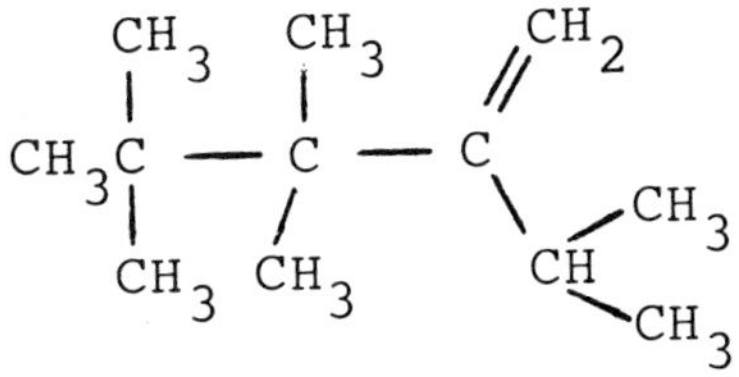

Olefin B

Although DDT (XXXVI) (Figure 20) does not react appreciably with ozone (25), we do know that DDE (XXXVII) is its principal breakdown product (43). It is known that $R_2CHCOOH$ will usually give rise to an epoxide:

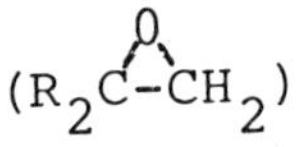

$$(R_2C \overset{O}{\triangle} CH_2)$$

when the R's are strong electron donating groups (14). There is a good possibility, therefore, that DDE will form an epoxide (XXXVIII), although this has not yet been shown (Figure 13).

M. Bacterial and viral cell products

In the course of ozonolysis certain forms of life that exist in the water will be destroyed, leaving various cell constituents behind as possible reactants with ozone.

1. Polyunsaturated fatty acids

For this reason, let us consider the ozonolysis of fatty acids. An experiment was conducted in which methyl oleate [18:1], methyl linoleate [18:2] and methyl linolenate [18:3] were ozonized at an ozone concentration of 1.5 ppm (44).

This is a reasonable ozone concentration, as two effectively operating water ozonization plants have reported that they use between 0.6 and 1.8 ppm (24,50). The ozonolysis of 18:1 yielded cis- and trans-methyl oleate ozonides and various unidentified compounds, while 18:2 reacted much faster and completely (44). The products included cis- and trans- forms of all the possible mono- and diozonides, many unidentified products and 12.6% of the initial ester was accounted for as malonaldehyde ($CHO-CH_2-CHO$) (44). The oxidation of 18:3 was extremely rapid and the yield of malonaldehyde was even higher than for 18:2 (44).

One concern here is the production of malonaldehyde. Research has shown that malonaldehyde is mutagenic on frameshift mutants with normal excision repair, in certain Salmonella typhimurium strains (36). There is also a concern with the possible production of malonaldehyde in the lungs of workers at ozonization plants, as a result of reactions with polyunsaturated fatty acids present in cell membranes (36).

Shamberger, et al. (45A) have studied the carcinogenicity of malonaldehyde. These investigators applied 0.25 ml of acetone with 12 or 6 mg malonaldehyde, 2.5 mg glycidaldehyde or 0.125 mg 7,12-dimethylbenz[a]anthracene (DMDA) to the shaved backs of 55-day old female Swiss mice. At three weeks a 27-week treatment was begun in which 0.25 ml of 0.1% croton oil in acetone was applied to their backs five days per week. The tumor incidence did not vary for those mice receiving 6 and 12 mg of malonaldehyde, and the overall tumor incidence was DMBA (95%), malonaldehyde (52%) and glycidaldehyde (40%).

It was noted that other studies have shown that various antioxidants may decrease the potency of some carcinogens. A daily application of 0.36 mg of malonaldehyde for 48 weeks to mice of the same characteristics as those mentioned above resulted in no tumor formation, which may indicate the existence of a threshold level.

A single application of 0.01% methyl cholanthrene, 0.01% benzo[a]pyrene and 0.01% DMBA to 55-day old female Swiss albino mice resulted in increased levels of malonaldehyde in the mouse skin.

The formation of malonaldehyde and the oxidation curves indicate that ozonolysis is occurring through direct cleavage of double bonds, and not by the normal autoxidation processes (normal oxidation of 18:2 yields no malonaldehyde) (44). The normal autoxidation processes involve hydroperoxide formation at the carbons α to the double bonds, with subsequent radical formation, resonance with double bond migrations, and formation of various products, including dimeric and polymeric cyclic peroxides (48).

It is known that in aqueous solution the Criegee zwitterion (Figure 2, IX) can be solvated by water to form peroxides. The peroxides formed can then catalyze the usual peroxidation reaction (44), or any of many other possible reactions with other organic compounds present in the water.

2. Purines

Ozonolysis of bacteria and virus life forms can be expected to leave small amounts of DNA and RNA bases in the water. The products of ozonolysis of these bases is not known, but it is possible that various N-oxides would result. Some purine N-oxides and certain of their derivatives have been found to have oncogenic properties (10,11,47). It has been shown that for 3-hydroxyxanthine (XXXIX) and guanine 3-N-oxide (XL) (Figure 14):

"Parallel titrations at 1.0, 0.5, and 0.1 mg/week for 26 weeks, administered subcutaneously, in female Wistar rats show that, for these conditions, the 50% tumor incidence doses lie between the two lower dose levels, or between a total of 2 and 10 mg of the free bases" (47).

The activity of adenine 1-N-oxide (XLI) is not as great as that of 3-hydroxyxanthine or guanine 3-N-oxide, but is considered to be one of the "significantly oncogenic purine N-oxide derivatives" (47). The lack of oncogenic properties of some purine N-oxides was shown by the low incidence of tumors which resulted from massive doses of 6-mercaptopurine-3-N-oxide (XLII) (47).

IV. Conclusions

The minute amounts formed of some of these oxidized organic products and their (quite often) short half-lives may mean that many of the potential public health problems are insignificant. The public health significance of the formation of these organic compounds, in these low amounts, as related to the need for water and sewage treatment, has not been established. Nevertheless, these findings are reported here to point out the possibilities of producing toxic compounds in potentially toxic amounts by ozonizing water and wastewaters.

LITERATURE CITED

A. Backer, J.J., Chem. Weekblad, 36:214, 1939.

1. Bailey, P.S., Bach, S.S., and Ashton, J.B., "Initial attack of ozone on an unsaturated system," in "Ozone Chemistry and Technology", 143-148, American Chemical Society, Washington, D.C., 1959.

2. Bailey, P.S., "The reactions of ozone with organic compounds," Chem. Rev., 58:939, 1958.

3. Ibid., 958-960.

4. Ibid., 975.

5. Ibid., 977.

6. Ibid., 979-980, 984.

6a. Bartlett, P.D. and Stiles, M., "Highly branched molecules. IV. Solvolysis of the p-nitrobenzoate of tri-t-butylcarbinol and some of its homologs," J. Am. Chem. Soc. 77:2806-2814 (1955).

7. Bean, E.L., "Ozone production and costs," in "Ozone Chemistry and Technology", 430-442, American Chemical Society, Washington, D.C., 1959.

8. Belew, J.S., "Ozonization," in "Oxidation, Techniques and Application in Organic Synthesis", Robert L. Augustine, ed., Vol. I, 259-335, Marcel Dekker, Inc., New York, N.Y., 1969.

9. Broude, N.E., and Buniovskii, E.I., "Reaction of glyoxal with nucleic acid components. V. Denaturation of DNA under the action of glyoxal." Biochem. Biophys. Acta., 294(3):378-384, 1973.

10. Brown, G.B., Sugmura, K., and Cresswell, R.M., "Purine N-oxides. XVI. Oncogenic derivatives of xanthine and guanine," Cancer Res., 33:1113-1118, 1973.

11. Brown, G.B., Teller, M.N., Smullyan, I., Birdsall, N.J.M., Lee, T.-C., and J.C., Stöhrer, G., "Correlations between oncogenic and chemical properties of several derivatives of 3-hydroxyxanthine and 3-hydroxyguanine., Cancer Res., 33:1113-1118, 1973.

12. Buell, G.C., and Mueller, P.K., "Toxicity of ozone, a supplemental review," AIHL-Report No. 18. State of Calif., Dept. of Public Health, Div. of Laboratories, Air and Industrial Hygiene Laboratory, Berkeley, Calif., 1965.

13. Cerkinsky, S.N. and Trahtman, N., "The present status of research on the disinfection of drinking water in the USSR," Bull. World Health Org., 46(2):277-283, 1972.

14. Criegee, R., "Products of ozonization of some olefins," in "Ozone Chemistry and Technology", 133-135, American Chemical Society, Washington, D.C., 1959.

15. Dillemuth, F.J., Skidmore, D.R., and Schubert, C.C., "The reaction of ozone with methane," J. Phys. Chem., 64:1496-1499, 1960.

16. Egyud, L.G., and Szent-Györgyi, A., "Cell division, SH, ketoaldehydes and cancer," Proc. Natl. Acad. Sci. U.S., 55(2):388-393, 1966.

17. Ibid., 388.

18. US/EPA, "Suspected carcinogens in drinking water," Interim Report to Congress, Appendix I. Washington, D.C., 1975.

19. Ibid., Appendix VII.

20. Faith, W.L., and Renzetti, N.A., "Sixth technical progress report," Report #30, 47-52, Air Pollution Foundation, San Marino, Calif., 1960.

21. Flinn, J.E., and Reimers, R.S., "Expanded drinking water contamination," in "Development of Predictions of Future Pollution Problems", Section XII, 110, Office of Research and Development, US/EPA, Washington, D.C.

22. Goldstein, E., Eagle, M.C., and Hoeprich, P.D., "Influence of ozone on pulmonary defense mechanisms of silicotic mice," Arch Environ. Health, 24:444-448, 1972.

23. Goldstein, E., Tyler, W.S., Hoeprich, P.D., and Eagle, C., "Adverse influence of ozone on pulmonary bactericidal activity of murine lung," Nature, 229:262-263, 1971.

24. Guinvarc'h, P., "Three years of ozone sterilization of water in Paris," in "Ozone Chemistry and Technology", 416-429, American Chemical Society, Washington, D.C., 1959.

25. Hoffmann, J., and Eichelsdörfer, D., Vom Wasser, 38:197-206, 1971.

26. Hueter, F.G., and Fritzhand, M., "Oxidants and lung biochemistry, a brief review," Archives of Internal Medicine, 128:51, 1971.

27. Kabler, P.W., et al., Position paper on: chlorine, reaction with nitrogenous groups, and effects on DNA, U.S. Public Health Service, 1959.

28. Kinman, R.N., "Water and wastewater disinfection with ozone: a critical review," CRC Critical Reviews in Environmental Control, 1975.

29. Kotin, P., and Falk, H.L., "Organic peroxides, hydrogen peroxide, epoxides, and neoplasia., Radiation Research, Supplement 3:193-211, 1963.

30. Leighton, P.A., and Perkins, W.A., "Photochemical secondary reactions in urban air," Report No. 24, 59-90, Air Pollution Foundation, San Marino, Calif., 1958.

31. Long, L., Jr., "The ozonization reactions," Chem. Rev., 27:437-493, 1940.

32. Lowndes, M.R., "Ozone for water and effluent treatment, Chemistry and Industry (London), 34:951-956, 1971.

33. Maggiolo, A., and Blain, E.A., "Ozone oxidation of sulfides and sulfoxides," in "Ozone Chemistry and Technology", 200-201, American Chemical Society, Washington, D.C., 1959.

34. Milas, N.A., and Nolan, J.T., Jr., "Some abnormal ozonization reactions," in "Ozone Chemistry and Technology", 136-139, American Chemical Society, Washington, D.C., 1959.

35. Mosher, W.A., "Structural relationships in addition of ozone to double bonds," in "Ozone Chemistry and Technology", 140-142, American Chemical Society, Washington, D.C., 1959.

36. Mukai, F.H., and Goldstein, B.D., "Mutagenicity of malonaldehyde, a decomposition product of peroxidized polyunsaturated fatty acids," Science, 191:868-869, 1976.

37. Anon., "Ozone treatment of potable water," PCI Ozone Corp., W. Caldwell, N.J., 1974.

38. Pate, C.T., Atkinson, R., and Pitts, J.N., Jr., "The gas phase reaction of ozone with a series of aromatic hydrocarbons," J. Am. Environmental Science and Health, Part A, Vol. A-11(1):1-10, 1976.

39. Razumovskii, S.D., Kefeli, A.A., and Zaikov, G.E., Zh. Org. Khim., 7(10):2044-2081, 1971. (Russ.)

40. Anon., "Recycled water poses disinfectant problem," Chem. & Engr. News, 45-46, Sept. 3, 1973.

41. Roberts, J.D., Stewart, R., and Caserio, M.C., "Organic Chemistry: Methane to Macromolecules", 279, W.A. Benjamin, Inc., Menlo Park, Calif., 1971.

42. *Ibid.*, Ch. 20.

43. *Ibid.*, 597.

44. Roehm. J.N., Hadley, J.G., and Menzel, D.B., "Oxidation of unsaturated fatty acids by ozone and nitrogen dioxide," Arch. Environ. Health, 23(2):142-148, 1971.

45. Schubert, C.C., and Pease, R.N., "The oxidation of lower paraffin hydrocarbons. I. Room temperature reaction of methane, propane, n-butane, and isobutane with ozonized oxygen," J. Amer. Chem. Soc., 78:2044-2048, 1956.

45a. Shamberger, R. J., Andreone, T.L. and Willis, C.E., "Antioxidants and cancer. IV. Initiating activity of malonaldehyde as a carcinogen. J. Natl. Cancer Inst. 53(6):1771-78, 1974.

46. Stokinger, H.E., "Factors modifying toxicity of ozone," in "Ozone Chemistry and Technology", 360-369, American Chemical Society, Washington, D.C., 1959.

47. Sugiura, K., Teller, M.N., Parham, J.C., and Brown, G.B., "A comparison of the oncogenicities of 3-hydroxyxanthine, guanine 3-N-oxide, and some related compounds," Cancer Res., 30:184-188, 1970.

48. Swern, D., "Primary products of olefinic autoxidations," in "Autoxidation and Antioxidants", I:1-54, W.O. Lundberg, ed., Interscience Publishers, Div. of John Wiley and Sons, New York, N.Y., 1961.

49. Szent-Gyorgyi, A., "Studies in growth," Proc. Robert A. Welch Found., Conf. Chem. Res., 8:167-76, 1965.

50. Torricelli, A., "Drinking water purification," in "Ozone Chemistry and Technology", 453-465, American Chemical Society, Washington, D.C., 1959.

51. Van Middelem, C.H., "Fate and persistence of organic pesticides in the environment", 228-249, American Chemical Society, Washington, D.C., 1966.

52. Zelac, R.E., Cromroy, H.L., Bolch, W.E., Jr., Dunavant, B.G., and Bevis, H.A., "Inhaled ozone as a mutagen. II. Effect on the frequency of chromosome aberrations observed in irradiated Chinese hamsters," Environmental Research, 4:325-342, 1971.

FIGURE 1

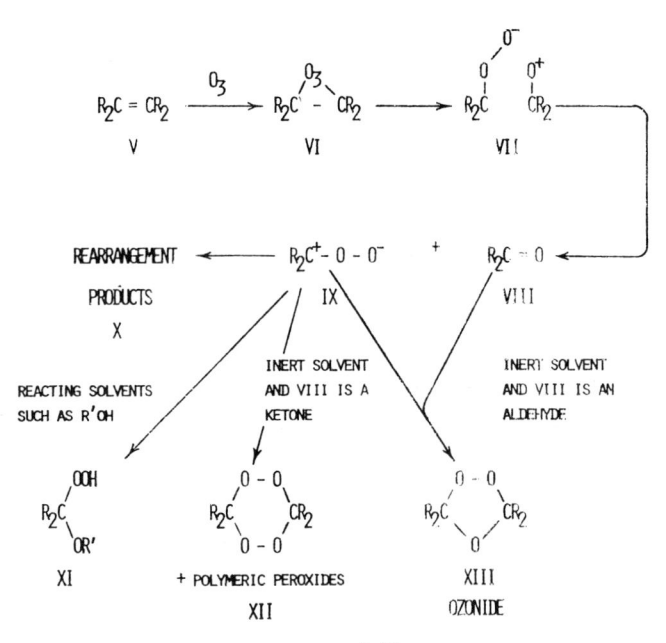

FIGURE 2 (1)

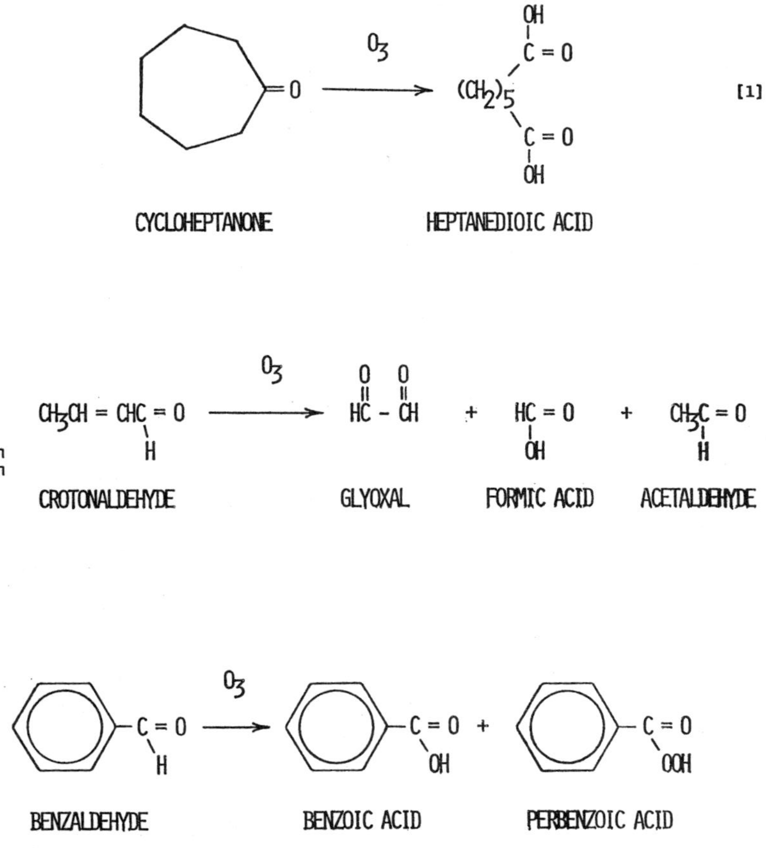

[1]

$$\text{cycloheptanone} \xrightarrow{O_3} \underset{\text{HEPTANEDIOIC ACID}}{\underset{OH}{\overset{C=O}{|}}-(CH_2)_5-\underset{OH}{\overset{C=O}{|}}}$$

CYCLOHEPTANONE

[2]

$$CH_3CH=CHC\underset{H}{\overset{=O}{|}} \xrightarrow{O_3} HC\overset{O}{-}C\overset{O}{H} + HC\underset{OH}{\overset{=O}{|}} + CH_3C\underset{H}{\overset{=O}{|}}$$

CROTONALDEHYDE   GLYOXAL   FORMIC ACID   ACETALDEHYDE

[3]

$$\text{BENZALDEHYDE} \xrightarrow{O_3} \text{BENZOIC ACID} + \text{PERBENZOIC ACID}$$

BENZALDEHYDE   BENZOIC ACID   PERBENZOIC ACID

[4]

$$CH_3CHC\underset{H}{\overset{=O}{|}} \xrightarrow{O_3} CH_3CHC\underset{OH}{\overset{=O}{|}} + CH_3CHC\underset{OOH}{\overset{=O}{|}}$$
$$\underset{CH_3}{} \quad\quad \underset{CH_3}{} \quad\quad \underset{CH_3}{}$$

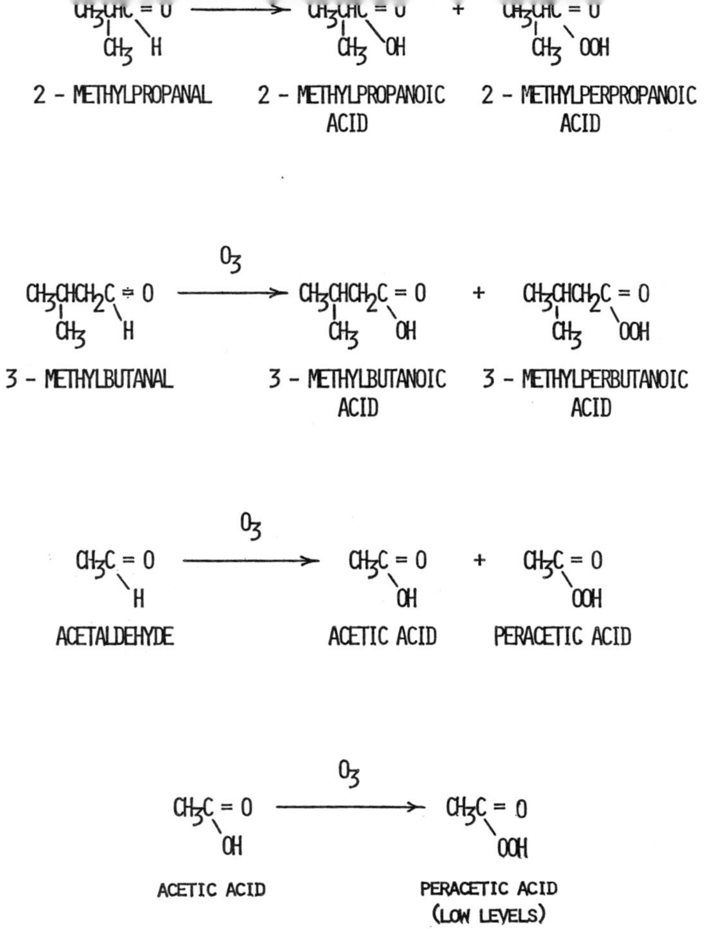

2 - METHYLPROPANAL   2 - METHYLPROPANOIC ACID   2 - METHYLPERPROPANOIC ACID

[5]

$$CH_3CH_2C\underset{H}{\overset{=O}{|}} \xrightarrow{O_3} CH_3CH_2C\underset{OH}{\overset{=O}{|}} + CH_3CH_2C\underset{OOH}{\overset{=O}{|}}$$
$$\underset{CH_3}{} \quad\quad \underset{CH_3}{} \quad\quad \underset{CH_3}{}$$

3 - METHYLBUTANAL   3 - METHYLBUTANOIC ACID   3 - METHYLPERBUTANOIC ACID

[6]

$$CH_3C\underset{H}{\overset{=O}{|}} \xrightarrow{O_3} CH_3C\underset{OH}{\overset{=O}{|}} + CH_3C\underset{OOH}{\overset{=O}{|}}$$

ACETALDEHYDE   ACETIC ACID   PERACETIC ACID

[7]

$$CH_3C\underset{OH}{\overset{=O}{|}} \xrightarrow{O_3} CH_3C\underset{OOH}{\overset{=O}{|}}$$

ACETIC ACID   PERACETIC ACID (LOW LEVELS)

Figure 2 (cont'd)

Figure 2 (cont'd)

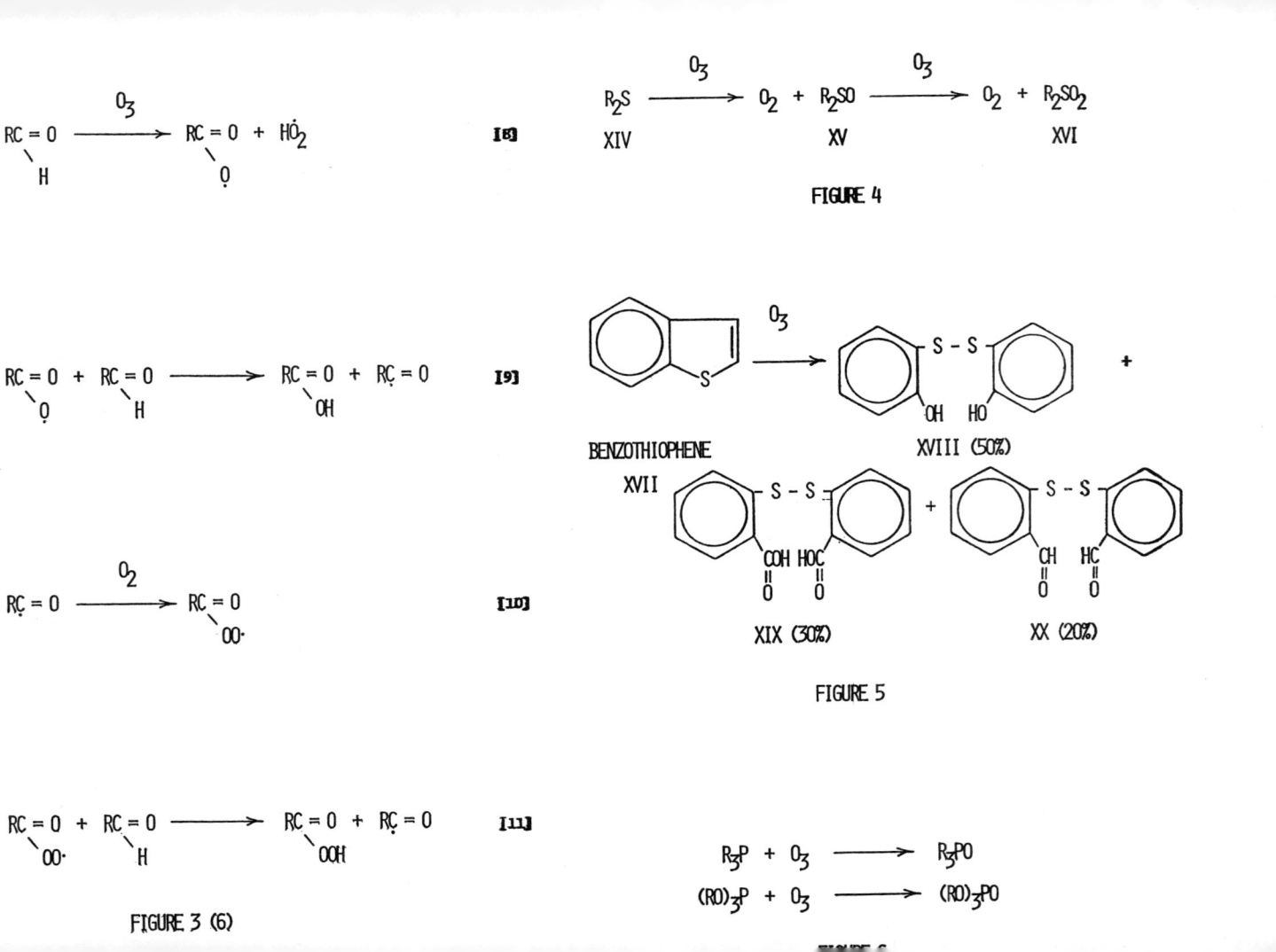

[8]

$$RC=O \xrightarrow{O_3} RC=O + H\dot{O}_2$$
$$\ |\phantom{\xrightarrow{O_3}}\ |$$
$$\ H\phantom{\xrightarrow{O_3}RC=O+H\dot{O}_2}\dot{O}$$

[9]

$$RC=O + RC=O \longrightarrow RC=O + R\dot{C}=O$$
$$\ |\ |\ |$$
$$\dot{O}HOH$$

[10]

$$R\dot{C}=O \xrightarrow{O_2} RC=O$$
$$\phantom{R\dot{C}=O\xrightarrow{O_2}RC}\ |$$
$$\phantom{R\dot{C}=O\xrightarrow{O_2}RC}O\dot{O}$$

[11]

$$RC=O + RC=O \longrightarrow RC=O + R\dot{C}=O$$
$$\ |\ |\ |$$
$$O\dot{O}HOOH$$

FIGURE 3 (6)

$$R_2S \xrightarrow{O_3} O_2 + R_2SO \xrightarrow{O_3} O_2 + R_2SO_2$$
XIV XV XVI

FIGURE 4

BENZOTHIOPHENE $\xrightarrow{O_3}$ XVIII (50%) + XIX (30%) + XX (20%)

XVII

FIGURE 5

$$R_3P + O_3 \longrightarrow R_3PO$$
$$(RO)_3P + O_3 \longrightarrow (RO)_3PO$$

-56-

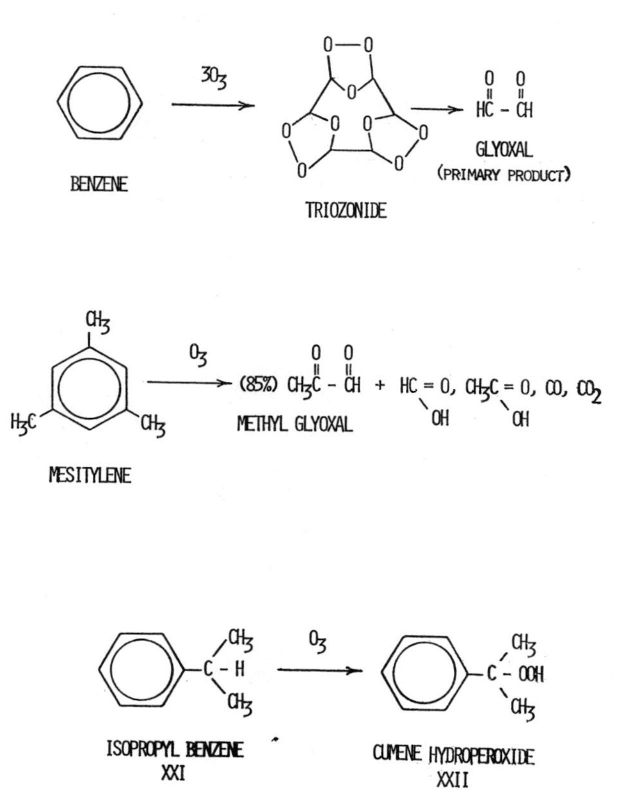

-57-

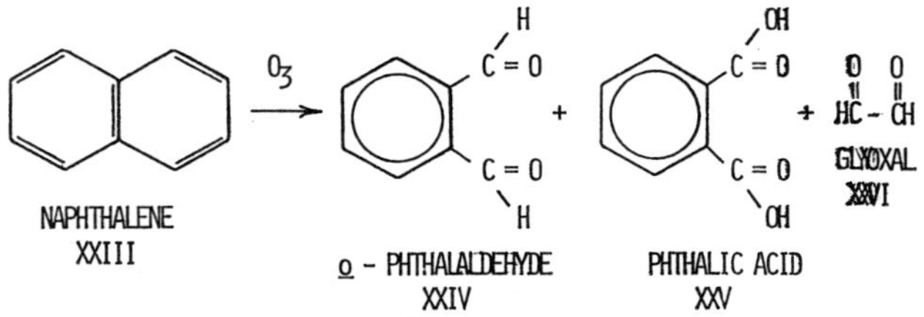

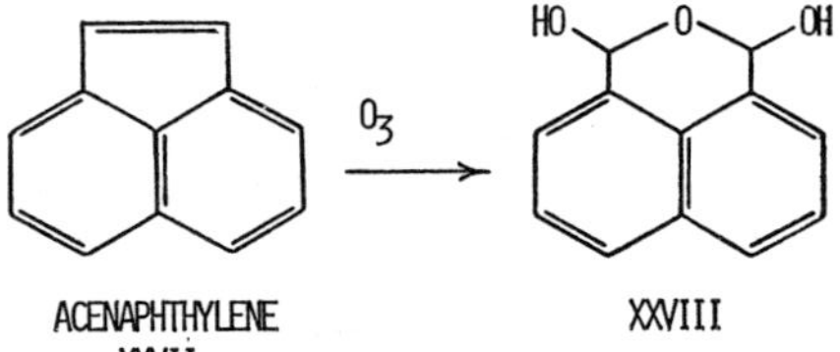

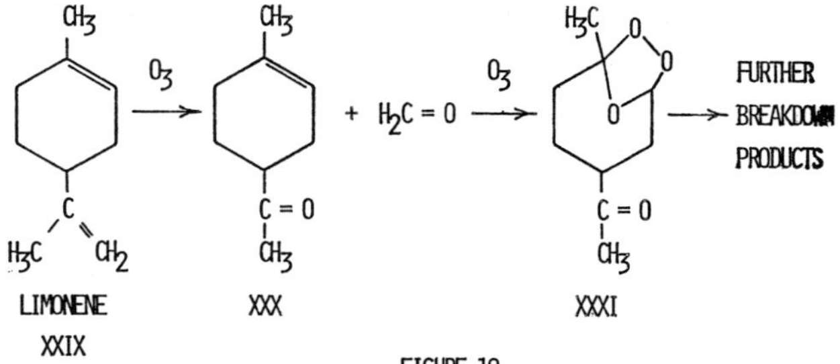

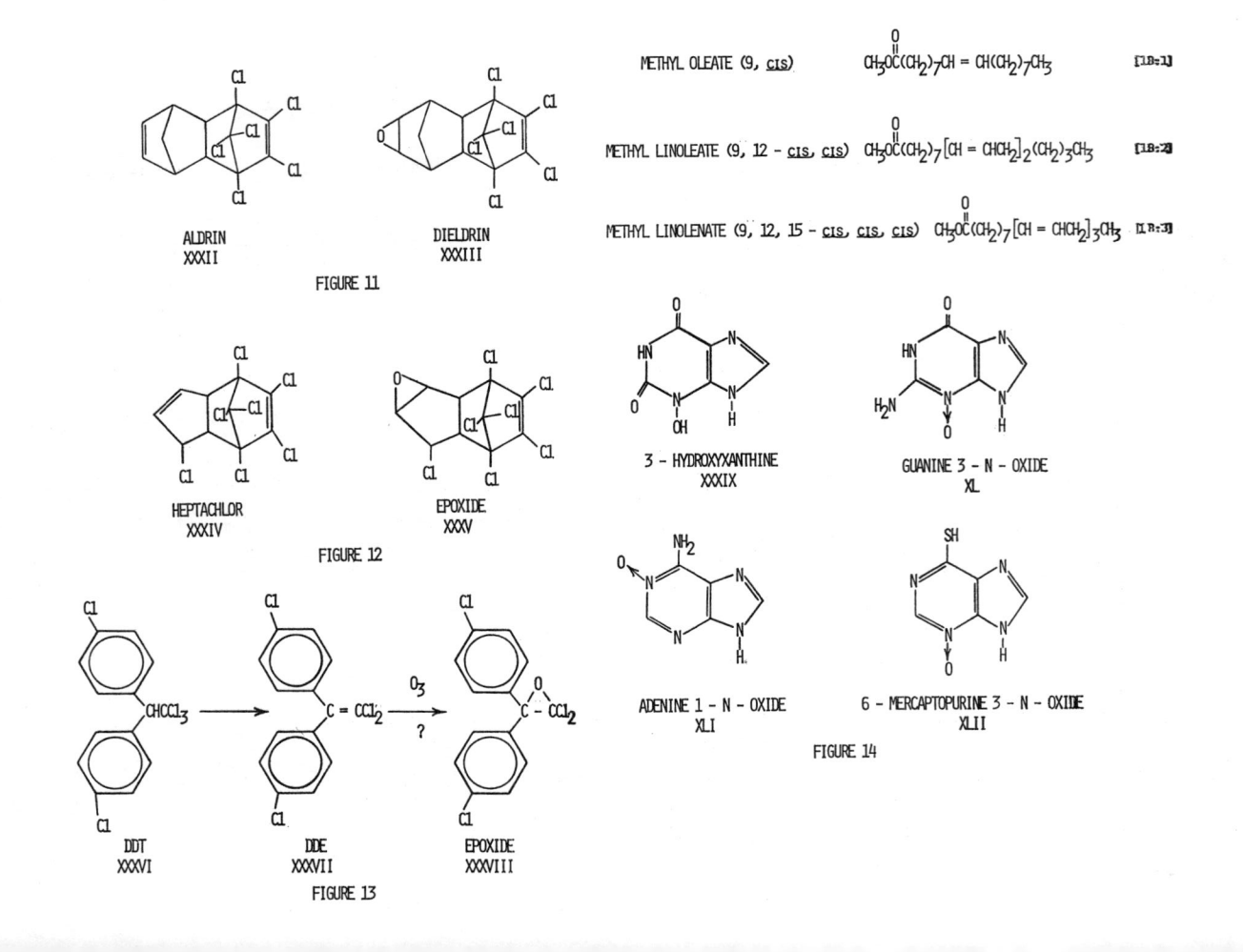

METHYL OLEATE (9, cis)   CH$_3$OCC(CH$_2$)$_7$-CH=CH(CH$_2$)$_7$-CH$_3$ [LR-1]

METHYL LINOLEATE (9, 12 - cis, cis)   CH$_3$OCC(CH$_2$)$_7$[CH=CHCH$_2$]$_2$(CH$_2$)$_3$-CH$_3$ [LR-2]

METHYL LINOLENATE (9, 12, 15 - cis, cis, cis)   CH$_3$OCC(CH$_2$)$_7$[CH=CHCH$_2$]$_3$-CH$_3$ [LR-3]

GUANINE 3 - N - OXIDE
XL

6 - MERCAPTOPURINE 3 - N - OXIDE
XLII

3 - HYDROXYXANTHINE
XXXIX

ADENINE 1 - N - OXIDE
XLI

FIGURE 14

DIELDRIN
XXXIII

EPOXIDE
XXXV

FIGURE 11

ALDRIN
XXXII

HEPTACHLOR
XXXIV

FIGURE 12

EPOXIDE
XXXVIII

DDE
XXXVII

DDT
XXXVI

FIGURE 13

-58-

OZONE'S RADICAL AND IONIC MECHANISMS OF

REACTION WITH ORGANIC COMPOUNDS IN WATER

by

Allison Maggiolo

Bennett College
Greensboro, North Carolina 27420

Abstract: Depending upon the pH, metals present, types of organic compounds and their relative concentrations in organic wastewaters, ozone may usually react by several mechanisms at the same time. A rationale is discussed regarding when and why one of the several possible paths may be tailored to predominate so as to utilize a practical minimum amount of ozone for aqueous pollution abatement and to minimize the potential formation of toxic post-ozonized breakdown products, such as epoxides.

Introduction

The reaction mechanisms operative during ozonation of organic compounds in very dilute wastewater solutions as well as the resulting products, usually are different from those reported in the literature. This is because organic chemical reactions reported in the literature normally are conducted in non-aqueous solvents. Thus, one notes in the literature that a particular compound, fairly concentrated in an organic solvent, when ozonized has been reported to yield an ozonide, peroxide, (their polymers) and in rare cases an epoxide. However, these particular types of products have never been reported when the same starting compound is ozonized in very dilute aqueous solutions of about 100 ppm (2 parts of compound to 10,000 parts of water). Under these conditions there is no doubt that all this water will, in almost every case, participate and be an initial part of the ozonized intermediate. Water may participate in the reaction by an <u>ionic</u> and/or a <u>free radical</u> mechanism depending on the pH or the presence of redox metal ion salts (such as iron, manganese, copper etc.).

Ionic Mechanism

Ozone reacts ionically with the unsaturated bond in organic compounds to give an ozonized intermediate which immediately reacts with water.

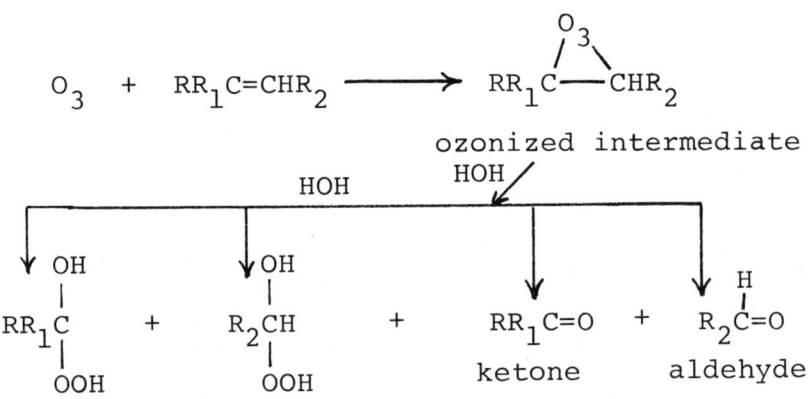

$$O_3 + RR_1C=CHR_2 \longrightarrow RR_1C\overset{O_3}{-}CHR_2$$

ozonized intermediate

HOH / HOH

hydroxyhydroperoxides:

of ketones: $RR_1C(OH)(OOH)$

of aldehydes: $R_2CH(OH)(OOH)$

$RR_1C=O$ ketone + $R_2C(H)=O$ aldehyde

The ionic mechanism with water is favored by one or more of the following conditions:

(1) pH<8
(2) no redox metal ions present (Fe^{++}, Fe^{+++}), (Mn^{++}, Mn^{+++}), (Cu^+, Cu^{++}) etc.;
(3) The compounds being attacked by ozone must have olefinic or acetylenic bonds or contain aromatic unsaturation that are somewhat reactive to electrophilic reagents (Friedel-Crafts halogenation, nitration, sulfonation, etc).

The ozonized water adducts as shown above are mixtures of hydroxyhydroperoxides and an aldehyde (if the ozonized unsaturated carbon atoms have a hydrogen attached) or a ketone (if they had no hydrogens attached). In the presence of all the excess oxygen accompanying the ozone and water the hydroxyhydroperoxides and aldehydes readily will react further while the stable ketones will tend to remain. The hydroxyhydroperoxides may react ionically in several ways shown below by the ionic conditions above, and later as discussed under "Free Radical Mechanisms" in several other ways.

$$R_1COOH + H_2O \longleftarrow R_1\underset{OOH}{\overset{OH}{CH}} \rightleftharpoons HOOH + R_1CHO$$

acid   hydrogen peroxide   aldehyde

$$2R_1CHO + O_3 \longrightarrow R\overset{O}{\underset{}{C}}OOH + RCOOH$$
                         peracid   acid

$$R_1CHO + O_2 \longrightarrow R_1\overset{O}{\underset{}{C}}OOH$$

$$R_1CHO + HOOH \longrightarrow R_1COOH \text{ (acid)}$$

$$R\overset{O}{\underset{}{C}}OOH + RCHO \longrightarrow 2RCOOH$$

If there is a hydrogen on the carbon atom attached to the hydroxy and hydroperoxide group it may rearrange directly to its acid (1), or be in equilibrium with its aldehyde and hydrogen peroxide as shown above. The aldehydes formed may react with ozone, oxygen or the hydrogen peroxide. With ozone the aldehyde will be oxidized to the peracid and acid (2); with oxygen to the peracid; and with hydrogen peroxide to the acid. Also, these peracids are readily reactive with any aldehydes to give acids. Thus the hydroperoxides and aldehydes are readily changed into acids. However, if there is no hydrogen on the carbon attached to the hydroxy and hydroperoxide group, the ozonized intermediate will revert to its ketone, which is stable, and the resulting hydrogen peroxide, which is found, will be used up in oxidizing some aldehyde, other susceptible organic compound, or break down to give hydroxyl free radicals.

In those rare cases in which ozone has been reported to ozonize highly hindered double bonds to yield <u>epoxides</u> directly, it must be acutely remembered that the solvent used was not water (3,4,5). First of all the reaction rate of ozone with these hindered double bonds is, relatively, much slower than with unhindered double bonds; thus ozone probably will attack something else in actual industrial or municipal wastes. More importantly, if ozone does react with the hindered double bond of the compound, the presence of about 10,000 parts of water (HOH) will allow the HOH to participate in the reaction to give either the glycol or its resulting aldehyde, but <u>no</u> epoxide. In fact, there has not yet been reported in the literature any epoxy formation under these overwhelmingly dilute aqueous conditions. In addition, it is believed that any <u>standard</u> epoxy-forming agents reacting under these highly dilute aqueous conditions would also form a glycol instead of an epoxide with either unhindered or hindered double bonds.

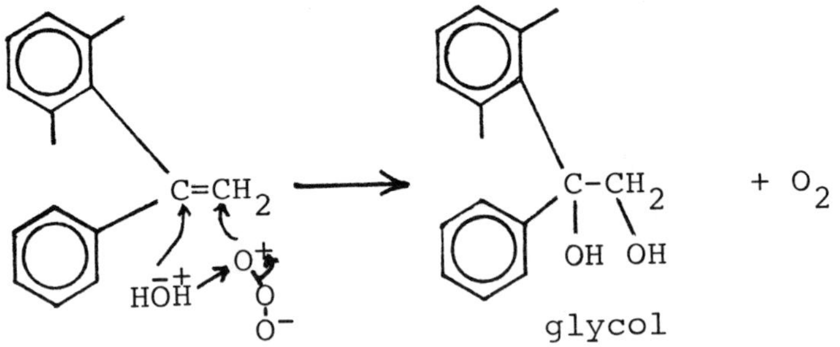

Free Radical Mechanism

Instead of ozone attacking both carbons of the double bond, cleaving the compound at the double bond to give acids and ketones as shown above, ozone may attack in several ways by free radical pathways with no cleavage at all. The free radical attack of organic compounds in water is favored at pH below 9, and in the presence of redox metal ions, oxygen, peroxides and sunlight.

Hoigné has shown that ozone tends to react with HOH at pH above 8 and especially above pH 9 to give, almost quantitatively, equal amounts of hydroxyl radicals (·OH) and hydroperoxide radicals (·OOH) instead of reacting with the double bond of aromatic compounds (6). With olefinic double bonds the reaction is not so clear-cut, in that ozone will react both with the olefinic double bond as well as with the water.

$$O_3 + HOH \xrightarrow{pH<9} \underset{\text{hydroxyl radical}}{\cdot OH} + \underset{\text{hydroperoxide radical}}{\cdot OOH}$$

Of the two radicals the (·OH) is extremely reactive and will attack any carbon having an active hydrogen or a double bond indiscriminately as compared to ozone, which is slower and selective for active double bonds. The hydroperoxide radical (·OOH) will react in the same general way as the hydroxyl radical but at a slower rate. These radicals react with organic compounds by addition to double bonds or by abstracting an active hydrogen to give organic free radical compounds. These organic free radicals can react in several ways to yield hydroxy and carbonyl compounds as shown later below.

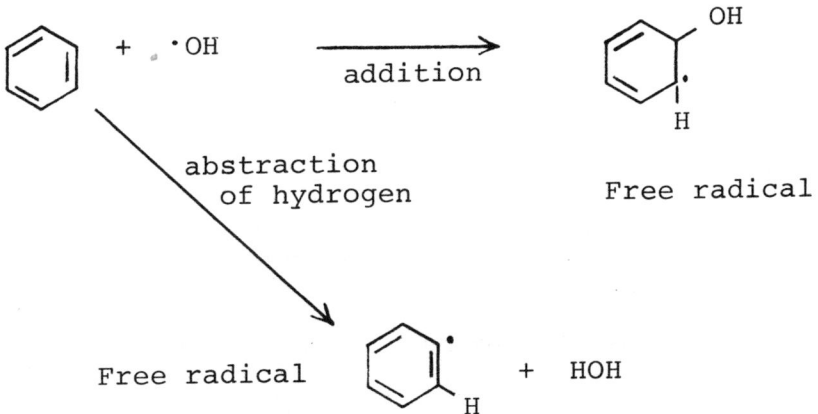

Oxygen is present at 20 and 50 times the amount of ozone in ozonized air or ozonized oxygen, respectively. Oxygen will participate in the reactions not only by oxidizing the aldehydes (formed upon ozonation) to acids, but will also propagate organic free radicals that react further with other molecules, or with some more oxygen to form active peroxy radicals.

$$O_2 + RH \xrightarrow{\text{(1) Metal ions, (2) Sunlight, (3) } O_3} R\cdot + \cdot OOH$$

$$R\cdot + R'H \longrightarrow R'\cdot\cdot + RH$$

$$R'\cdot\cdot + O_2 \longrightarrow R'OO\cdot$$

$$R'OO\cdot + RH \longrightarrow R'OOH + R\cdot$$

The hydroperoxide formed will break down to aldehydes, ketones and alcohols.

$$\underset{H}{\overset{H}{RCOOH}} \longrightarrow \underset{H}{\overset{H}{R\text{-}CO\cdot}} + \cdot OH \text{ (active hydroxyl-radical)}$$

$$\longrightarrow RCH_2OH + RH$$
$$\text{alcohol}$$

$$RCH_2OOH \longrightarrow RCHO + HOH$$
$$\text{aldehyde}$$

$$RCH_2R'OH \longrightarrow RCOR' + HOH$$
$$\text{ketone}$$

In the presence of iron, manganese or copper salts or other redox metals usually found in industrial wastewaters, the following type of reaction takes place to give alkoxy and peroxy radicals which will react further as shown above.

$$Fe^{++} + ROOH \longrightarrow Fe^{+++} + (OH)^- + RO\cdot$$
$$\text{alkoxy radical}$$

$$Fe^{+++} + ROOH \longrightarrow Fe^{++} + H^+ + ROO\cdot$$
$$\text{peroxy radical}$$

<u>Typical Examples of Free Radical vs Ionic Ozonation</u>

Niegowski (7) has shown that much less ozone is needed to destroy phenol at pH 12 than at pH 7. It took 1000 ppm of ozone to destroy 99% of 600 ppm of phenol at pH 12, while 2000 ppm of ozone was required to destroy 99% at pH 7. Thus, in this case the free radical hydroxyl formation plus the excess oxygen was more efficient than the ionic ozone cleavage of a double bond.

Hay et al. (8) have shown that ozone in the presence of cobaltous acetate reacts as a continuous free radical initiator of oxygen oxidation of p-methoxytoluene in acetic acid solvent to give exclusively p-methoxytoluic acid. There was no ionic attack at, or cleavage of, the aromatic double bonds.

$$CH_3O-\bigcirc-CH_3 \xrightarrow[Co^{++}/Co^{+++}]{O_3/O_2} CH_3O-\bigcirc-COOH$$

Plakidin et al. (9) have shown that they could eliminate the breaking down of the very susceptible (to ozone) condensed aromatic rings in 2,2'-dibenzanthronyl with an ozone-oxygen mixture in the presence of manganese sulfate in aqueous sulfuric acid. They obtained an excellent yield of the di-phenolic (hydroxy) compound, 16,17-dihydroxyviolathrone.

2,2'-dibenzanthronyl   16,17-dihydroxyviol-
                                                              anthrone

Conclusions

Ozonation of unsaturated aliphatic or aromatic compounds in the presence of <u>large excesses</u> of water (as in municipal or diluted industrial wastes) will <u>not</u> yield ozonides, peroxides or epoxides. Instead, water will participate in the reaction, as will excess oxygen (which accompanies the ozone), to yield mostly acids and ketones along with smaller amounts of alcohols.

However, if a particular industrial <u>aromatic</u> waste to be ozonized should have a pH above 9 and/or contain metal redox salts (such as iron, manganese and copper), then a certain amount of hydroxyaromatic (phenolic) structures will form. These would tend to be toxic, unless further oxidized by additional ozone or by acclimated biological activated waste treatment.

ACKNOWLEDGEMENTS

The aid extended by Dr. Rip G. Rice in suggesting this presentation, supplying key references and preparing this manuscript for publication is deeply appreciated.

Also, the numerous references and up to date background furnished by Drs. Phillip C. Singer and J. Donald Johnson, of the School of Public Health, University of North Carolina at Chapel Hill, N.C. was invaluable.

LITERATURE CITED

(1) Criegee, R., Ann. der Chemie, 583:12 (1953).

(2) Düll, H., "Ozone Reaction and Ozonides", Translation, Inaugural-Dissertation (1933), Albert Ludwigs University.

(3) Bartlett, P.D., and Stiles, M., J. Am. Chem. Soc. 77:2806 (1955).

(4) Kompa, G., and Roschier, R.H., Ann. 470:129 (1929).

(5) Fuson, R.C. <u>et al</u>., J. Am. Chem. Soc. 66:1274 (1944).

(6) Hoigné, J., "Identification and Kinetic Properties of the Oxidizing Decomposition Products of Ozone in Water and Its Impact on Water Purification", Proc. Second Intl. Symposium on Ozone Technology, R.G. Rice, P. Pichet & M.A. Vincent, Editors, Intl. Ozone Inst., Cleveland, Ohio (1976) 271-282.

(7) Niegowski, S.J., "Destruction of Phenols by Oxidation with Ozone", Indl. & Engr. Chem. 45:632-634 (1953).

(8) Hay, A.S., et al., J. Org. Chem. 25:616-7 (1960).

(9) Plakidin, V.L., et al., "Catalytic Oxidation of 2,2'-Dibenzanthronyl with Ozone-Oxygen Mixture", Kinetika i Kataliz 3(2):292-295 (1962) (Russian).

BROMOFORM PRODUCTION BY OXIDATIVE BIOCIDES IN MARINE WATERS*

G.R. Helz and R.Y. Hsu
Department of Chemistry
University of Maryland
College Park, MD 20742

and

R. M. Block
Chesapeake Biological Laboratory
University of Maryland
Solomons, MD 20688

Abstract

Experiments with natural estuarine water which has been treated with chlorine or with ozone establish that bromoform is produced in both cases. This apparently results from oxidation of bromide in seasalt followed by substitution of the oxidized bromine into organic matter. When the two oxidants are applied so as to produce comparable residuals, the bromoform yield from each is quite similar. Negligible amounts of chlorine-containing haloforms

---

*Paper not presented at Workshop, but included in Proceedings because of the close relationship of its subject matter to the formation of bromo-containing trihalomethanes and other bromo-organic compounds found in water and wastewater.

are produced. Dechlorination with $Na_2S_2O_3$ after an average contact time of 20 minutes does not reduce bromoform yield, indicating that the organic substitution reactions reach completion in this length of time. The results suggest that little reduction in haloform yields can be achieved by replacing chlorine with ozone in seawater applications and that to eliminate haloform production through dechlorination will require that the chlorination-dechlorination cycle be complete in a period probably shorter than 100 seconds.

Introduction

Recent research has shown that as much as 1-2% of the chlorine which is used for disinfection and other biocidal purposes during the treatment of drinking waters and wastewaters becomes bound to carbon in chloro-organic compounds (Jolley, 1974; Bellar et al., 1974). A number of such compounds, of which chloroform is usually the most prominent, have now been identified. Based on an annual usage in the U.S. of 56×10^{10} g of chlorine for water treatment (Klingman, 1975) and assuming 1% conversion of this to chloroform, the inadvertent and dispersive generation of chloroform in this manner would exceed 5% of the industrial production in the United States. Because there is some concern about the environmental impact and human health effects of the chloro-organic products, many scientists, engineers and public health officials are actively exploring the feasibility of using other biocides, such as ozone, in water treatment.

In coastal regions, electric power plants are among the largest and most rapidly growing users of chlorine. Because of the high biologic productivity of coastal marine waters, it is usually necessary to chlorinate these waters on a frequent or even continuous basis prior to passage through power plant cooling systems in order to prevent settling of fouling organisms such as barnacles and mussels (White, 1972). The present paper presents initial results from a study aimed at determining whether haloform compounds are produced in marine waters as they are in drinking waters, and whether replacement of chlorine by ozone will prevent this problem.

Methods and Results

Natural estuarine water was used in all experiments. Chlorinated, ozonated, and untreated control solutions were examined in parallel experiments, run simultaneously so as to insure that prior to treatment the water quality was the same in each case. For the chlorine experiments, Ca(OCl)$_2$ stock solution (1.06 equiv/l) was metered at 0.5 ml/min into a 3.8 l/min stream of estuarine water which flowed continuously through an 80 l glass tank; thus the dosage was 139 m-equiv/l or about 5 ppm Cl. Turnover time of water in this tank was roughly 20 minutes. The system was allowed to run for several hours until a steady state level of oxidant in the tank was observed. Then a sample was collected in a glass bottle and sealed so that no air-space remained in the bottle.

A second sample was obtained at the same time, but dechlorinated with Na$_2$S$_2$O$_3$ before sealing. Next, a dozen White Perch (<u>Morone americana</u>) were introduced into the tank; after 30 minutes two additional water samples (one of which was quenched with Na$_2$S$_2$O$_3$) were taken and sealed. In the ozone experiments, the above procedure was conducted in a similar fashion except that the flowing water was treated with the oxygen-ozone gas effluent from a Welsbach T-816 ozonator; dose was adjusted to give approximately the same oxidant level as in the chlorine testing tank. Turnover time of water in the ozone tank was 8 minutes. With the controls, all procedures, such as taking of a quenched and an unquenched sample and testing both in the presence and absence of fish, were followed.

The samples were analyzed for volatile halo-organics 18-24 hours after collection. Forty ml of each sample was rapidly transferred into a 50 ml vial and sealed with a septum cap. These vials were equilibrated in a water bath at 60°C, after which 1.0 ml of the head space gas was injected into a gas chromatograph equipped with a 6 ft Tenax GC column and with a Hall Electrolytic Conductivity Detector (Tracor Inc.) operated in the halogen mode. Quantification was accomplished with the aid of a digital integrator. Precision, based on repeated tests with standards, was about ±3%. Identification of products was based solely on retention time, although during earlier developmental work on the analytical procedure using estuarine samples, identifications were confirmed by mass spectrometry. Because the headspace technique and the selective detector used in these experiments discriminate against all compounds other than volatile halogenated organics, the likelihood of error in the identifications is very small.

As shown in Table 1, bromoform concentrations in the neighborhood of 20 µg/l were found in all the samples which had been treated with hypochlorite or ozone. No gas chromatographic peaks belonging to other compounds were observed in any samples. Furthermore, bromoform was not detectable in the controls. This indicates that bromoform is produced by the oxidants and is not introduced by sample handling or contamination.

In both the hypochlorite and the ozone tests, introduction of fish lowered the steady state oxidant level and the amount of bromoform observed. In the unquenched samples, the yield of bromoform is essentially the same for both hypochlorite and ozone. Furthermore, in the hypochlorite experiments, quenching with $Na_2S_2O_3$ had little effect on the yield.

Discussion

The virtual absence of any chlorine-containing haloforms contrasts with results from drinking waters where $CHCl_3$ normally predominates and $CHBr_3$ is the least abundant of the four possible chloro-bromo haloforms (Symons, et al., 1975). It is well known that chlorination of marine waters results in oxidation of dissolved bromide to hypobromous acid (Carpenter and Macalady, 1976; Eppley et al., 1976; Duursma and Parsi, 1976; etc.). Similarly, ozone appears to generate hypobromous acid in seawater (Blogoslawski, et al., 1976). Theoretical calculations based on thermodynamic and kinetic data suggest that when normal seawater (pH 8) containing 840 µ-mol/l of bromide is treated with a typical chlorine dose of perhaps 100 µ-equiv/l, reduction of free chlorine species by bromide should be virtually quantitative and should reach 99% completion in 10 seconds or so (Sugam and Helz, 1977, and in preparation).

The almost total absence of substituted chlorine in the products of our experiments therefore implies that the substitution reactions which lead to haloform production are much slower than the conversion of free chlorine species to oxidative bromine species. On the other hand, the virtually identical haloform yields in the quenched and unquenched chlorine samples indicate that the substitution reaction must be fast compared to turnover time of water in the experimental tank (i.e., 20 minutes, or roughly 1000 seconds), because this is the average time between addition of chlorine and addition of $Na_2S_2O_3$ to the quenched sample. Thus, these observations bracket the order of magnitude of the time needed to complete the halogen substitution at n x 100 seconds.

It is important to distinguish between the substitution reactions discussed above and the haloform production reactions. Laboratory studies of haloform production from chlorinated solutions of fulvic or humic acids, believed to be the sources of carbon in the haloforms, suggest that this production requires hours to reach completion (Stevens, et al., 1976; Rook, 1976 and 1977). Furthermore, Kopfler, et al. (1976) and Nicholson, et al. (1977) have provided evidence for the existence of a long lived halo-organic intermediate which decays to release haloforms upon heating. Thus the substitution of halogen atoms into organic structures in water and the release of haloform molecules from those organic structures appear to be separate processes with quite different rates. The latter process apparently does not require the presence of an oxidant residual. This implies that in seawater, at least, it would be difficult to limit haloform production by instituting dechlorination practice at power plants unless the entire chlorination-dechlorination cycle can be accomplished in a time span somewhat less than 100 seconds.

The excellent agreement in bromoform yield from ozone and chlorine is somewhat surprising. In the case of the unquenched samples, the yield of bromoform is the same within experimental error. For the quenched samples, slightly less bromoform was obtained from ozone than from chlorine, but this probably reflects simply the shorter turnover time of water in the ozone-treated tank. It would seem from these data that replacement of chlorine by ozone as an antifoulant in marine power plants offers little advantage so far as minimizing volatile halo-organics is concerned, unless it can be shown that ozone is biocidally more effective than chlorine so that it can be used at lower doses.

The chief purpose of placing fish in the test tanks after the first set of samples had been collected was to determine if reaction of the oxidants with excretory products, body slimes or tissue produced a large increase in the haloform yield or in the Cl/Br ratio in the products. A large enhancement in yield could make haloforms potential toxic factors in bioassay experiments intended to assess the acute toxicity of chlorine and ozone to fish. However, the data indicate small decreases in bromoform content in the tanks when fish are present. A small decline in the residual oxidant is also observed. One possible explanation is that fish produce compounds which react rapidly with oxidative biocides, but which do not generate volatile halo-organic compounds. Ammonia, urea, and other nitrogenous products are likely candidates. Probably they compete with the naturally occurring bromoform precursors for the available halogen and thus lower bromoform yield. An alternate hypothesis for the decrease in bromoform may be uptake of this compound by the fish, either through absorption into body slimes or transport into the body via the mouth or gills.

The total yield of volatile halo-organic material in these experiments is relatively small compared to results with drinking waters. On a molar basis, 0.4% of the chlorine dose becomes converted into bromoform. At present, several hypotheses can be offered to account for this. For example, the humic acids, which may be the most important haloform precursors, tend to be flocculated and removed from estuarine water (Sholkovitz, 1976), and thus the yields may be limited by the availability of appropriate precursors. Alternatively, competing reactions may deprive available haloform precursors of reactive halogen. Finally, it is possible that the organic intermediates may be more stable than those produced in fresh waters and thus may be slower in releasing haloforms. This last alternative could have important biological consequences if the intermediates are toxic.

Conclusions

Our results suggest the need for many additional biological and chemical experiments, some of which we are now pursuing. However, several conclusions can be drawn at this time:

1. Haloform production can occur in marine waters as well as in fresh waters.

2. Bromine is almost exclusively the halogen substituent in haloforms generated in marine waters. This is consistent with theory and experiments which show that Br^- in seasalt is readily oxidized by chlorine and ozone.

3. In our experiments, comparable oxidant residuals from ozone and chlorine yielded comparable amounts of bromoform. Thus there may be no advantage to replacing chlorine with ozone in seawater applications if the main purpose is to minimize haloform concentrations.

4. The substitution reactions which ultimately lead to bromoform production apparently reach completion in n x 100 seconds, so if dechlorination is to be used to limit haloform production in marine waters, the chlorination-dechlorination cycle must be completed in a time period shorter than this.

Acknowledgments

This work was sponsored by the U.S. Environmental Protection Agency, Grants No. R803839 02 and R804683010, as well as by the Maryland Power Plant Siting Program. The technical assistance of Steven R. Gullans and Leonard B. Richardson is gratefully acknowledged.

LITERATURE CITED

Bellar, T. A., Lichtenberg, J. J. and Kroner, R. C. (1974) The occurrence of organohalides in chlorinated drinking waters. J. Am. Water Works Assoc., 66, 703.

Blogoslawski, W., Farrell, L., Garceau, R. and Derrig, P. (1976) Production of oxidants in ozonized seawater. In: R.G. Rice, P. Pichet and M.A. Vincent (Eds.) Proc. Sec. Internat. Symp. on Ozone Tech. Intl. Ozone Inst., Cleveland, Ohio, p. 671-681.

Carpenter, J. H. and Macalady, D. L. (1976) Chemistry of halogens in seawater. In: R. L. Jolley (Ed.) Proc. Conf. on the Environmental Impact of Water Chlorination, Oak Ridge Nat. Lab. CONF-751096, p. 177.

Duursma, E. K. and Parsi, P. (1976) Persistence of total and combined chlorine in seawater. Netherlands J. Sea Res., 10, 192.

Eppley, R.W., Renger, E. H. and Williams, P. M. (1976) Chlorine reactions with seawater constituents and the inhibition of photosynthesis of natural marine phytoplankton. Estuarine and Coastal Mar. Sci., 4, 147.

Jolley, R. L. (1974) Determination of chlorine-containing organics in chlorinated sewage effluents by coupled ^{36}Cl tracer-high resolution chromatography. Environ. Lett., 7, 321.

Klingman, L. L. (1975) pers. comm., U.S. Bur. Mines.

Kopfler, F. C., Melton, R. G., Lingg, R. D. and Coleman, W. E. (1976) GC/MS Determination of Volatiles for the National Organics Reconnaissance Survey (NORS) of Drinking Water. In: L. H. Keith (Ed.) Identification and Analysis of Organic Pollutants in Water, Ann Arbor Science Publishers, Ann Arbor, Michigan.

Nicholson, A. A., Meresz, O. and Lemyk, B. (1977) Determination of free and total potential haloforms in drinking water. Anal. Chem., 49, 814.

Rook, J. J. (1976) Haloforms in drinking water. J. Am. Water Works Assoc., 68, 168.

Rook, J. J. (1977) Chlorination reactions of fulvic acids in natural waters. Environ. Sci. and Tech., 11, 478.

Sholkovitz, E. R. (1976) Flocculation of dissolved organic and inorganic matter during the mixing of river water and seawater. Geochim. et Cosmochim. Acta., 40, 831.

Stevens, A. A., Slocum, C. J., Seeger, D. R. and Robeck, G. G. (1976) Chlorination of organics in drinking water. In: R. L. Jolley (ed.) Proc. Conf. on the Environmental Impact of Water Chlorination, Oak Ridge Nat. Lab. CONF-751096, p 85.

Sugam, R. and Helz, G. R. (1977) Speciation of chlorine produced oxidants in marine waters: Theoretical aspects. Chesapeake Sci., 18, 113.

Symons, J. M., Bellar, T. A., Carswell, J. K., Demarco, J., Kropp, K. L., Robeck, G. G., Seeger, O. R., Slocum, C. J., Smith, B. L. and Stevens, A. A. (1975) National organics reconnaisance survey for halogenated organics. J. Am. Water Works Assoc., 67, 634.

White, G. C. (1972) Handbook of Chlorination, Van Nostrand Reinhold Co., New York, 744 pp.

Table 1. Yields of bromoform ($CHBr_3$) in estuarine water* treated with chlorine (as OCl^-) and ozone.

	Fish Absent	Fish Present
Hypochlorite (μ-equiv/l)	23.8	21.2
Bromoform Yield (μg/l)		
$Na_2S_2O_3$ Quench	24.2	20.1
No Quench	25.8	19.8
Ozone (μ-equiv/l)	23.2	21.8
Bromoform Yield (μg/l)		
$Na_2S_2O_3$ Quench	19.0	17.0
No Quench	26.6	19.0
Controls		
$Na_2S_2O_3$ Quench	<1	<1
No Quench	<1	<1

* Chemical characteristics of the water are: salinity = 14 g/kg, pH = 8.0, T = 15°C, NH_3 < 0.1 mg/l, initial dissolved oxygen = 11 mg/l (dissolved oxygen in ozonated tank = 17.7 mg/l).

OZONATION OF SEAWATER*

Robert S. Ingols
Principal Research Scientist, Retired
Engineering Experiment Station
Georgia Institute of Technology,
Atlanta, Georgia 30332

When ozone has been bubbled through pure water, its residual persists with a half-life of 20 minutes. On the other hand, Blogoslawski (1976) reports that an oxidizing residual persists in seawater for many hours. He further quoted Helz (1976) at the Chicago Forum on Disinfection With Ozone (June 1976) to indicate that bromophenols are produced when phenol is added to ozonized seawater. Pichet and Hurtubise (1976) have studied ozonized bromine-containing water and conclude that it reacts as a dilute bromine solution.

In spite of the evidence for the presence of bromine in ozonized seawater, Johnson (1976) expressed doubt that the (postulated) hypobromous acid or its ion would persist at pH 8.0 rather than disproportionate to the bromate ion.

---

*Paper not presented at Workshop, but included in Proceedings because of the close relationship of its subject matter to the formation of bromo-containing trihalomethanes and other bromo-organic compounds found in water and wastewater.

This paper is a brief discussion of the aqueous chemistry of bromine and observations of ozonized seawater.

Theoretical Considerations

Because seawater and deionized water may have different pH values and will have different buffer concentrations, it is necessary to know what happens to bromine at various pH values. Bromine disproportionates in water to form hydrobromic and hypobromous acids according to the equation:

$$Br_2 + H_2O = H^+ + Br^- + HOBr \quad (K = 1.6 \times 10^{-9}) \quad (1)$$

A plot of the pH against completion of the reaction is given by the left hand curve in Figure 1 for 0.1 millimolar concentration (8.0 mg/l Br^-).

Hypobromous acid also ionizes as shown in equation (2) and as plotted in the right hand curve of Figure 1 at 0.1 millimolar concentration.

$$HOBr = H^+ + BrO^- \quad (K = 2.06 \times 10^{-9}) \quad (2)$$

From the equations of Figure 1, it would seem that bromine behaves similarly to chlorine in disproportionating to two acids and in the ionization of hypobromous acid. There is a real difference, however, for the hypobromite ion disproportionates to bromide and bromate as shown in the following equation:

$$3BrO^- = 2Br^- + BrO_3 \quad (K = 10^{15}) \quad (3)$$

The equilibrium constant would indicate that any hypobromite ion should completely disproportionate to the bromate ion at all pH values where ionization of hypobromous acid occurs. This equation explains Johnson's (1976) dilemma as expressed at the Chicago meeting. At trace concentrations, however, the first step in the overall reaction is slow according to Radford (1965).

Sodium hypobromite solutions are also forbidden by the above equation. Actually, if bromine is added to a strong alkali, a direct reaction occurs according to the equation:

$$3Br_2 + 6OH^- = 5Br^- + BrO_3 + 3H_2O \quad (4)$$

The bromate anion as part of bromic acid is so stable that it has no bactericidal properties. Its inertness explains Johnson's perplexity about the persistent bactericidal property of the "ozone" residual in seawater. While a strongly acid solution can induce bromates to liberate iodine from iodide, no reaction will occur at neutral pH values. Apparently, 10^{-4} molar hypobromite ion is not subject to immediate disproportionation, as the equilibrium constant indicates, for too many have observed an oxidizing residual with time.

Attempts to find an ultraviolet absorption spectrum of the bromine system in water which would compare with the chlorine system were unsuccessful.

Observations with Seawater

Independent analytical procedures for bromine would help in deciding the fate of ozone in seawater. The element bromine can be extracted with carbon tetrachloride and identified by the reddish-brown color of the organic solvent.

When ozonated seawater was shaken with carbon tetrachloride, no color change occurred until a drop of concentrated hydrochloric acid was added to the water. At the lower pH, elemental bromine formed which then dissolved in the organic solvent. The reversal of the ionization of hypobromous acid to free bromine is good evidence that bromate was not formed.

The formation of bromophenol as reported by Helz (1976) is further evidence for the presence of hypobromous acid and not bromic acid or bromate ion under the conditions of observation of ozonated seawater.

Conclusions

It is concluded that bromides in seawater are oxidized to bromine or hypobromous acid or its anion by ozone, depending on the pH of the medium. A 0.1 millinormal bromide ion concentration in seawater is low enough in concentration to retard the disproportioning formation of bromates and explains the observed persistent "ozone residual", in spite of the equilibrium reaction constant of disproportionation, which indicates that the hypobromite ion should not persist at any pH.

LITERATURE CITED

Radford, (1966) "Oxyacids of Bromine and Their Salts," a chapter in Jolles, Zvi E., <u>Bromine and Its Compounds</u>, Academic Press, New York, N.Y.

Johnson, J. D., Forum on Disinfection with Ozone, (June 1976), E. Fochtman, R. G. Rice, and M. E. Browning, Eds., Intl. Ozone Inst., Cleveland, Ohio (1977), p. 301, 304.

Blogoslawski, W. J., quotes Helz, George, on formation of bromophenols, Forum on Disinfection with Ozone, (June 1976), E. Fochtman, R. G. Rice, and M. E. Browning, Eds., Intl. Ozone Inst., Cleveland, Ohio (1977), p. 297-298.

Anon., <u>Handbook of Physics and Chemistry</u> (1965), Chemical Rubber Co., Cleveland, Ohio

Ingols, R. S., R. H. Fetner and W. H. Eberhardt, (1959), "Determining Ozone in Solution," in Ozone Chemistry and Technology, ACS, Advances in Chemistry Series, 21:102 (1959).

Pichet, P., and C. Hurtubise, (1976) "Reactions of Ozone in Artificial Seawater," Proc. Sec. Intl. Symposium on Ozone Technology, P. Pichet, R. G. Rice & M. A. Vincent, Eds., Intl. Ozone Inst., Cleveland, Ohio (1976), p. 664-669.

$$HOBr = H^+ + BrO^- \quad (K - 2.06 \times 10^{-9}) \qquad (2)$$

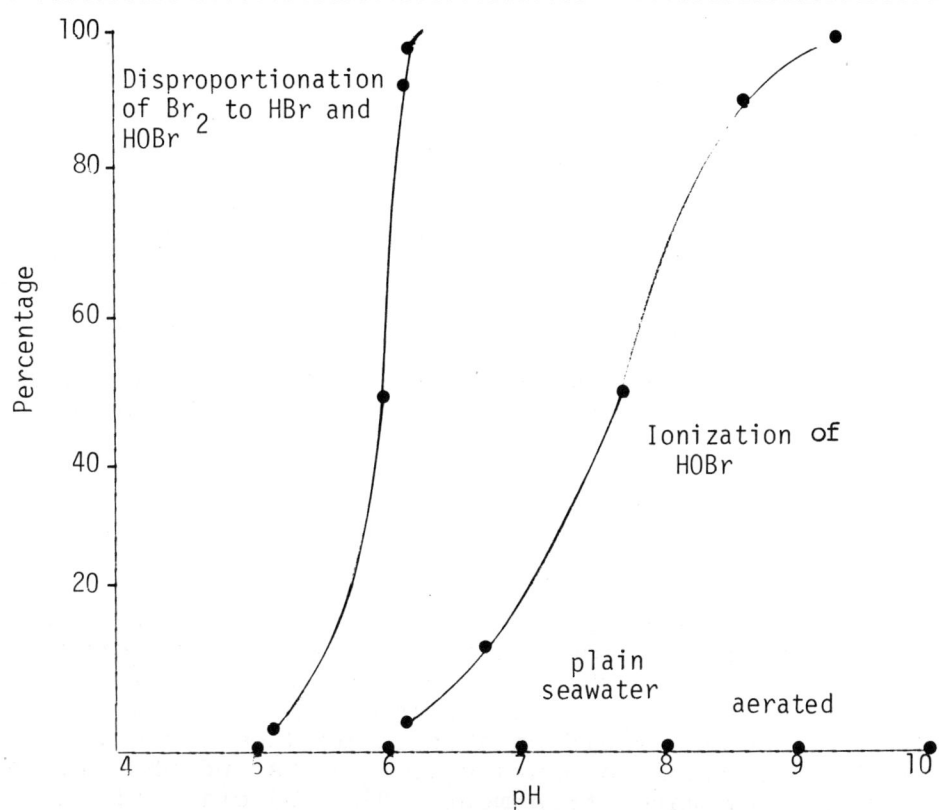

Figure 1. Percent of disproportionation of bromine gas and izonation of hypobromous acid in a 10^{-4}M(16 mg/l Br_2)

BIOCHEMICAL ASPECTS OF THE TOXICITY INVOLVED WITH OZONE ORGANIC OXIDATION PRODUCTS IN WATER

Ph. Hartemann, J.C. Block, M. Maugras and J.M. Foliguet

Laboratoire d'Hygiène et de Recherche en Santé Publique;
INSERM, Unité de Cancérologie Expérimentale U.95
Plateau de Brabois, 54500 Vandoeuvre-lès-Nancy, France

ABSTRACT

The use of chlorine or ozone as disinfectant in water treatment processes may produce some discrete organic compounds when some organic material is present in the water. It is important to know if these compounds would be able to produce toxicological, carcinogenic or mutagenic effects and what is the best way for water treatment. The objective of this paper is to compare ozone and chlorine treated waters containing respective organic compounds from the view of their toxicological properties. The classical bioassays using bacteria or algae are not sensitive enough for this purpose. We describe here a new approach to determine the toxicity of water for the mammals using short-term biochemical tests for the study of hepatic detoxification in mice.

INTRODUCTION

Water disinfection which generally involves the use of chlorine or ozone produces some complex chemical reactions with the organic compounds present in the water before treatment. The chemistry of disinfectants and their reactions with interfering substances are now being understood better. However, it is still very hard to choose between ozone and chlorine for the treatment of waters containing particular organic compounds. Both oxidizing agents have advantages as well as disadvantages.

It has been seen that with chlorine, which has been used for a long time, besides disinfection other benefits can result, such as color removal. Chlorine also suppresses the unwanted biological growths because of the possibility to maintain the desired antimicrobial residual concentration in water. However, it has become known recently that chlorine gives rise to some toxic halogenated organic compounds which may also act as carcinogens and mutagens (1,2,3).

On the other hand, the ozone treatment can destroy bacteria and virus and is efficient against some organics. However, it is not known whether ozone can also produce some toxic or carcinogenic ozonated organic compounds. We know that ozonated organic compounds are more biodegradable than the parent compounds. Figure 1 depicts an experiment where ozonation was done on water containing large amounts of earth extracts. The ratio of the Biochemical Oxygen Demand in 5 days (BOD 5) to the Chemical Oxygen Demand (COD) increases from 0.07 to 0.28 after 40 minutes of ozonation. For this final value some bacterial strains are able to cause biodegradation of the ozonized organic compounds, after adaptation. The ozonized organics can now be halogenated efficiently. Thus, the ozonated water shows an increasing chlorine demand (4,5,6).

It is well known that ozone, contrary to chlorine, does not remain in water after treatment. For this reason, in many parts of Europe, the ozonated waters are subjected to an additional chlorination to prevent a microbial proliferation in water pipes. ROOK has recently reported that the levels of haloforms in water treated with this dual disinfection system are half of those obtained with use of chlorine alone (7).

Thus, we think that there is no duality between chlorine and ozone. Since it is very difficult to assess the efficiency of a water treatment by study of only a few parameters, thus the most promising solution would be perhaps the consecutive use of chlorine treatment after ozonation. However, many questions have been raised concerning what happens to organic materials when they are ozonized and then chlorinated. More information also is needed on the health significance of these low concentrations.

A first approach to this problem is to identify the sources, the distribution and the chemical transformations of a variety of organic chemicals with analytical methods. The second one, which will be presented in this report, is to study directly samples of water containing some organics treated by chlorine or ozone in view of their toxicity. This toxicity may be appreciated by short-term or long-term bioassays using bacteria, algae, cell cultures of human or animal origin. The use of mammals like mice or rats is most interesting, since it permits a prudent extrapolation for the eventual effects on human health.

The intensity and duration of the action of most drugs and foreign compounds in mammals is determined in large part by their rate of metabolism. The biotransformations can occur in the intestine, in the lung, the kidney or the skin; by far the greatest number of these biochemical reactions are carried out in the liver. The central step in the metabolism of most of the foreign substances involves an oxidation reaction mediated by a complex of enzymes containing cytochrome P 450.

We will describe here only short-term toxicity experiments on mice involving the study of the hepatic detoxification enzymes in liver by measure of the level of two biochemical parameters: cytochrome P 450 in liver microsomes and transaminases in serum.

DESCRIPTION OF THE WATER SAMPLES USED FOR THE EXPERIMENTS

We used for these experiments two types of water containing organic compounds:

- distilled water containing a detergent: dodecyl sodium benzene sulfonate (10 mg/l)
- distilled water slightly yellow colored by addition of 10 ml/l from an extract obtained by incubation of compost (5 kg) in water (20 l) at pH 11 during 48 hrs, centrifugation, successive filtrations up to 0.45 µ and sterilization by autoclaving.

These samples are treated by ozone or chlorine.

Ozone is produced by a laboratory ozonator (Degremont) using air. The ozone is bubbled in two flasks equipped with a glass filter for dispersion. The first bottle contains the sample of water and the second one is used to retain the ozone which has not reacted. The quantities of ozone are determined by the iodometric method (8).

Chlorination is conducted by addition of commercial chlorine solution which contains 152 g/l of active chlorine. This solution is added to 200 ml of water sample in 500 ml glass stoppered flasks, and after manual agitation the bottles are stored in the dark for 30 minutes. Determination of chlorine is made by the iodometric method using thiosulfate (8).

The efficiency of the treatments is observed by the survey of some distinctive parameters as shown in Table 1 and Figure 2. Table 1 shows the decrease of the residual color and the residual organic carbon in the water sample containing earth extract treated by ozone and Figure 2 shows the increase of the chlorine demand of the same sample. The results are similar during the treatment of the water containing detergent; the residual specific group and the residual organic carbon decrease while the chlorine demand increases.

Thus it appears evident in this experiment that the improvement of the quality of the water samples (color, specific group,...) does not correspond to an ultimate transformation of the responsible organic substances, since after 45 minutes of contact the residual organic carbon remains higher than 50% in both cases. We could also point out that the chlorine demand measured 18 hours after ozonation is significantly lower than that measured 30 minutes after ozone treatment, principally with earth extract. An explanation of this phenomenon could be a new polymerisation of the humic acids some time after the treatment.

So, it is clear that both treatments are able to produce in the water samples "new" organic compounds whose eventual toxicity should be studied by suitable assays.

Short-term toxicity tests using unicellular organisms

Two assays were conducted to compare the immediate toxicity of water containing detergent or earth extract treated by either ozone or chlorine. They are short-term toxicity tests, using microscopic organisms, first a bacteria: Pseudomonas fluorescens, second an algae: Chlamydomonas variabilis. These tests were developed for large amounts of organics in wastewater and are not suitable for our experiments, since they are not sensitive enough.

The test involving Pseudomonas measures the number of surviving bacteria after 4 hours contact. The results are expressed in median tolerance limit (TL 50), which is the level that gives 50% survival.

The test involving Chlamydomonas measures the inhibition of flagellar movements. The results are expressed as Inhibition Dose 50, which gives 50% inhibition of algae movement.

These tests, performed with water samples containing earth extract, show that the water treated with ozone as described above, has no toxic effect on these microorganisms. Contrary to ozone, the same tests using chlorine-treated water show a marked toxicity. But it is difficult to know if this dramatic increase of toxicity is due to the residual chlorine or to the compounds resulting from the addition of chlorine to earth extract organics. The question of the toxicity of the residual chlorine to microorganisms would certainly be discussed in other reports, but this effect prevents a good comparison of ozone and chlorine-treated water. Therefore, we have developed biochemical short term tests of toxicity on mammals.

DESCRIPTION OF A BIOCHEMICAL TEST OF TOXICITY INVOLVING HEPATIC CYTOCHROME P 450 DETERMINATION

This test uses mice of pure genetic origin who

receive by intraperitoneal injections the water samples previously described. We used this type of administration for this first approach to the problem because it takes less time to obtain the same level of organics in the blood than orally and it avoids the digestive absorption which would complicate the phenomenon by the presence of the digestive tract barriers.

The biotransformation of drugs and other foreign compounds in the liver is accomplished by several enzyme systems linked to the membranes of the endoplasmic reticulum of the cells. There are only a few biochemical reactions involved in the hepatic metabolism of foreign compounds: oxidations, reductions, hydrolyses and conjugations.

Oxidation accounts for most of the transformations, since there are so many ways in which a compound can be oxidized, such as alkyl chains or aromatic rings, including polycyclic hydrocarbons. The key enzyme of these oxidases is cytochrome P 450, a complex of protein and haeme. Like hemoglobin, the cytochromes serve to bind oxygen and deliver it to oxidize its substrate. Cytochrome P 450 gets its designation from the fact that in its reduced form it binds carbon monoxide and then absorbs light at 450 nanometers. The amplitude of the absorbance peak of a microsomal suspension obtained by homogenization and centrifugation of liver cells is the basis of the quantitative studies.

The system is quite susceptible to numerous environmental agents which stimulate the synthesis of cytochrome P 450. More than 200 steroid hormones, drugs, insecticides, carcinogens and other foreign chemicals are now known to stimulate this system in experimental animals and many of them have been shown to do the same thing in man. On the other hand, some other substances inhibit cytochrome P 450, like heavy metals (9). The complete procedure is depicted in Figure 3 and requires the following steps:

- Treatment of mice

The samples diluted with the same quantity of a 1.8% (w/v) NaCl aqueous solution are injected in mice (1 ml of solution by intraperitoneal route) every day for three days. For each experiment we used at the minimum five mice for statistical treatment of the results. After the last injection, the animals are given only water for 12 hours. They are killed by cervical dislocation, the blood is collected by orbital sinus and the livers are removed and washed with isotonic cold KCl solution.

- Extraction of microsomes and measurements of the level of cytochrome P 450 according to OMURA and SATO (10)

All the manipulations were conducted at 4°C. Each liver is separately homegenized with a Potter homogenizer in isotonic KCl solution. The extracts were centrifuged at 10,000 g for 20 minutes for elimination of cellular debris, nuclei and mitochondria. The supernatants are again centrifuged at 100,000 g for 90 minutes; the supernatants containing cytosol and soluble enzymes may be used for certain enzymatic determinations and the pellets were placed into suspension in 0.1 M tris buffer, pH 7.4, containing 20% glycerol.

The level of cytochrome P 450 was measured by recording the following spectrum: (control cuvette) suspension of microsomes reduced with dithionite against the same suspension in which CO was bubbled for 20 seconds (sample cuvette). Determination of the level was made using 9,100/mM x cm as extinction coefficient for the absorption at 450 and 490 nm and related to the protein concentration determined by the method of Lowry (11).

- Enzymatic activity measurements

Some enzymatic activities may be measured to determine the level of some detoxification hepatic enzymes on the liver extracts (hydroxylases, demethylases or reductases). On the serum the importance of the release of hepatic enzymes due to the liver necrosis may be determined. However, in this paper are reported only the determinations of two enzymatic activities; Seric Glutamate Oxaloacetate Transaminase (SGOT) from hepatic origin and Seric Glutamate Pyruvate Transaminase from cardiac and muscular origin (SGPT).

These enzymatic activities are measured with test combination kits (Boehringer) on centrifuged blood obtained from the killed mice. The principle of the method is to obtain first the transaminase reaction which produces oxaloacetic acid (for the S.G.O. Transaminase) or pyruvic acid (for the S.G.P. Transaminase). These compounds are transformed in a second step into 2,4-dinitrophenylhydrazones which are determined by the colorimetric method (absorption at 540 nm). Results are expressed in units per liter in conformance with the I.F.C.C.

The determination of the ratio SGOT/SGPT provides good information on the liver necrosis. The higher the ratio is, the higher will be the necrosis.

- Interpretation of the results

Three consequences are to be studied on the level of cytochrome P 450:

. Level of cytochrome is decreasing: this decrease can be explained by the inhibition of the synthesis of cytochrome P 450 or more probably by a destruction of the cytochrome. In both cases the compound which produces a decrease of the level of cytochrome P 450 must be considered as very toxic. This type of toxicity can also be shown by measuring the ratio of the activities of the two transaminases SGOT/SGPT. If this ratio increases in regard to the control, that means that liver necrosis occurs.

. Level of cytochrome is increasing: this increase indicates an induction of the haemoprotein. High levels permit the liver to detoxify the compound responsible for the induction. So, that means that the compounds which induce the cytochrome P 450 are toxic but the liver can metabolize them. In this case, the value of the ratio SGOT/SGPT must remain quite the same as the control.

Level of cytochrome is unchanged: that means either that the compound is not toxic at the concentration tested or that there is a balance between a destruction and an induction of cytochrome, probably caused by two different compounds present in the sample. In the first hypothesis the transaminase ratio remains unchanged, contrary to the second one in which a liver necrosis must occur and gives an increase of the ratio.

RESULTS AND DISCUSSIONS

All experiments reported here were conducted with injection of water to mice of the strain XVII. Distilled water treated with chlorine or ozone does not induce any change in the level of cytochrome P 450 or in the ratio of the seric transaminases.

The results obtained with the water containing earth extract (E.E.) are reported in Figure 4. Mice treated with E.E., chlorinated E.E. or ozonated E.E. have a lower level of cytochrome P 450 than the control animals. At this dose Earth Extract is toxic for the liver and causes a necrosis, since the ratio of the transaminases increases to an important extent. Although we do not note any significant action of chlorination or ozonation on the liver necrosis, there is a statistically significant difference between the levels of cytochrome in animals receiving water containing E.E. treated with ozone and chlorine. The ozonation seems to render the sample a little less toxic than those containing E.E. untreated or E.E. treated with chlorine, but this difference is small.

On the other hand, the results obtained with the waters containing the detergent and reported in Figure 5, clearly show a dramatic change in the level of cytochrome P 450 in the sample treated with ozone. The water containing dodecyl sodium benzene sulfonate causes a slight decrease of the cytochrome level, which is not very significant according to statistical tests. The treatment of this detergent with chlorine causes a significant increase of the cytochrome level, which is similar to the control level. In this case, the destruction of P 450 by liver necrosis, pointed out by the increase of the ratio of the transaminases, is balanced by the induction of a new synthesis of the haemoprotein. The action of ozone treatment on the detergent is very significant, since the level of cytochrome increases and the ratio of the transaminases shows a slight necrosis. This experiment gives a good example for the induction of the cytochrome P 450 system.

Thus, the ozonation probably partially transforms dodecyl sodium benzene sulfonate into organic compounds(s) which is able to induce the hepatic system of detoxification and, for this reason, is certainly less toxic than the parent compound. The chlorination also has an effect on this detergent but this effect is less significant. In this case, the simultaneous use of analytical methods could provide very important and useful information (in the field of toxicology).

The biotransformation of drugs and other foreign compounds in the liver is accomplished by several enzyme stages that can metabolize a wide variety of toxic agents. Although the cytochrome P 450 is the key enzymatic system of biological oxidation, other enzymes such as hydroxylase (aniline hydroxylase) or demethylase (ethyl morphine and aminopyrine) must be studied to have a more complete survey of this metabolic pathway. Reductions, conjugations, hydrolysis reactions also are of great importance, and particularly N.A.D.P.H. cytochrome C reductase, which acts in one of the two electron transfer chains coupled with the working of the cytochrome P 450 mechanism, will give good complementary information.

Another advantage of the measure of cytochrome levels is to give more information on the type of induction. Then the polycyclic hydrocarbons are metabolized by a mixed function oxidase system (aryl hydrocarbon hydroxylase, A.H.H.) which requires N.A.D.P.H. and molecular oxygen, but the terminal oxidase is somewhat different from that which is induced by other drugs. The catalytic properties of the cytochrome are changed as are its spectral properties, with an absorbance maximum at 448 nm rather than at 450 nm. The polychlorobiphenyls are also able to induce the formation of this cytochrome P 448.

It would be of great interest to find an induction of cytochrome P 448 in animals receiving water samples containing organics treated either by chlorine or ozone. This would suggest the possibility for the organic compounds produced by the water treatment to have carcinogenic properties.

The study of the induction of microsomal liver enzymes by different dose levels could be established for water samples artificially polluted with organic compounds and treated in a laboratory plant or for real drinking water after suitable concentration. This could permit determination of a no-effect level in this system in comparison with the levels of organic compounds found by the aid of the analytical methods. A similar result was recently reported for organochlorine pesticides given in the diet of rats during 2 weeks (12). A no-effect level was established for each pesticide and compared with the no-effect levels from the literature based on histopathological abnormalities in the liver in long term experiments. In general, good agreement was found, except for some compounds for which a much lower no-effect level was found in the biochemical short-term test.

CONCLUSION

The cytochrome P 450 system gave us interesting results. This system is the site of many competitive interactions of a wide variety of foreign substances which will be found in drinking water. This system is widely found in living systems: microorganisms, plants, insects and mammals. The modifications of this system according to different xenobiotics actually seem to be similar. Although the measure of levels of hydroxylating cytochrome gives interesting results, it must be made in conjunction with other determinations. Particularly, analytical methods are necessary to determine which compound formed after water treatment is responsible for the measured modifications. Also, mutagenicity tests involving bacteria like that developed by AMES on <u>Salmonella-typhimurium</u> may certainly contribute to a better knowledge of the toxicological, carcinogenic and mutagenic properties of the drinking waters treated with chlorine or ozone. From the preliminary results presented here, it is impossible to set a conclusion from the view of the role of ozonation in the production of toxic organic compounds. These experiments are now in progress.

REFERENCES

(1) J.J. Rook, "Formation of haloforms during chlorination of natural waters". Water Treat. Exam. 23(2):234-243 part 2 (1974).

(2) R.H. Harris, "The implication of cancer causing substances in Mississippi River waters". Environmental Defense Fund, Washington, D.C. (1974).

(3) B. Dowty, "Halogenated hydrocarbons in New Orleans drinking water and blood plasma". Science 187:75-77 (1975).

(4) R. Buydens and G. Fransolet, "L'action de l'ozone sur le chlore, le bioxyde de chlore et le chlorite contenus dans les eaux traitées". Tribune Cebedeau 24:4-6 (1971).

(5) Y. Richard, "Interférences chlore-ozone dans le traitement de l'eau". Sec. Intl. Symp. on Ozone Technology, R.G. Rice, P. Pichet & M.A. Vincent, Editors. Intl. Ozone Inst., Cleveland, Ohio (1976) pp. 169-180.

(6) J.C. Block, M. Morlot and J.M. Foliguet, "Problèmes liés à l'évolution du caractère d'oxydabilité de certains corps organiques présents dans l'eau traitée par l'ozone". T.S.M. l'Eau, 71: 29-34 (1976).

(7) J.J. Rook, "Haloforms in drinking water". J. Am. Water Works. Assoc. 68(3):168-172 (1976).

(8) "Standard Methods for the Examination of Water and Wastewater". Am. Public Health Assoc. Inc., 13th Edition, 271 (1971).

(9) A. Kappas and A. Alvares, "How the liver metabolizes foreign substances". Scientific American, 232:22-31 (1975).

(10) T. Omura and R. Sato, "The carbon monoxide binding pigment of liver microsomes. I. Evidence for its haemoprotein nature". J. Biol. Chem. 239:2370-2378 (1964).

(11) O.H. Lowry, N.J. Rosenrough, A.L. Farr and R.J. Randall, "Protein measurement with the folin phenol reagent". J. Biol. Chem. 193:265-275 (1951).

(12) E. Den Tonkelaar and G. Van Esch, "No-effect levels of organochlorine pesticides based on induction of microsomal liver enzymes in short-term toxicity experiments". Toxicology 2:371-380 (1974).

DISCUSSION

Joe Cotruvo: In the slides where there were comparisons of the different levels of cytochrome P 450 induction, the differences between the tests usually seemed to be in the range of 1 or 2 nanomoles per milligram. It was a change of from 7 to 6 or from 5 to 4. Can you give some kind of a qualitative indication of what that means in terms of the effect on the liver? Is it a substantial change to go from 7 to 6, or merely something that is measureable?

Hartemann: With this level of organic compound injected into the animals it is _very_ significant. As you know, it is necessary in the Ames mutagenicity test with Salmonella to use quite the same microsomal preparation after induction by phenobartitol, Arochlor, or other carcinogenic compounds. We give very high levels of organic compounds to rats or mice to observe more important changes. Perhaps we are able to obtain four times more cytochrome P-450 after injection of Arochlor than in the control animals, but we use very low levels in our experiments. We inject only water, 0.5 milliliter of water, in one animal.

I did not have time to present results obtained with other enzymatic determinations, because cytochrome P-450 is not the only enzyme involved in the detoxification process. However, the results are the same as with cytochrome B-5 or some detoxification enzymes.

EARTH EXTRACT + OZONE

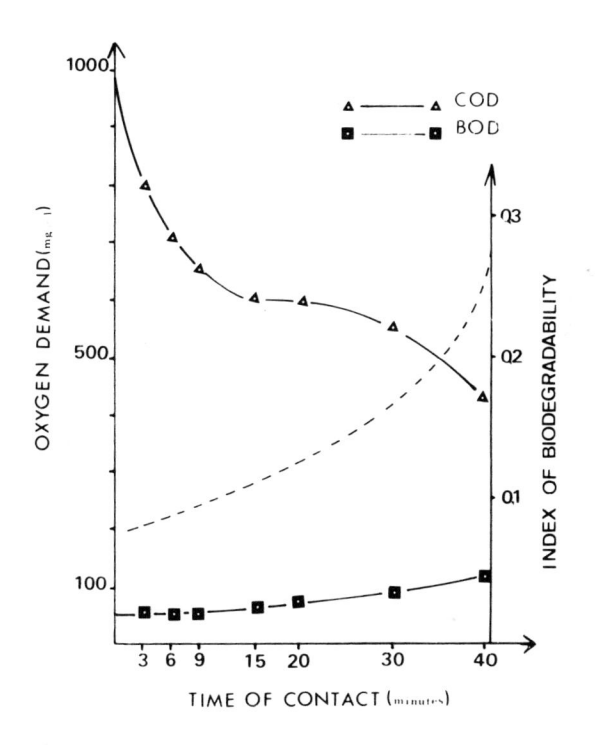

Fig 1. Ozonation of a water containing earth extract.

CONTACT TIME min.	EARTH EXTRACT		DETERGENT	
	Residual color %	Residual organic carbon %	Residual specific group %	Residual organic carbon %
0	100	100	100	100
5	60	75	96	96
15	40	70	82	66
30	20	62	78	60
45	4	53	69	55

Table 1. Evolution of some parameters during the treatment of the water sample by ozone.

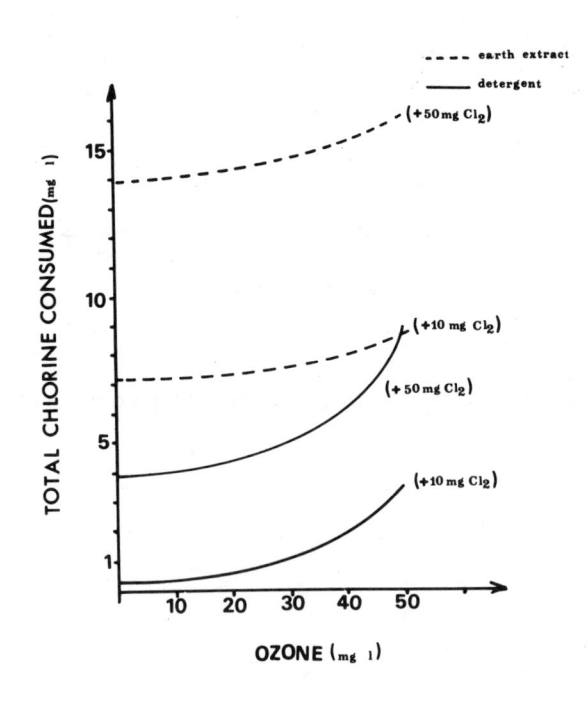

Fig 2. Evolution of the chlorine demand after ozonation.

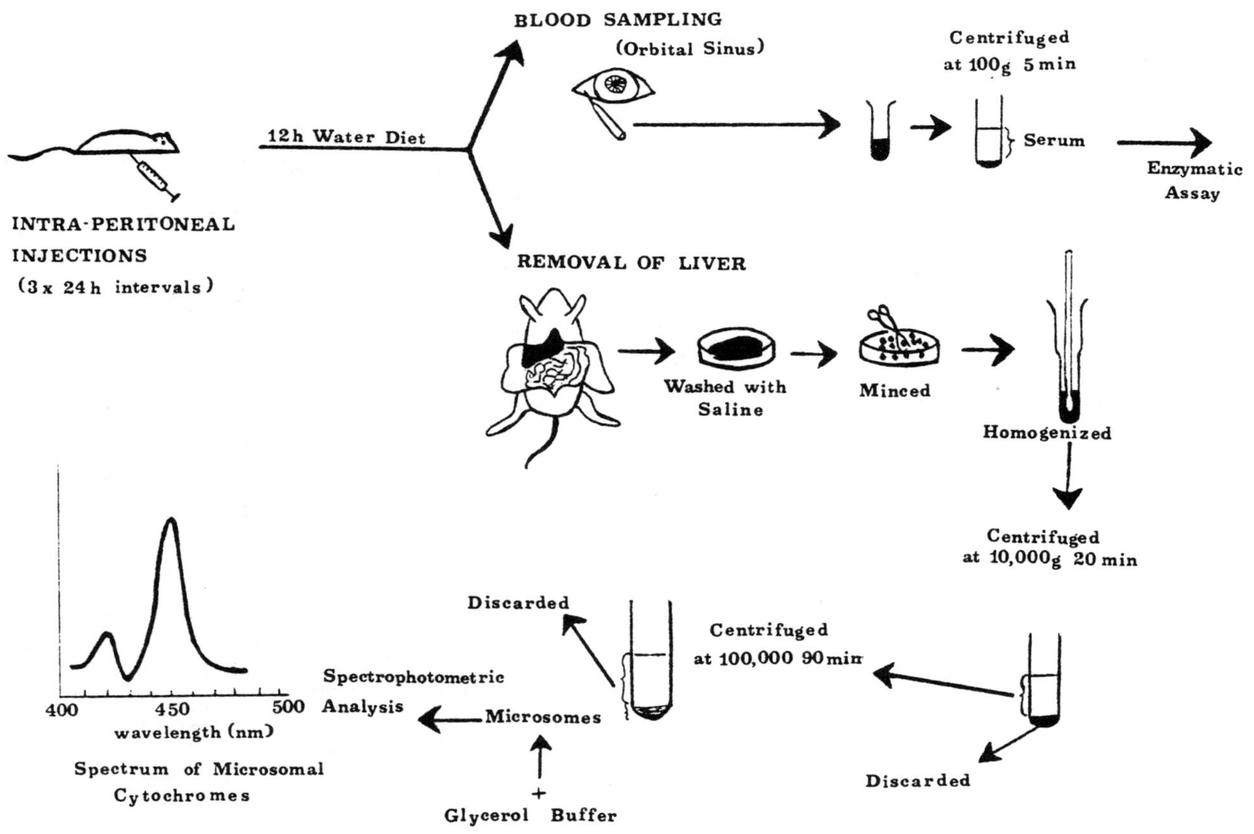

Fig 3. Description of the biochemical test of toxicity.

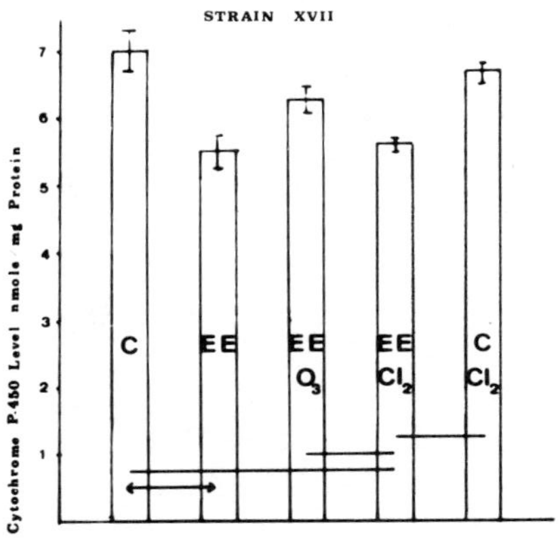

C Control; EE Earth Extract (10mg/l); EE Cl₂ Earth Extract with Chlorine (10mg Cl₂/l); EE O₃ Earth with Ozone (10mg O₃/l); C Cl₂ Water with Chlorine (10.6mg Cl₂/l)

Fig 4. Effect of an earth extract on the cytochrome P450 in mice liver

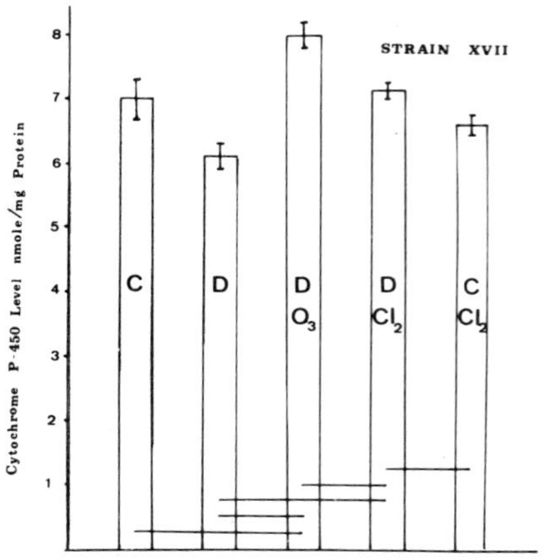

C Control; D Detergent; D DO₃ Detergent with Ozone (10mg D/l + 10mg O₃/l); DCl₂ Detergent with Chlorine (10 mg D/l+ 10mg Cl₂/l); C Cl₂ Water with Chlorine (10.6 mg Cl₂/l)

Fig 5. Effect of detergent on the cytochrome P450 in mice liver.

EFFECT OF OZONE ON HOSPITAL WASTEWATER CYTOTOXICITY

by

Riley N. Kinman, Janet Rickabaugh,
Victor Elia, Kevin McGinnis, Terrence Cody,
Scott Clark and Robert Christian

University of Cincinnati, Cincinnati, Ohio 45221

An Examination of Products Formed by Ozonation In A Wastewater Reuse System

Disinfection of wastewaters is one of the most complex tasks assigned to the environmental engineer, scientist, and treatment plant operator. Organisms that must either be killed or inactivated are surrounded by a variety of chemical compounds which compete for the disinfecting agent. Each compound has its own specific chemistry in aqueous solution. Each wastewater has a pH, buffer capacity, chemical content, and physical quality peculiar to the domestic and industrial contributions in the collection system and the treatment units ahead of the disinfection step. These treatment units always function at varying degrees of efficiency. Sometimes they spill solids which encapsulate and protect organisms from the disinfecting agent. Sometimes products remain which exert greater than expected demands for the disinfectant. Many times short circuiting occurs in the treatment units, causing wastewater that is not adequately treated to reach the disinfection step. When all of these possibilities are considered, plus the changes in temperature and flow which occur, disinfection of wastewater on a continuous basis appears to be an impossible task.

The ideal wastewater disinfectant should have the characteristics shown in Table 1.

Table 1

Characteristics of the Ideal Wastewater Disinfectant

1) Soluble in Water
2) Bactericidal
3) Cysticidal
4) Viricidal
5) Sporicidal
6) Readily Available
7) Easy to Apply
8) Measureable in Wastewater
9) Chemically Weak
10) Relatively Inexpensive
11) Reaction Products that are Non-Toxic, Non-Mutagenic and Biodegradable
12) Chemistry in Aqueous Solution is Known

No presently used disinfectant possesses all of these ideal characteristics. Chlorine has been criticized, because of the toxicity of some of its end products to aquatic life (1) and the potential for harm to humans from chlorinated compounds formed in the disinfection process (2,3,4,5). Because of these problems associated with the use of chlorine, ozone has been proposed as a replacement or supplement for chlorine. Ozone was used as one of the unit processes in a hospital wastewater reuse system in these studies. This paper describes an investigation of the reaction products formed during ozonation of this particular wastewater and the changes in cytotoxicity which occurred.

Oxidation Of Organic Compounds - Ozone Versus Chlorine

Chlorine has been used as a chemical oxidant in water and wastewater treatment for many years. Black and Christman reported in 1963 (6) that Miami, Florida had used chlorine for more than 15 years to reduce color values to less than 10 units. They realized then that the end products of chlorine oxidation of color by the practice of heavy or super-chlorination did not result in the simple end products of CO_2 and H_2O. They found that ozone was more effective than either chlorine or chlorine dioxide for color removal. They were not able to isolate pure compounds from degradative oxidation of the fulvic acids, but they postulated that soluble mixtures of complex carboxylic acids were produced.

Chlorine in wastewater treatment functions both as an oxidizing agent and as a chlorinating agent. Ozone functions as an oxidizing agent. Table 2 depicts the characteristics of an ideal chemical oxidant. Any consideration of substitution of ozone for chlorine should take into account the quantities and compounds produced by each chemical along with their respective problems.

Table 2

Characteristics Of An Ideal Chemical Oxidant

1. Powerful, Non-Specific Oxidant
2. Readily Available; Soluble In Water; Easy to Apply
3. Chemistry Well Known; Reaction Products Non-Toxic and Biodegradable
4. Measurable in Wastewater
5. Relatively Inexpensive

In recent years attempts have been made to determine the products and their concentrations which result when chlorine or ozone is applied to water containing organic compounds. There is not time to cite all the work, but some important studies were as follows: Burttschell et al. (7) studied the chlorination of phenol and indicated the following step-wise formation of various chlorinated phenols. Initially phenol was chlorinated to form either 2- or 4-chlorophenol. The 2-chlorophenol was chlorinated to form either 2,4- or 2,6-dichlorophenol while 4-chlorophenol formed 2,4-dichlorophenol. Both 2,4- and 2,6-dichlorophenol were chlorinated to form 2,4,6-trichlorophenol. The 2,4,6-trichlorophenol reacted with aqueous chlorine to form a mixture of non-phenolic oxidation products. Many of the above products are "suspect" in terms of human health.

Spaeth (8) studied the chlorination of five coal-tar derivatives in water and showed that there were reaction products formed. Chlorination of biphenyl, fluorene, naphthalene, acenaphthene and pyrene produced chlorinated naphthalenes and chlorinated acenaphthenes in chlorine demand-free water and in untreated Ohio river water. Substitution products formed were confirmed by GC-MS techniques. Reactivity of the five coal-tar derivatives examined followed the order: biphenyl < fluorene < naphthalene < acenaphthene < pyrene.

Eisenhauer (9) studied the ozonation of phenol. He found very little of the phenol converted completely to carbon dioxide and water within the contact time and concentration of ozone which he used. Dibasic acids such as oxalic and maleic were some of the products formed in these studies.

Gould and Weber (10) studied the oxidation of phenols by ozone. Virtually complete removal of phenol and its aromatic oxidation products required 4 to 6 moles of ozone for each mole of phenol originally present. This much ozone would not be used in a disinfection process, but might be used in a chemical oxidation process. They found that glyoxal, glyoxylic acid and oxalic acid remained at this point. They believed that subsequent dilution of the discharged effluent would reduce the levels of these organics in the receiving water to acceptable concentrations.

Rook (2,5) studied the formation of halomethanes in chlorinated water. He found that chloroform and other chlorinated methanes were formed during chlorination of natural waters for disinfection.

Jolley (11) identified some seventeen compounds formed in the sewage chlorination process. The yield of organic chlorinated compounds was not very great. The important fact is that many trace chlorinated organics may be formed in the chlorination of water containing organic compounds. Little is known concerning the overall health significance of these chlorinated compounds.

Bellar et al. (3,4) studied the levels of halogenated organics in drinking waters that were chlorinated. They concluded that chloroform, bromodichloromethane, dibromochloromethane and bromoform were present in most U.S. drinking water supplies and that these compounds resulted from the addition of chlorine to the waters for disinfection purposes.

These few studies and others indicate that many compounds may be formed during the chlorination of a wastewater. The number of compounds and their respective concentrations are a function of many complex variables. Chlorine species, compounds being chlorinated, temperature, pH, presence or absence of catalysts, chlorine dose, contact time, etc., all affect the reactions which will occur. Both oxidation and halogenation may occur. Little is known concerning the health significance of these chlorinated compounds in the concentrations that have been measured at this point in time. Some of the compounds (chloroform is an example) have been found to be carcinogenic in laboratory animal studies. Many compounds have not been studied for their carcinogenic, mutagenic or other toxic properties.

Oxidation products formed by the use of ozone in water and wastewater treatment have not been identified to any great degree. Even less is known concerning actual concentrations of these oxidation products in the treated waters. Much pressure has been placed on the U.S. Environmental Protection Agency to set limits on the trace chlorinated organics found in drinking water, even though the concentrations that have been found are in the parts per billion (μg/l) range.

No such pressure has been applied with respect to ozone, probably because much ozone literature describes the reaction products of ozone disinfection to be carbon dioxide and water. This may not be true. Many oxidation products can be produced when ozone is added to water containing organic compounds. Whether the oxidation products from ozone are more harmful or less harmful than the reaction products from chlorine is not known at this time. Additional research concerning toxicity, carcinogenicity, mutagenicity and other health effects is necessary after the compounds have been identified and quantified.

This presentation includes the evaluation of certain ozonated organic compounds that would be present in a hospital water reuse system. In these studies ozone was used to treat synthetic military hospital wastewater. The five volatile organic compounds that were in the wastewater were methanol,

ethanol, isopropanol, acetone and acetic acid. These compounds and their concentrations are listed in Table 3. The fate of these compounds was studied after treatment with ozone. In addition, studies were carried out to determine if ozone caused a decrease in toxicity for this wastewater or an increase in toxicity. These studies indicated that the ozonated wastewater was more toxic than the unozonated wastewater. A description of the test cell system used for evaluation of cytotoxicity is given in the Appendix.

Table 3

Composition Of Synthetic MUST Permeate

```
Ethanol     -------------  185 mg/l
Methanol    ------------   100 mg/l
Isopropanol ---------       10 mg/l
Acetone     -------------   10 mg/l
Acetic Acid ---------       90 mg/l
```

In this paper a mixture of these 5 compounds is called an 8X synthetic permeate.

Background of This Project

A few years ago this laboratory initiated the development of a test system for assessing the potability of water. The system established was an _in vitro_ mammalian cell culture system. The Bioassay Method was initially applied to the examination of waters produced by a direct reuse system, which is being developed by the U.S. Army for treatment of the wastewater from mobile surgical hospitals. These hospitals are referred to as Medical Units, Self-contained, Transportable: (MUST). The processes employed in this treatment plant include ultrafiltration, reverse osmosis and ozonation combined with ultraviolet radiation. A cell culture bioassay system is currently used to study the effectiveness of each treatment stage in removing toxic components from wastewaters and also to investigate the possibility of toxic products being formed as a result of ozonation.

Early in the project a sample of reverse osmosis permeate (RO-Permeate) was obtained from investigators at AirResearch, Inc. who were working on the development of the pilot plant. This permeate was produced from a synthetic hospital waste. The formula for this synthetic waste was derived from a survey of the waste components of an Army surgical hospital.

The MUST Direct Reuse Plant is designed to treat raw wastes from five sources within the hospital. These sources are: the operating room, the kitchen, the laboratory, the X-ray developing lab and the shower. The permeate received from AirResearch was prepared from a synthetic waste containing all components at 8 times the average daily concentration estimated to be in the waste stream. This RO-Permeate was examined by gas chromatography-mass spectroscopy. The five volatile components shown in Table 3 were identified.

Ozone Treatment of MUST Permeate

The synthetic permeate was treated with ozone in the ozonation system shown in Figures 1 and 2. The ozone contact unit was constructed entirely of plexiglas and the system was operated on a continuous flow basis. The water to be treated was placed in the feed tank and heated to approximately 45°C. The pH was adjusted to 9.5 with sodium hydroxide. Then the ozone feedwater was pumped into the contact unit. No attempt was made to control the temperature of the water in the contact unit. Ozone, produced by a Welsbach (Model T-816) ozone generator, was introduced at the bottom of the unit. The concentration of ozone in the gas stream was approximately 50 mg/l (or 3.8% by weight).

The ozone contact unit is shown in detail in Figure 2. Ozone was introduced to the contact unit through the spargers shown. The unit had a total liquid capacity of 70 liters (20 gal) with a capacity of 6.2 liters (1.7 gal) for each column. Feedwater was introduced at the top left and exited at the top right. The water flowed in an alternating co-current-countercurrent direction, relative to the ozone gas stream.

Ozonation was initiated after the unit had completely filled with feedwater. After 87 liters (23 gal) of water had been pumped through the unit, steady state conditions were assumed and sample collection was begun. After an 8 liter volume had been collected from the exit port of the contact unit, 100 ml samples were taken for analysis. Sample residence times were estimated from the average flow rate between sampling times. The operating parameters for the ozonation of the synthetic permeate are shown in Table 4.

Table 4

Operating Conditions For Ozonation Of Synthetic Permeate

Ozone Generator

Operating Voltage	118 volts
Operating Power	180 watts
Oxygen Flow Rate	6.2 l/min @ 8.7 psi

Feedwater

Flow Rate	5-20 gal/hr
Temperature	47-55°C
Ozonation Times	18-193 min

Feedwater was pumped through the contact unit at a rate of 5-30 gal/hr. Feedwater temperature was 47-55°C, and ozonation times ranged from 18 to 193 minutes.

Significant reduction in total organic carbon (TOC) and chemical oxygen demand (COD) were achieved as shown in Figure 3. The TOC was reduced approximately 28% while the COD was reduced 31%.

Chromatographic analyses of the ozonated samples were carried out using a Hewlett Packard (Model 402) Gas Chromatograph equipped with a flame ionization detector. The samples were analyzed by direct injection of 5 µl of sample. The analyses were carried out isothermally at 120°C using a Porapak-Q column.

The fates of methanol, ethanol and acetone are shown in Figure 4. Methanol concentration was reduced from 100 mg/l to about 30 mg/l. This represented a 70% reduction. Ethanol concentration was reduced from 175 mg/l to approximately 8 mg/l, a 95% reduction. Acetone concentration, on the other hand, increased from 10 mg/l to about 16 mg/l, which represented a 50% increase. Most of the change in concentration of these components occurred during the first 90 minutes of ozonation.

During gas chromatographic analysis of the ozonated samples an additional peak was observed which had a retention time only slightly greater than that of methanol. This peak was tentatively identified as acetaldehyde by comparison to the retention time of pure acetaldehyde. Confirmation of this was achieved by using different chromatographic conditions and comparing the retention times. Final verification was achieved by GC-MS.

The effects of ozonation on the concentrations of acetic acid, isopropanol and acetaldehyde are shown in Figure 5. Acetic acid concentration increased from 90 mg/l to 130 mg/l during the first hour of ozonation; this represents a 45% increase. The acetic acid concentration then decreased to 100 mg/l. Isopropanol concentration decreased by at least 90% during the first hour, from an initial concentration of 10 mg/l to below the detection limit of 1 mg/l. During this time interval, acetaldehyde reached a concentration of approximately 35 mg/l.

Comparison of the actual total organic carbon values and the total organic carbon values calculated from the concentration of the five major components and acetaldehyde as determined by GC are shown in Figure 6. After one hour there was a 30% difference, which remained approximately constant for the balance of the ozonation period.

Changes in Cytotoxicity

Cell toxicity data indicated an increase in cytotoxicity for this permeate upon ozonation. Figure 7 shows a plot of average protein production for a 10% dilution with various ozonation periods. Ozonation for 30 minutes caused the permeate to increase in cytotoxicity; additional ozonation caused a further increase in cytotoxicity. Ozonation for 90 minutes caused the cytotoxicity to decrease slightly. Ozonation for 120 minutes returned the cytotoxicity to a value near that of the control. An 18% dilution is shown in Figure 7, also with an ozonation period of 120 minutes.

Several dilutions of the permeate are shown in Figure 8, all of which were ozonized 30 minutes. Effects of dilution are clearly shown, Greater concentrations of permeate caused greater cytotoxicity with the same applied dose of ozone. Studies are continuing to better define the cytotoxicity effects of the applied ozone.

Summary

Ozone may not oxidize the organics found in water or wastewater completely to carbon dioxide and water. Other intermediate oxidation products may be formed. Cytotoxicity of this particular wastewater increased upon initial ozonation and decreased with prolonged ozonation. Further research is needed to clarify the significance of these observations with respect to ozone oxidation of organics and ozone disinfection of water and wastewater.

Acknowledgements

This research was supported by the U.S. Army Medical Research and Development Command through contract number DADA-17-73-C-3013.

Appendix

Method for Testing Toxicity of Water Samples
With Cultured Mammalian Cells

A. Preparation of Water Sample

1. Water sample (1 liter) is filtered under pressure through a 0.2µ Millipore filter.

2. Osmolality of the water sample is measured by the freezing point depression method.

   2a. If osmolality of sample is less than 10 milliosmoles/kg water, go to Section 3.

   2b. If osmolality of sample is $\leq$ 10 milliosmoles/kg water, go to Section 4.

3. Dissolve prepackaged dry powdered MEM Eagle's Minimum Essential Medium with Hank's Salts) in 750 ml of water sample, add 0.35 g $NaHCO_3$, and 50 ppm Gentamicin Reagent Solution. When dissolution is complete, add sample water to make up 1000 ml of solution. Adjust pH to 7.0 with NaOH or HCl if necessary. The osmolality of the medium should not exceed 310 milliosmoles/kg water. (Go to Section 5).

4. If the osmolality of the water sample exceeds 10 milliosmoles/kg water, the NaCl content of the medium must be reduced to compensate for the osmotic pressure contribution from the sample. Calculate the amount of NaCl required to make a 300 milliosmoles/kg solution.

Example:

273	mosm/kg water are required from NaCl if osmolality of sample = 0
−32	actual osmolality of sample
241	milliosmoles still required
÷2	(each millimole NaCl is equivalent to 2 milliosmoles)
120.5	milliosmoles salt required to make up difference
X 58.45	(M.W. of NaCl)
7043.2	mg NaCl required to make 300 milliosmolal medium

Dissolve prepackaged dry powdered Eagle's Minimum Essential Medium with Hank's salts minus NaCl (must be special ordered from tissue culture supply house) in 750 ml of sample water. Add calculated quantity of NaCl, 0.35 g NaHCO$_3$ and 50 mg of Gentamicin. When dissolution is complete, add sample water to make a total volume of 1 liter. Adjust pH with NaOH or HCl, if necessary; the osmolality of this medium must be checked. Adjustment is not usually necessary, but may be to get into a range of 290-310 mosmoles/kg of water. Small adjustments may be made by adding small quantities of salt or sample water.

5. A control medium is prepared for carrying out simultaneous control growth experiments and diluting the water sample medium when determining concentration-related effects. Control medium is prepared in exactly the same manner as the water sample medium, except that double-distilled water is used as the makeup water instead of the water sample.

6. All media are sterilized by filtration through sterile 0.2µ membrane filters. All media should be prepared 24-48 hours before use, tested for sterility and refrigerated until used.

B. Preparation of Cultured Cells

1. Continuous cultures of L-cells (Clone 929) are maintained at 37°C in MEM containing 5% fetal calf serum (F.C.S.). The cells are passed every Tuesday and Friday. The medium is replaced the day before passage.

2. Confluent cells are used for preparation of subcultures to be used for the bioassay.

3. Fifty-thousand cells suspended in 2 ml of MEM containing 1% F.C.S. are added to sterile disposable glass test tubes and the tubes are placed horizontally, in racks, at 37°C where the cells become attached to the walls of the tube. The usual procedure is to prepare 12 racks of tubes, each rack containing 26 tubes. After 72 hours (the cells will have doubled at least twice), the cultures in the tubes are ready for addition of the test water media.

4. The growth medium is aspirated from the culture tubes and is replaced with 2 ml of test medium containing 1% F.C.S. The test media consist of the test water medium, half-log dilutions of the test medium, and a control group.

5. Before the addition of test water medium, one culture tube is removed from each rack for an immediate protein determination. These data show whether or not the cultures are growing satisfactorily and have approximately the same number of viable cells before starting the test experiment. Another culture tube is removed from each rack before the addition of test water medium. This tube contains the "zero" time or pretreatment culture. At 24, 48, 72 and 96 hours, 6 cultures are removed from each rack. The medium in the remaining cultures

is renewed after the 48-hour and 72-hour cultures are removed.

6. When culture tubes are removed, the medium is aspirated, the cell sheet is washed twice with phosphate-buffered saline, and the tubes are inverted to drain and air-dry.

7. The effect of the test water on the cultures is determined by measuring the differences in protein content of the air-dried tubes by the colorimetric Lowry technique*.

---

* Oyama, V. and H. Eagle, "Measurement of Cell Growth With a Phenol Reagent (Folin-Ciocalteaw)", Proc. Soc. Exptl. Biol. Med. 91:305 (1956).

LITERATURE CITED

1. "Proceedings of Wyoming Workshop on Disinfection of Wastewater and its Effect on Aquatic Life." Wyoming, Michigan. Oct. 30, 1974, U.S.E.P.A. et al.

2. Rook, J.J., "Formation of Haloforms During Chlorination of Natural Water." Water Treatment & Examination, 23(2):234 (1974).

3. Bellar, T.A., Lichtenberg, J.J., and Kroner, R.C., "The Occurrence of Organohalides in Chlorinated Drinking Water.", J. Am. Water Works Assoc. 66(12):703 (1974).

4. Symons, J.M., Bellar, T.A., Carswell, J.K., Demarco, J., Kropp, K.L., Robeck, G.G., Seeger, D.R., Slocum, C.J., Smith, B.L., and Stevens, A.A., "National Organics Reconnaissance Survey for Halogenated Organics", J. Am. Water Works Assoc. 67:634 (1975).

5. Rook, J.J., "Haloforms in Drinking Water," J. Am. Water Works Assoc. 68:168 (1976).

6. Black, A.P. and Christman, R.F., "Chemical Characteristics of Fulvic Acids." J. Am. Water Works Assoc. 55(6) (1963).

7. Burttschell, R.H., Rosen, A.A., Middleton, F.M. and Ettinger, M.B., "Chlorine Derivatives of Phenol Causing Taste and Odor", J. Am. Water Works Assoc. 51:205 (1959).

8. Spaeth, D.P., "The Chlorination of Coal Tar Derivatives in Water", Ph.D. Dissertation, Univ. of Cincinnati, Cincinnati, Ohio (1972).

9. Eisenhauer, H.R., "Ozonization of Phenolic Wastes." J. Water Poll. Control Fed. 40:1887 (1968).

10. Gould, J.P. and Weber, W.J., Jr., "Oxidation of Phenols by Ozone." J. Water Poll. Control Fed. 48:47 (1976).

11. Jolley, R.L., "Chlorination Effects on Organic Constituents in Effluents from Domestic Sanitary Sewage Treatment Plants." Oak Ridge National Laboratory Publication, O.R.N.L. - TNA - 4290 (October, 1973).

DISCUSSION

Fred Ranch, Nalco Chemical Company: Dr. Kinman, do you have any standard errors on the data that you have? I noticed that you didn't have the values for the protein presented. Do you think that L cells which are non-mammalian, that are growing rapidly, are representative of normal cells in a human? Finally, do you think these data have any realtionships to toxicity of this material in intact animals?

Kinman: You asked me three beauties. The statistics are being run, and I could have shown some of those. You have the standard deviations above and below the line and all the numbers are a little bit confusing when you have five dilutions and then five or six times of ozonization, but they are being run.

In terms of the cells, whether these represent exactly the behavior in the human body, I can't say. All I am saying is that it does give you a measure of how you are affecting these particular cells, and when you combine this with GC/MS techniques, you can at least begin to eliminate, hopefully, certain compounds that might be in there that would be contributing to the overall toxicity.

Now your third question is on the relationship of the L cells, whether this is the best kind of a cell. The cytotoxicity work is being carried on principally under Dr. Terry Cody, and his selection of this particular cell group is his own. The goal here was to develop a quick test that would predict whether a TOC value of 10 or 5 really means anything in terms of toxicity. Along with this work in the MUST program there is a lot of animal toxicology going on, and there is an attempt to relate the cell cytotoxicity with the animal toxicology, and then with what we know about human toxicology. And of course, there are some mightly large steps here, there is no question about it, but you have to start somewhere.

Dave Rosenblatt, U.S. Army Medical Bioengineering R&D Labs: Don't you think you run a risk in trying to do gas chromatography on some of these intermediates which may be peroxidic in nature? That is, you may destroy the compound in trying to analyze for them.

Kinman: The answer, of course, is yes. In the TOC we could not account for what should have been there in terms of TOC versus the GC compounds that we measured. I think there is a great danger, in answer to your question, yes.

One other point, however, you can follow this by mass spectroscopy, and hopefully pin down some things you can't do with GC alone.

Unknown: Have you measured hydrogen peroxide in your system after ozonation?

Kinman: No, we have not.

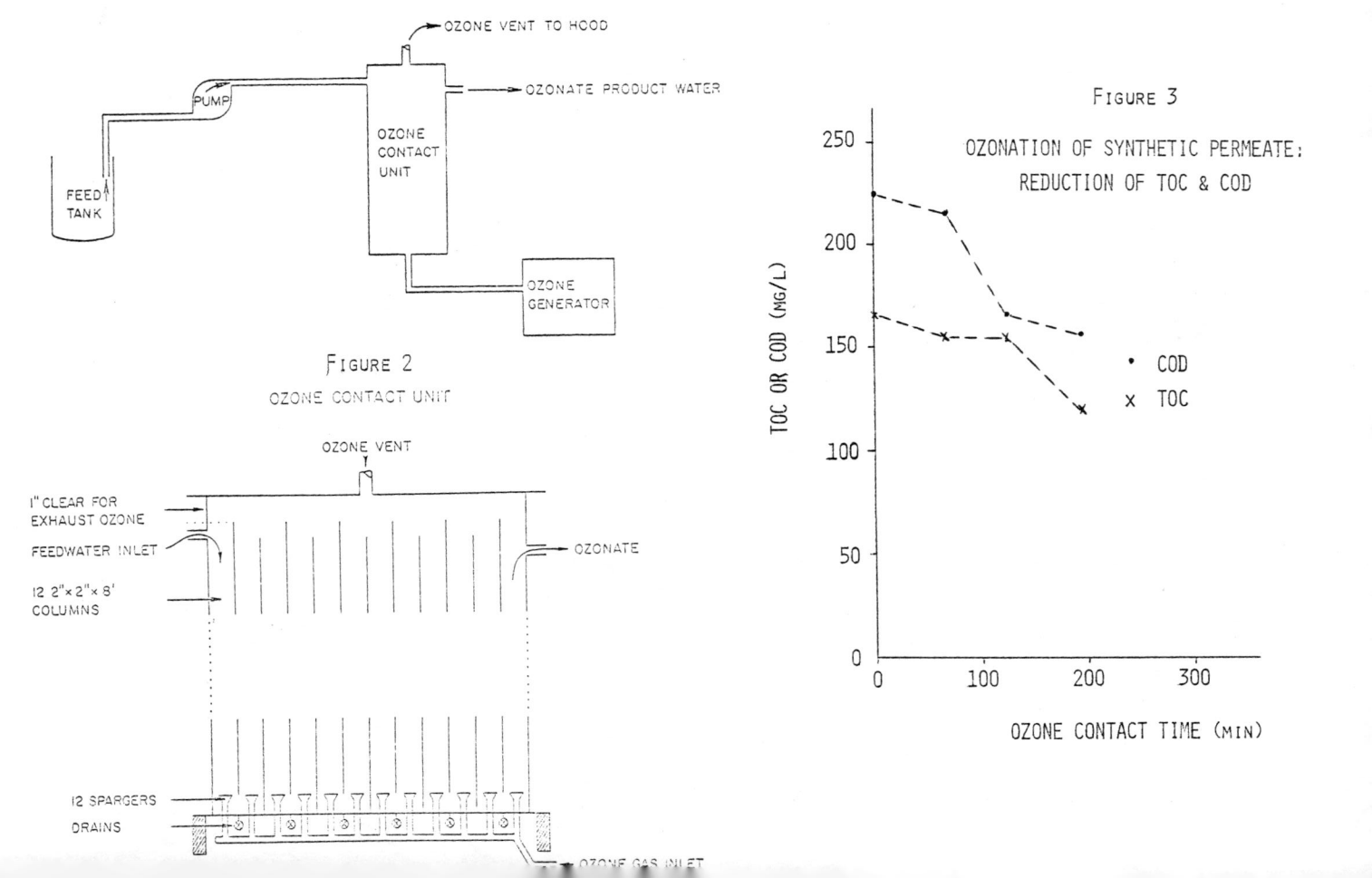

Figure 1 OZONATION SYSTEM

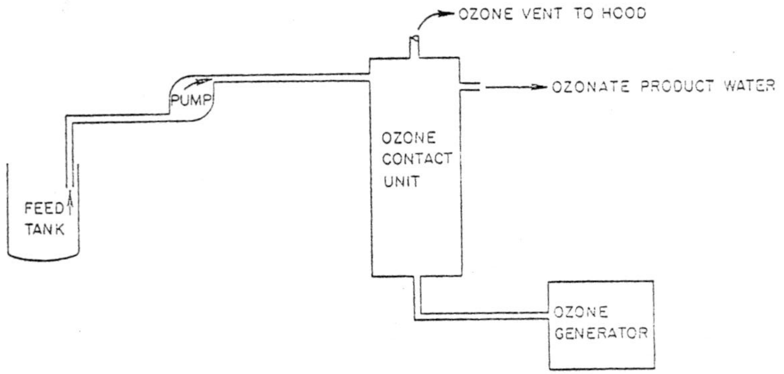

Figure 2
OZONE CONTACT UNIT

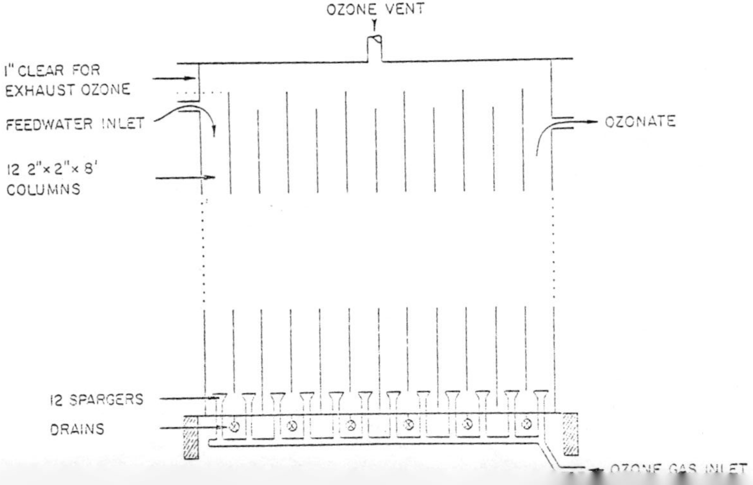

Figure 3
OZONATION OF SYNTHETIC PERMEATE:
REDUCTION OF TOC & COD

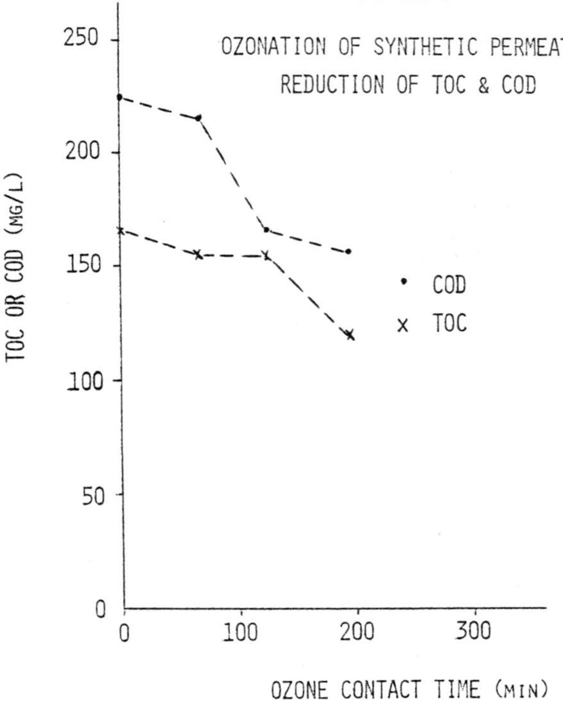

• COD
× TOC

-112-

FIGURE 4

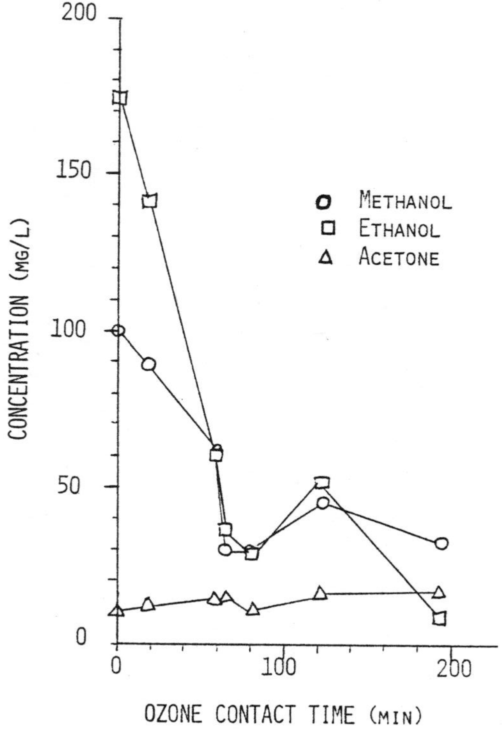

FIGURE 5

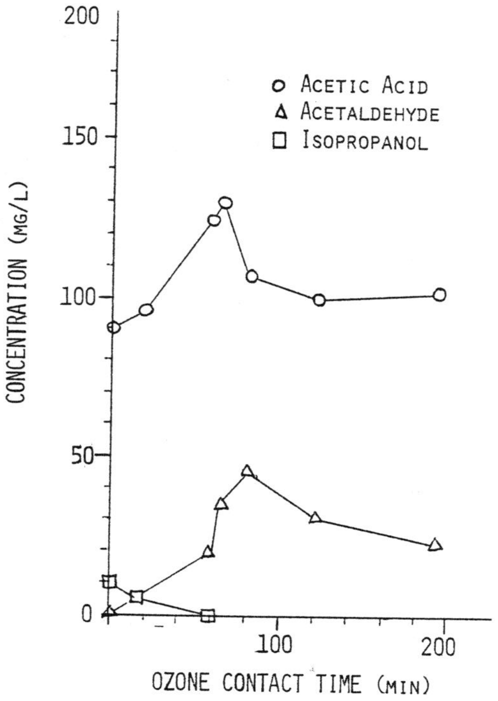

FIGURE 6

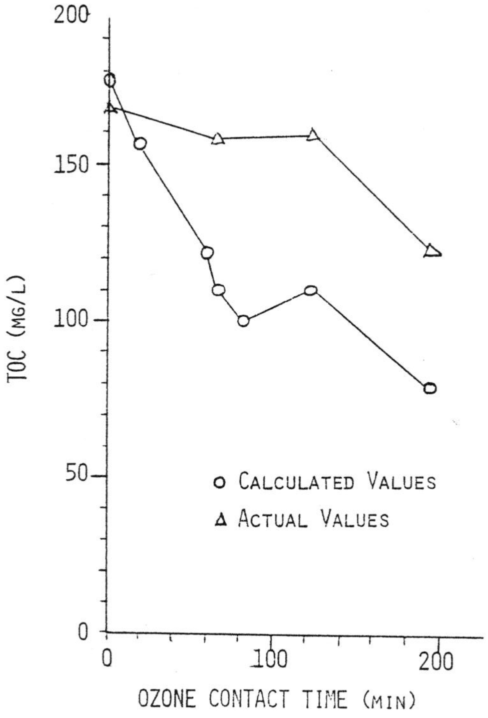

-113-

FIGURE 7 EFFECT OF OZONATION ON SYNTHETIC PERMEATE CYTOTOXICITY

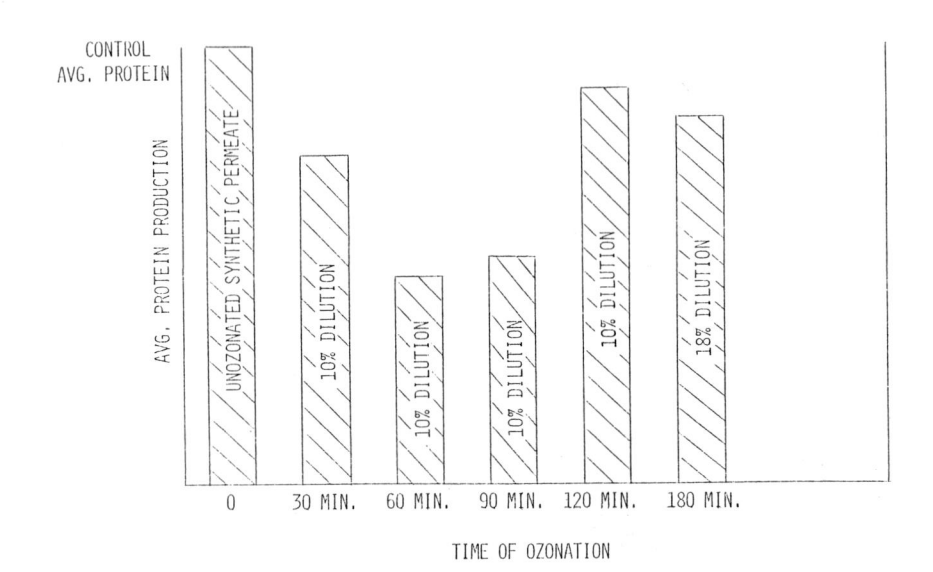

FIGURE 8 EFFECT OF OZONATION ON SYNTHETIC PERMEATE CYTOTOXICITY

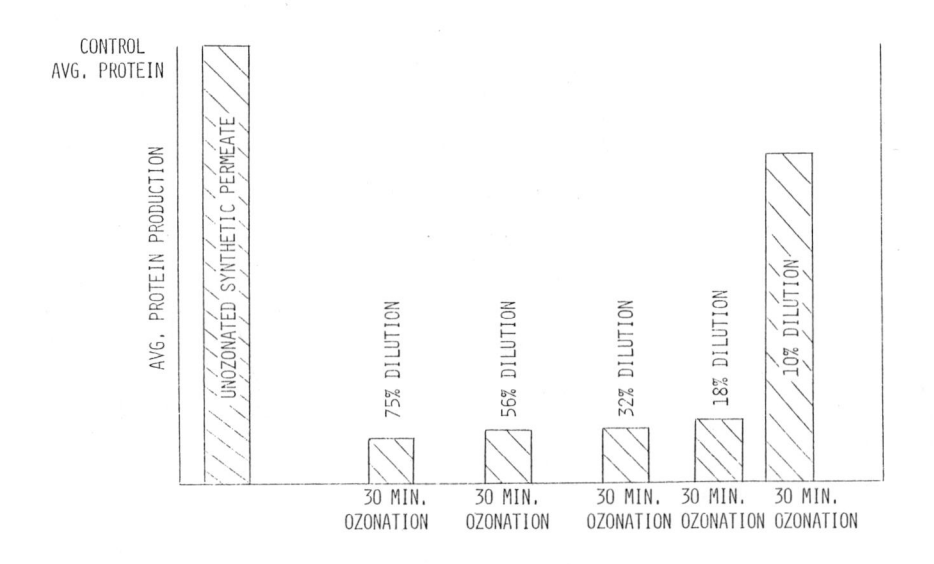

OZONE METHODS AND OZONE CHEMISTRY OF SELECTED
ORGANICS IN WATER*

1. BASIC CHEMISTRY

Ronald J. Spanggord and Vernon J. McClurg

Stanford Research Institute, 333 Ravenswood Ave.,
Menlo Park, California 94025

Introduction

The use of chlorine in water purification treatment has recently come under critical review due to the findings of low levels of chlorinated organics in drinking water. The potential biological hazard that these chemicals represent to man is unknown; however, some of these materials, such as $CHCl_3$, are known carcinogens in animals.

These findings have led to the search for alternatives to chlorination for water purification. Ozonation is one alternative and this process has been used in Europe for many years. However, the biological effects of the products of ozonation are unknown and should be evaluated before ozonation is considered as a replacement for chlorination in the United States.

In this study we have attempted to evaluate the potential biological hazard of ozonation through the use of rapid microbial bioassays of selected organic compounds subjected to ozonation conditions. These evaluations serve as a preliminary estimate of the production of biologically active compounds by use of the ozonation water treatment technique.

Chemistry

The objectives of the chemical evaluations were to provide aqueous solutions of selected organic compounds for mutagenic screening under pre- and post-ozonation conditions, and to identify organics formed in biologically active solutions. The chemistry of selected compounds is discussed in this report.

Chemicals Investigated

A listing of the chemicals investigated in this study appears in Table 1. Initial concentrations,

---

*Discussion follows subsequent paper.

percent reacted, ozonation time and concentration, pH change, and method of analysis of these chemicals are also listed.

In this study, our goal was to produce sufficient concentrations of intermediate components to evaluate them for potential mutagenicity by Ames testing. To achieve sufficient concentration, reaction times and reactant concentrations were not selected to duplicate water purification treatment. Maximum aqueous solubility of test chemicals or 1% solutions were used in this study.

Methods

Ozonations were performed in an all-Pyrex glass reactor, using a Welsbach Model T-408 ozone generator. The ozonation vessel is shown in Figure 1. Ozone was bubbled into the reactor from two all-glass frits, and a high speed overhead stirrer initiated efficient mass transfer action. Four glass baffles with 1/8-in. spacings from the reactor walls prevented solution vortexing.

Ozone concentrations were monitored by standard iodometric titrations (1), and component identifications were based on gas chromatographic/mass spectral (gc/ms) data and other spectrometric techniques.

Product Evaluation

Since very few of the test chemicals possessed mutagenicity after ozonation, product identifications were not extensive in most cases. However, we were able to identify and suggest some intermediates that resulted during the study. These intermediates are discussed below.

Oleic Acid. Oleic acid reacted readily with ozone and yielded classical ozonation products (Eq. 1)

[1] $C_8H_{17}CH=CH(CH_2)_7COOH + O_3 \longrightarrow$

$C_8H_{17}C(=O)OH$

$+$

$HC(=O)(CH_2)_7C(=O)OH$

$+$

$HOC(=O)(CH_2)_7C(=O)OH$

These products were identified by gc/ms as their trimethylsilyl esters. Intermediate ozonides or epoxides probably do not survive in the final pH (3.8) of the system.

Diethylamine. The ozonation of diethylamine produced two new products with molecular ions at m/e = 59 and m/e = 87. One compound was positively identified as acetaldoxime (I, m/e = 59).

[2] $CH_3CH_2\!\!>\!\!N-H + O_3 \longrightarrow CH_3CH=NOH$ (I) m/e = 59
CH_3CH_2
$+$
$CH_2=CHNCH_2CH_3$ m/e = 87
    $|$
    OH

The other product may be represented as Compound II or Compound III.

$CH_2=CHNCH_2CH_3$ $CH_3CH=\overset{+}{N}CH_2CH_3$
      $|$
(II) OH (III) $\underline{O}$

Compound III could possibly be an intermediate to acetaldoxime (Eq. 3) through an intramolecular elimination of ethylene.

[3] $CH_3CH=\overset{+}{N}\overset{CH_2}{\underset{O\;\;\;H}{\diagdown\!\diagup}}CH_2 \longrightarrow CH_3CH=N\!\!-\!\!OH + CH_2=CH_2$

N-ethylacetamide (m/e = 87), a potential product of the ozonation of diethylamine, was eliminated on the basis of its mass spectrum.

Several nitrogen function compounds were found to incorporate 16 mass units into the parent structure. Caffeine is one example (Eq. 4).

[4]

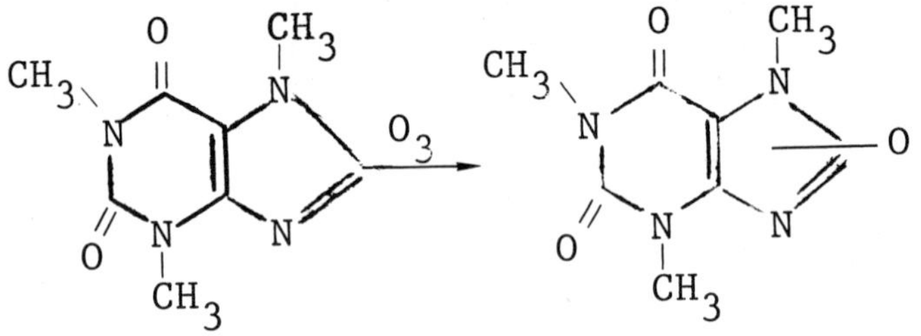

[Structure: caffeine + O₃ → oxidized caffeine derivative]

A reaction common to tertiary amines (3) is the formation of amine oxides (Eq. 5) and this reaction may be occurring at one of the tertiary amine sites in caffeine.

[5] $$R_3N + O_3 \longrightarrow R_3\overset{+}{N}-O^-$$

Diphenylhydrazine was ozonized as the hydrochloride salt. One product was observed in the gas chromatographic profile that indicated the incorporation of oxygen in the aromatic rings (Eq. 6).

[6]

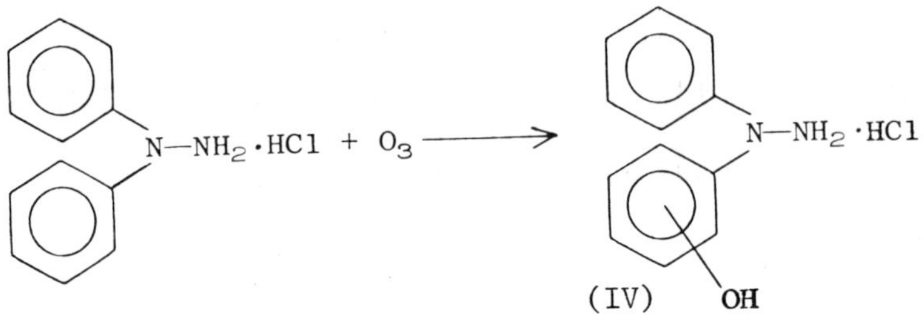

[Structure: diphenylhydrazine·HCl + O₃ → hydroxylated product (IV)]

Compound IV formed mono-, di-, and tri-TMS derivatives. Amine oxides have to be ruled out on the basis that a tri-TMS derivative would have a molecular ion at m/e = 417 while the tri-TMS derivative of Compound IV would have a molecular ion at m/e = 416 which was found. Ring hydroxylation is not unusual at the pH of the reaction mixture (2.9). The amine function is protonated at this pH and is probably nonreactive.

Under buffered conditions (pH 7), both ring- and nitrogen-hydroxylated derivatives of diphenylamine were found along with diphenylamine (Eq. 7).

[7]

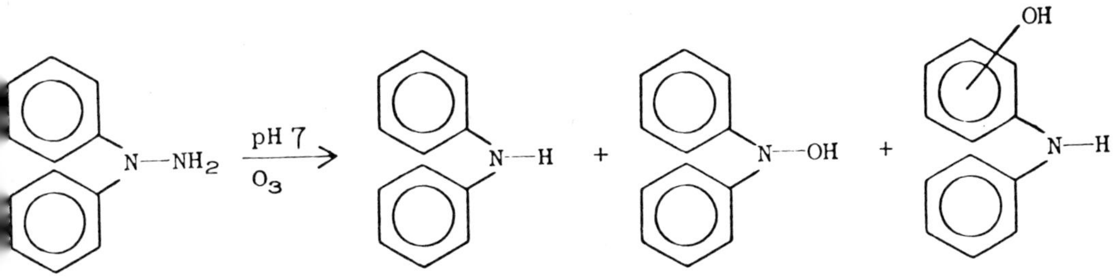

$$\text{Ph}_2\text{N-NH}_2 \xrightarrow[O_3]{pH\ 7} \text{Ph}_2\text{N-H} + \text{Ph}_2\text{N-OH} + (\text{HO-C}_6\text{H}_4)\text{PhN-H}$$

Phenol. Phenol produced two products on ozonation that were identified as catechol (V) and resorcinol (VI).

[8]

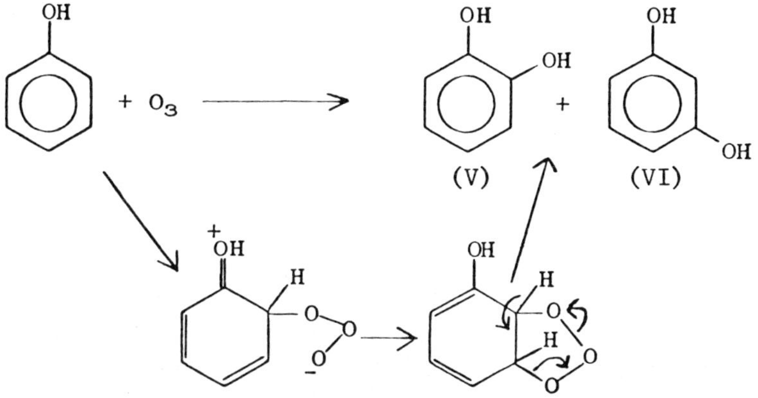

(V), (VI), (VII)

Catechol has been observed by other workers (4), but resorcinol has not. An intermediate such as (VII) may be responsible for both products. Compound VII could initially be formed by an electrophilic substitution of ozone on phenol followed by nucleophilic substitution on the ring. Two modes for the elimination of oxygen are possible to yield both catechol and resorcinol (Eq. 8).

Glycine. One product was found in the reaction of glycine with ozone. This compound formed a tetra-TMS derivative (m/e = 395) and might be represented as Compound VIII, indicating both electrophilic attack of ozone on the amine function and side chain oxidation.

$$^+NH_3CH_2\overset{O}{\underset{\|}{C}}-O^- + O_3 \longrightarrow HONHC\underset{OH}{\overset{O}{\underset{\|}{\overset{\|}{C}}}}H-OH \longrightarrow$$

(VIII)

$$TMSON\underset{OTMS}{\overset{TMS\ O}{\underset{\|}{\overset{\backslash\ \|}{C}}}}HC-OTMS$$

Ethanol. The reaction of ethanol with ozone was found to produce acetaldehyde and acetic acid (Eq. 9).

[9] $$CH_3CH_2OH + O_3 \longrightarrow CH_3C\overset{O}{\underset{H}{\diagdown}} + CH_3C\overset{O}{\underset{OH}{\diagdown}}$$

These products have been identified by other workers (5). However, in this study mutagenic activity was observed from ozonized ethanol solutions and this activity could not be attributed to acetaldehyde or acetic acid.

The ozonized ethanol solution was found to oxidize iodide to iodine, and liberate oxygen when treated with lead tetraacetate. The liberation of oxygen with lead tetraacetate is specific for hydroperoxides (6). Since glc and hplc profiles showed no other components besides acetaldehyde and acetic acid, thin-layer chromatography was used and plates were sprayed with potassium iodide to detect the hydroperoxide. A spot was detected which had an Rf identical to that of hydrogen peroxide. Also, a low temperature titration of the hydroperoxide with ceric ion, a method reported to distinguish alkyl hydroperoxides from hydrogen peroxide (7), was positive for hydrogen peroxide. However, biological tests on hydrogen peroxide (Ames test) indicated much less mutagenic activity than observed in ozonized ethanol.

The ozonation of ethanol was performed in D_2O to evaluate the nuclear magnetic resonance spectrum. During the course of the reaction, a doublet appeared at 1.3δ, a singlet at 5.0δ, and a quartet at 5.4δ, in addition to an acetic acid singlet at 2.1δ, and a singlet due to water at 4.8δ. Ethanol was removed by concentration under vacuum, and integration of the proton signals of the unknown indicated a ratio of 3:1:1. The ^{13}C-nmr spectrum of the mixture showed three signals for the unknown and two signals for acetic acid. Off-spin resonance of the ^{13}C-nmr signals indicated that the signals arose from methyl (40 ppm), methylene (110 ppm), and methine carbons (118 ppm). Based on this data as well as chemical shifts, the structure of the unknown is Compound IX.

$$CH_3CH(OOH)-OCH_2O-CH(OOH)-CH_3$$

(IX)

When acetaldehyde, formaldehyde, and hydrogen peroxide were mixed together, proton nmr signals identical to those of the ozonized ethanol solution were observed.

The formation of Compound IX appears to arise from the condensation of a reaction product of acetaldehyde and hydrogen peroxide with formaldehyde (Eq. 10).

[10] $$CH_3CHO + H_2O_2 \rightleftharpoons CH_3CH(OH)(OOH)$$

$$2\ CH_3CH(OH)(OOH) + CH_2O \longrightarrow CH_3CH(OOH)-OCH_2O-CH(OOH)CH_3$$

Formaldehyde could arise from the ozonation of the enol form of acetaldehyde (Eq. 11) or more likely by complex free radical processes on ethylhydroxyhydroperoxide (Eq. 12).

[11] $$CH_3CHO \rightleftharpoons CH_2=CH(OH) \xrightarrow{O_3} CH_2=O + O=CH(OH)$$

[12] $$CH_3CH(OH)(OOH) \longrightarrow CH_3CH(OH)(O\cdot) \longrightarrow [CH_3\cdot] \xrightarrow{O_2}$$

$$CH_3O_2\cdot \longrightarrow CH_3OH + CH_2O$$

LITERATURE CITED

1. Federal Register, Vol. 36, No. 84, Part II, p. 8195, April 30, 1971.

2. H. B. Henbest and M. T. Stratford, J. Chem. Soc. 711 (1964).

3. P. S. Bailey, et al. J. Org. Chem. $\underline{33}$, 2675 (1968).

4. H. R. Eisenhauer, J. Water Pollution Control Federation, $\underline{40}$, 1887 (1968).

5. O. S. Borsis and J. N. Kokavoaras, Zeit. Phys. Chem., $\underline{50}$, 160 (1968).

6. D. C. Criegee, Collect Czech. Chem. Comm. $\underline{33}$, 3468 (1968).

7. L. A. Bodouskopa, et al. Zh. Anal. Khim. $\underline{22}$, 1268 (1967), CA. $\underline{80}$, 2674.

ACKNOWLEDGEMENT

This work was performed under Contract 68-01-2894 funded by the U.S. Environmental Protection Agency, Offices of Planning & Management, Research & Development and Water Supply, and by the U.S. Army Research & Development Command.

Table 1

OZONATION OF SELECTED ORGANICS

Compound	Initial Conc.	% Reacted	Ozonation Time (min)	pH Initial	pH Final	Ozone Conc. (g/hr)	Methods of Analysis
2,4-Dinitrotoluene	80 ppm	55	17	8.0	3.5	6	GC
Benzoic Acid	3,622 ppm	14	60	4.3	3.9	6.8	GC
Ethanol	7,759 ppm	74	360	5.4	3.2	9.3	GC, tlc, nmr
Benzo(a)pyrene	5.3×10^{-4} ppm	100	60	6.8	5.9	5.5	HPLC, fluorescence detection
HMX	0.3 ppm	46	75	9.2	5.5	5.5	HPLC, uv detection
RDX	31 ppm	8	30	8.4	5.2	6.7	HPLC, uv detection
Oleic Acid	99 ppm	100	30	5.8	3.8	5.1	GC
Humic Acids	0.05% suspension	solubilization	20	7.1	3.2	5.1	GC, HPLC
Condensate water	40 ppm	88	120	7.2	3.6	4.5	GC
Diphenyl-hydrazine·HCl	8,190 ppm	34	2	2.9	2.2	6.7	GC, HPLC
Nitrilotriacetic Acid, Na3	10,450 ppm	50	60	11.0	9.7	5.7	UV, zirconium complex
Aldrin	0.005% suspension	100	30	7.5	4.4	5.7	GC
DDT	7×10^{-3} ppm	86	90	7.9	4.5	5.7	GC
Aroclor 1254	0.005% suspension	--	60	6.7	4.5	5.7	GC
Acetic Acid	12,630 ppm	3.5	360	2.7	2.7	6.2	GC

Table 1 (Concluded)

Compound	Initial Conc.	% Reacted	Ozonation Time (min)	pH Initial	pH Final	Ozone Conc. (g/hr)	Methods of Analysis
Glucose	10,000 ppm	100	30	7.2	3.6	4.2	GC
Urea	9,432 ppm	3.8	180	7.6	4.2	4.1	GC
Cholesterol	Saturated	--	30	6.5	4.6	4.5	GC
Benzidine·2 HCl	4,830 ppm	64	30	2.8	2.1	4.5	GC
Glycine	11,900 ppm	22	130	6.3	3.9	4.5	GC
Cysteine	9,622 ppm	18	32	5.0	2.8	4.5	GC
Benzene	510 ppm	~26	25	7.1	3.6	5.2	GC
Thymine	2,227 ppm	95	16	7.3	3.0	5.2	GC
Caffeine	9,660 ppm	61	60	7.1	4.9	6.2	GC
Diethylamine	8,040 ppm	37	60	11.1	11.1	6.2	GC
Phenol	9,140 ppm	41	110	6.9	2.4	3.8	GC
Hydroquinone	15,260 ppm	16	95	6.9	3.2	3.8	GC
Glycerol	7,240 ppm	100	300	5.1	2.8	3.9	GC
Alanyl glycyl glycine	462 ppm	0	45	5.3	3.8	6.0	UV

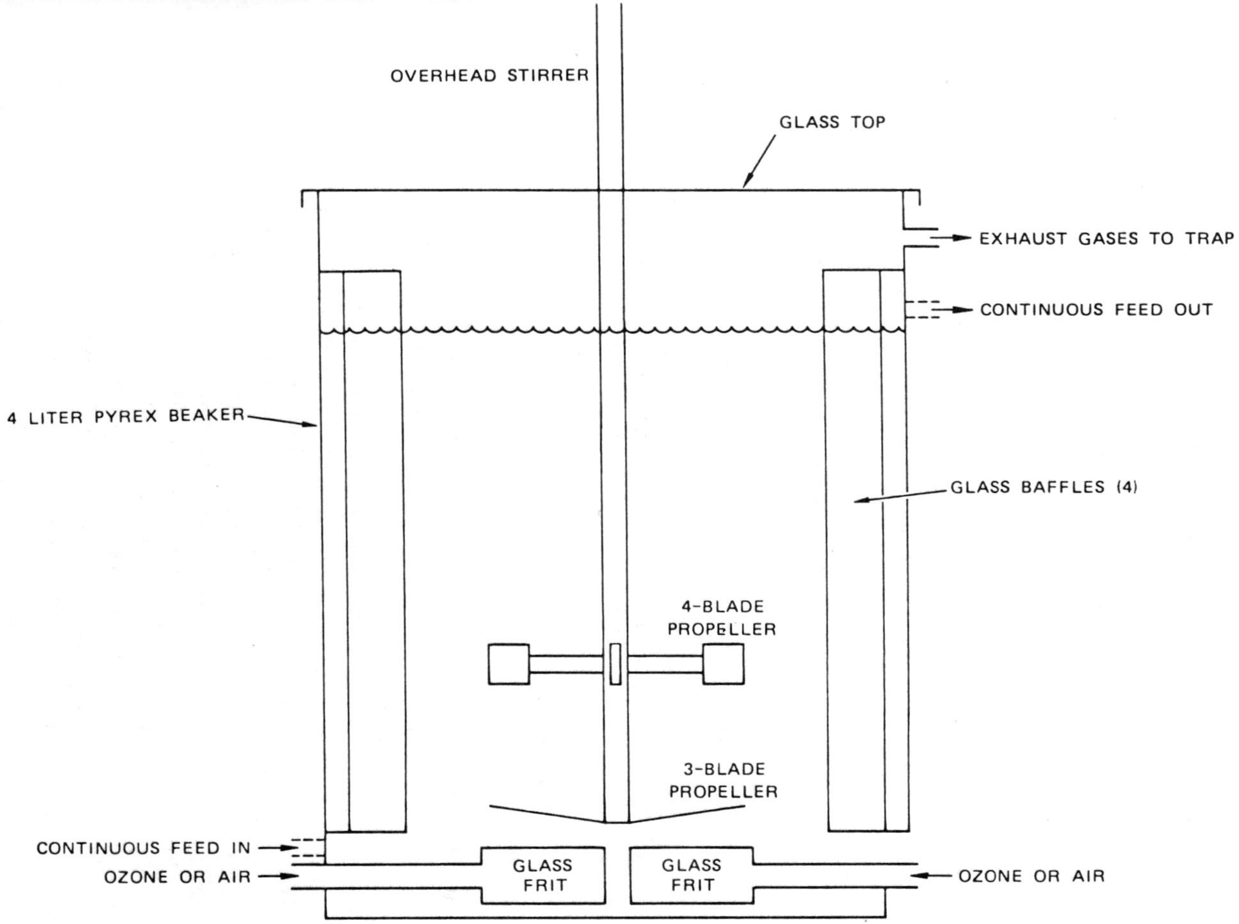

FIGURE 1 SCHEMATIC DIAGRAM OF OZONATION REACTION VESSEL

- 125 -

OZONE METHODS AND OZONE CHEMISTRY OF SELECTED
ORGANICS IN WATER

2. MUTAGENIC ASSAYS

Vincent F. Simmon, Sharon L. Eckford and Ann F. Griffin

Stanford Research Institute, 333 Ravenswood Avenue,
Menlo Park, California   94025

Introduction

The use of microbiological mutagenicity assays to assess the potential carcinogenic activity of organic compounds has gained increased acceptance because these assays are relatively inexpensive, rapid, and have shown a high correlation with known organic carcinogens. It has been found that a number of halogenated hydrocarbons are formed when water is disinfected by chlorination. Before an alternative disinfection process is chosen, it would be desirable to know what, if any, potentially hazardous products are formed. Therefore, these studies were undertaken to determine if mutagens are formed when selected organic carcinogens are ozonated. The chemicals ozonated and chemical analysis are presented in the accompanying paper.

Methods

1.   Mutagenic assays with Salmonella typhimurium

The Salmonella typhimurium strains used at SRI are all histidine auxotrophs by virtue of mutations in the histidine operon. When these histidine-dependent cells are grown on a minimal media petri plate deficient in histidine, only those cells that revert to histidine independence (his^+) are able to grow (mutants). The spontaneous mutation frequency of each strain is relatively constant, but when a mutagen is added to the agar the mutation frequency is increased 3 to 100 times.
We obtained our S. typhimurium strains from Dr. Bruce Ames of the University of California at Berkeley.[1-4] In addition to having mutations in the histidine operon, all the indicator strains have mutations in the lipopolysaccharide coat (rfa^-) and deletions that cover a gene involved in the repair of uv damage ($uvrB^-$). The rfa^- mutations makes the strains more permeable to large molecules, thereby increasing their sensitivity to these molecules. The $uvrB^-$ mutation decreases repair of some types of chemically damaged DNA and thereby enhances

sensitivity to some mutagenic chemicals. Strain TA1535 is reverted to histidine prototrophy (his$^+$) by many mutagens that cause base-pair substitutions. TA100 is derived from TA1535 by the introduction of the R factor plasmid pKM101.[5] The introduction of this plasmid, which confers ampicillin resistance to the strain, greatly enhances the sensitivity of the strain to some base-pair substitution mutagens. We have shown that mutagens such as benzyl chloride and 2-(2-furyl)-3-(5-nitro-2-furyl) acrylamide (known as AF2) can be detected in plate assays by TA100 but not by TA1535. The presence of this plasmid also makes strain TA100 sensitive to some frameshift mutagens-- eg., ICR-191, benzo(a)pyrene, alfatoxin B_1, and 7,12-dimethylbenz(a) anthracene. Strains TA1535 and TA1538 are reverted by many frameshift mutagens. TA1537 is more sensitive than TA1538 to mutation by some acridines and benzanthracenes but the difference is quantitative rather than qualitative. TA98 is derived from TA1538 by the addition of the plasmid pKM101, which makes this strain more sensitive to some mutagens.

All the indicator strains are stored at -80°C. For each experiment, an inoculum from frozen stock cultures is grown overnight at 37°C in a nutrient broth consisting of 1% tryptone and 0.5% yeast extract. After stationary overnight growth, the cultures are shaken for sensitivity to crystal violet. The presence of <u>rfa$^-$</u> mutation makes the indicator strains sensitive to this dye, whereas the parent strain, <u>rfa$^+$</u>, is not sensitive to the dye. However, the mutation is reversible, leading to the accumulation of <u>rfa$^+$</u> cells in the culture. Therefore, the cells must be tested routinely to ensure their sensitivity to crystal violet. Each culture also is tested by specific mutagens known to revert each test strain (positive controls).

Assays in Agar

To a sterile 13 x 100 mm test tube placed in a 43°C heating block, we add in the following order:

(1) 2 ml of 0.6% agar*
(2) 0.05 ml of indicator organisms
(3) 0.5 ml of metabolic activation mixture (optional)
(4) 250 µl of a solution of the test chemical

For the negative controls, we use steps (1) (2), and (3) (optional) and 50 µl of the solvent used for the test chemical.

---

* 0.6% agar contains 0.05 mM histidine and 0.05 mM biotin.

This mixture is stirred gently and then poured onto minimal agar plates.† After the soft agar has set, the plates are incubated at 37°C for 2 days. The number of his⁺ revertants are counted and recorded. Some of the revertants are routinely tested to confirm that they are his⁺, require biotin, and are sensitive to crystal violet (rfa⁻).

2. Mutagenic assays with Saccharomyces cerevisiae

The yeast S. cerevisiae D3 is a diploid heterozygous for a mutation in an adenine-metabolizing enzyme.[6] Cells homozygous for this mutation produce a red dye when grown on medium containing adenine. Adenine-requiring homozygotes can be generated from the heterozygotes by mitotic recombination. Many mutagens increase the frequency of mitotic recombination. Mitotic recombination is indicated by the development of colonies with red pigmentation, and the degree of conversion to this pigmented colony indicates the mutagenicity of a compound or its metabolite.[7]

The in vitro yeast mitotic recombination assay is conducted in a 0.067 M phosphate buffer suspension containing the chemical and a metabolic activation system (optional). The suspension is incubated at 30° for 4 hours. After incubation, the sample is diluted serially in sterile saline and plated on tryptone-yeast agar plates. Plates of a 10^{-5} and 10^{-3} dilution are incubated for 2 days at 30°C, followed by 2 days at 4°C to enhance the development of the red pigment indicative of adenine-negative homozygosity. To detect red colonies or red sectors, we scan the plates with a dissecting microscope at 10 x magnification. Plates of a 10^{-5} dilution are incubated for 2 days at 30°C for determination of the total number of colony-forming units.

3. Metabolic Activation—Aroclor 1254-Stimulated Metabolic Activation System

Some carcinogenic mutagens (e.g., dimethylnitrosamine) are inactive unless they are converted to their active form by being metabolized. Ames has described the metabolic activation systems we use.[8,9] Adult male rates (200 to 250 g) are given single 500-mg/kg intraperitoneal injection of a polychlorinated biphenyl (Aroclor 1254)[9]. Four days after the injection, the animals' food is removed. On the fifth day, the rats are killed and the liver homogenate prepared.

---

† Minimal agar plates consist of 15 g of agar, 20 g of glucose, 0.2 g of $MgSO_4 \cdot 7 H_2O$, 2 g of citric acid monohydrate, 10 g of K_2HPO_4, and 3.5 g of $NaHNH_4PO_4 \cdot H_2O$ per liter.

The livers are removed aseptically and placed in preweighed, sterile glass beakers. The organ weight is determined, and all subsequent operations to the metabolic activation step are conducted in an ice bath. The organ is washed in an equal volume of cold, sterile surgical scissors in three volumes of 0.15 MKCl, and homogenized with a Potter-Elvehjem apparatus. The homogenate is centrifuged for 10 minutes at 9000 x $\underline{g}$, and the supernatant is removed and stored in liquid nitrogen. To the postmitochondrial supernatant are added $MgCl_2$, KCl, glucose-6-phosphate, TPN, and sodium phosphate (pH 7.4).

The metabolic activation system for each experiment consists of (for 10 ml):

(1) 1.0 ml postmitochondrial supernatant
(2) 0.2 ml of $MgCl_2$ (0.4 M) KCl (1.65M)
(3) 0.05 ml of glucose-6-phosphate (1 M)
(4) 0.4 ml of NADP (0.1 M)
(5) 5.0 ml of sodium phosphate (0.2 m pH 7.4)
(6) 3.35 ml of H_2O.

4. Chemicals

The chemicals tested and the ozonation of them are described in the previous paper. In the assays with $\underline{S.\ typhimurium}$ 0.25 ml (or a dilution with the same volume) was added in each assay. In the $\underline{S.\ cerevisiae}$ assay, 1 ml of the test chemical before and after ozonation (or a dilution) was added to 1.5 ml of buffer, yeast cells and S9 mix (when present).

Results
───────

Oleic acid, diethylamine, glycine, 2,4-dinitrotoluene, benzoic acid, benzo(a)pyrene, HMX, RDX, humic acids, TNT condensate water, aldrin, DDT, Aroclor 1254, acetic acid, glucose, urea, cholesterol, cysteine, benzene, thymine and caffeine were not mutagenic prior to or after ozonation at the concentrations tested. Some of these chemicals (e.g., benzo(a)pyrene) are known to be mutagenic; however, not at the concentrations used in these assays.

1. Ethanol

Prior to ozonation, the ethanol assay was not mutagenic. After prolonged ozonation (4-6 hours) mutagenic activity was observed. There was a slight (~ 2-fold) increase in the number of revertants in TA100, and a greater than 20-fold increase in the number of mitotic recombinants in $\underline{Saccharomyces}$. Metabolic activation significantly decreased mutagenic activity. Acetaldehyde was considered as the possible mutagen, but when tested with Saccharomyces it was not mutagenic. Preliminary experiments indicated that hydrogen peroxide was weakly mutagenic and toxic in TA100 and in Saccharomyces.

However, the mutagenic activity we observed with H_2O_2 was far less (1/100) than expected. Ethyl hydroperoxide with the appropriate oxidizing potential was mutagenic, but far more toxic than ozonated ethanol. A mixture of H_2O_2, acetaldehyde ethanol and acetic acid in proportions approximating their concentrations after extended ozonation of ethanol had approximately the same mutagenic activity as ozonated ethanol, but was quite toxic.

In another reconstruction experiment, acetaldehyde and H_2O_2 were added to form ethylhydroxy hydroperoxide which was mutagenic, but more toxic than ozonated ethanol.

A mixture of acetaldehyde, formaldehyde, and H_2O_2 gave a product identical in mutagenic activity and in an NMR spectrum to ozonized ethanol.

2. 1,1-Diphenylhydrazine

Prior to ozonation, diphenylhydrazine was toxic, but not mutagenic in the Salmonella assays. After ozonation, it was weakly positive in Salmonella TA1535 and TA100 when it was exhaustively ozonated (60 minutes), but was not mutagenic in experiments in which ozonation was limited to less than 100%. Diphenylhydrazine was not mutagenic in assays with Saccharomyces.

3. Nitrilotriacetic acid (NTA)

NTA was not mutagenic in Salmonella pre- or post-ozonation. In one experiment there was a slight increase in mutagenic activity in S. cerevisiae D3 after ozonation.

4. Phenol

Phenol was not mutagenic prior to ozonation. After ozonation, toxic products were present and the solution was weakly mutagenic in some assays with S. cerevisiae D3.

5. Benzidine

The known carcinogen/mutagen (when metabolically activated) was mutagenic in Salmonella when metabolically activated prior to ozonation. Mutagenic activity was observed both in the presence and absence of metabolic activation after ozonation. This result suggests that oxidation of the amine group (e.g., to a hydroxylamine or nitro group) may have occured as a result of ozonation. Prior to ozonation, benzidine was not mutagenic in Saccharomyces even in the presence of a metabolic activation system. However, after ozonation, mutagenic activity was observed. These results reinforce the observations in Salmonella that direct acting mutagenic products(s) are formed when benzidine is ozonized.

LITERATURE CITED

1. J. McCann, E. Choi, E. Yamasaki, and B. N. Ames, Proc. Nat. Acad. Sci. USA, 72, 5135 (1975).

2. B. N. Ames, E. G. Gurney, J.A. Miller and H. Bartsch. Proc. Nat. Acad. Sci. USA, 68, 3128 (1972).

3. B. N. Ames, F. D. Lee, and W. E. Durston. Proc. Nat. Acad. Sci. USA, 70, 782 (1973).

4. B. N. Ames, E. W. Durston, E. Yamasaki, and F. D. Lee. Proc. Nat. Acad. Sci. USA, 70, 2281-2285 (1973).

5. J. McCann, N. E. Spingarn, J. Kobori, and B. N. Ames. Proc. Nat. Acad. Sci. USA 72, 979-983 (1975).

6. F. K. Zimmermann and R. Schwaier. Mol. Gen. Genet., 100, 63 (1967).

7. D. J. Brusick and V. W. Mayer. Environ. Health Perspectives, 6, 83 (1973).

8. L. D. Kier, E. Yamasaki, and B. N. Ames. Proc. Nat. Acad. Sci. USA, 71, 4159 (1974).

9. B. N. Ames, J. McCann, and E. Yamasaki. Mut. Res., 31, 347 (1975).

ACKNOWLEDGEMENT

This work was performed under Contract 68-01-2894 funded by the U.S. Environmental Protection Agency, Offices of Planning & Management, Research & Development and Water Supply, and by the U.S. Army Research & Development Command.

DISCUSSION OF SPANGGORD AND SIMMON PAPER

Ed Greenberg, FMC Corporation: What are the lifetimes of any of the peroxides or hydroperoxides in the culture media when you ran the Ames test? You've got aerobic organisms which also produce catalase and probably metallic ions which will also catalyze the decomposition. What do you expect are the lifetimes of the hydroperoxide and the peroxides?

Spanggord: Let me say something first about the stability of that hydroperoxide that we found. It has very good stability. You can leave it on the shelf for two months, which we did, and rerun the assay and still find the mutagenic response. It's a highly stable compound, at least at pH 3 of the reaction mixture, which began with purified water. However, you are correct that metal ions and other substances would affect the stability in a more typical water.

Simmon: This may explain why it appeared to be more mutagenic in yeast assay systems on the Ames plate. With these exposures, in buffer for a period of 4 hours, the cells are not growing, whereas in the Ames assay the chemical is incorporated into the top Agar, and as you point out, this is a medium for growth, and there may be inactivation of the compound. In fact, there was much less mutagenic activity found in the Ames procedure than in the yeast procedure.

Greenberg: Were you able to see any chemical reaction of this peroxide with the medium?

Simmon: No. There was no visible reaction.

Unknown: I would like to know the concentration of the organics and the ozone dosage.

Spanggord: For which compound?

Unknown: For ethanol, acetic acid, etc.

Spanggord: Usually they would run about 6.3 g/hr of ozone. These were high doses, and their contact times varied, but I think for benzidine we went 60 minutes.

Unknown: What was the amount of benzidine?

Spanggord: Ethanol was 7,000 parts per million. I think benzidine was also at a high level since it was present as its hydrochloride salt --- I don't have the data right here, but I think it was in the area of 7 or 8 thousand parts per million.

Simmon: These data will all appear in the paper, along with those tables which were shown here.

Walter Blogoslawski, National Marine Fisheries, Milford, Connecticut: When you say 6.4 or 6.5 grams per hour do you mean applied dose or dissolved dose of ozone?

Spanggord: That is applied dose.

Blogoslawski: Then you don't have a dissolved dose?

Spanggord: No.

Bob Sievers, University of Colorado: You mentioned that positive tests for certain strains infer that you are dealing with polynuclear aromatic hydrocarbons, in the case in the Cincinnati water. Have you made any effort to identify which of the compounds these might be specifically?

Simmon: We know from the tests which have been run on pure compounds what classes of compounds are detected by each of the strains. And routinely, when we test any compound or any concentrate, it's on all of the strains, both in the presence and absence of the activation system. Now, those particular fractions, petroleum ether, diethyl ether and acetone, are currently being studied at EPA here in Cincinnati, and GC/Mass Spec and other techniques are being used to identify what is in each of those different fractions of the concentrate. So hopefully, that information will be available and will be published in the near future.

FRENCH METHODS FOR EVALUATING TOTAL MICROPOLLUTANT LOAD IN WATER, AND FOR DETERMINING ITS TOXICITY

Michel Rapinat

Compagnie Générale des Eaux
52 rue d'Anjou
75008 Paris, France

Evaluation of the quality and toxicity of water is a general problem encountered by everyone having a responsibility in the field of health and water distribution.

As a result of the highly diverse sources of pollution, the number of substances liable to be present in trace quantities in water is extremely high.

In the field of inorganic chemistry, the substances to be determined are relatively few in number, test methods are relatively simple, and results are obtained rapidly. This makes it possible to carry out a specific test for a given compound.

Organic compounds, on the other hand, are extremely varied; each year, a multitude of new products comes into use: their biodegradability, metabolic and toxic effects are often poorly understood. Furthermore, the corresponding measurement methods are time-consuming and difficult.

These factors make it essential, in this context, to have available global or overall methods for evaluating the quality and toxicity of a given water.

1 General problems presented by global methods

1.1 Interpretation of results. By the term micropollutant we mean a substance which is capable, even at very low dosage, of producing harmful or toxic effects on living organisms, on man in particular.

A major stumbling block in global methods is the problem of taking into account, not only micropollutants, but also a large number of substances which have no harmful effect on man, and which may even be beneficial to him.

At an overall level, correlation between quantity and quality, or between dosage and effect, is thus very difficult.

As a result, global measurements must be supplemented by specific analyses similar to those which are available for detergents, pesticides, and aromatic hydrocarbons, so as to make the results obtained more specific, and designate the substances to be eliminated.

However, when the nature of the micropollution is not subject to major changes, the use of global methods provides a good indication of quality level.

In particular, the results obtained can be useful in the following:

-- surveillance over time of changes in quality and toxicity of water intended for human consumption, or of surface and subterranean water sources used for the preparation of drinking water;

-- comparison of the effectiveness of new treatment procedures.

Furthermore, only global methods make it possible to evaluate the toxicity effects of interaction among combinations of pollutants. This is an extremely important factor, inasmuch as studies have shown that the toxicity of a mixture is commonly less than the sum of the individual toxicities of its components.

1.2 Concentration: the basic technique in current global methods. The total micropollutant load in a given water is, as a rule, extremely low. A value of several ppm constitutes in itself a high figure.

Thus, the efforts of research workers have been directed into two different channels:

-- the development of extremely sophisticated concentration procedures, to reduce experimental error and obtain representative concentrates;

-- the development of techniques of sufficient sensitivity to be usable directly, with no preliminary concentration step.

The first channel has already given rise to certain interesting results. The appearance of new techniques gives the second high promise for the future. In particular, mention may be made of organic carbon measurement in evaluating total micropollutant load.

In recent years, French laboratories have done much work on these themes: the aim of this paper is to:

-- present the principle of the most widely used concentration methods, with their advantages and defects;

-- give some results obtained by use of global techniques for evaluating total micropollutant load;

-- discuss a certain number of studies dealing with toxicity.

2 Concentration procedures

2.1 <u>Qualities desired in a concentration technique</u>.
The aim of concentration techniques is to obtain extracts that are as representative as possible of the initial samples. More particularly, the techniques in use must:

-- produce an extract containing all substances present in the water, in ionic or molecular form. Special attention must be paid to volatile products, which are liable to evaporate or be entrained;

-- avoid changing the nature of substances. In particular, heating of samples must be limited, since high temperatures can lead to decomposition or to recombination;

-- introduce as little contamination as possible, as a result of sample contact with equipment, reagents, solvents and adsorbents. This point must be examined with great care, when extracts are to be used for tests of toxicity, since contaminants can modify response to such tests;

-- restrict the concentration of inorganic compounds, which may interfere with toxicity testing;

-- provide the mass of extract necessary for subsequent testing: weighing, identification of families of compounds, toxicity testing.

In addition, certain practical criteria may be important, if routine operating tests are desired:
-- volume of sample to be taken
-- time required for analysis
-- simplicity of operations performed
-- operating precautions required.

2.2 <u>Principal techniques used</u>. Most of these methods involve an initial extraction step, followed by a concentration step. Extraction may be performed:

-- with a solvent;

-- by adsorption on activated carbon or a suitable resin, followed by solvent elution;

-- by inert gas entrainment, fixation on activated carbon, and finally solvent elution.

The choice of solvent is a delicate matter, which has a great influence on the extraction performance of the method used. A good solvent should have the following characteristics:

-- lack of selectivity, so as to take into account the greatest possible number of substances;

-- low solubility in water, facilitating subsequent separation;

-- low boiling point, to restrict changes resulting from heating and from evaporation of volatile products;

-- freedom from contamination of extracts intended for toxicity testing;

-- convenience in handling, including low toxicity and non-flammable nature.

Naturally, there is no universal solvent. It is necessary to select the solvent best suited to the application intended.

Chloroform is one of the most widely used solvents, as a result of its qualities, including limited selectivity, convenience in use, and low contamination of extracts for toxicity testing. However, its relatively high boiling point, 62°C at atmospheric pressure, makes it an unsuitable solvent for use with products of high volatility.

Cyclohexane also has desirable properties. It is suitable for extracting volatile substances. Unfortunately, it is not usable for toxicity testing.

Other solvents can also be used, including methanol, butanol, ethanol, and others.

Resin or activated carbon adsorption is a method which makes it possible to handle large volumes of water, since the adsorption columns are installed in the treatment plant. However, a certain degree of selectivity may be noted, depending on the adsorbent.

Concentration may be performed by the following:
-- evaporation at atmospheric pressure
-- vacuum evaporation
-- vacuum sublimation
-- freezing
-- freeze-drying.

The tendency in selecting one of these methods has been toward elimination of sample heating, so as to avoid decomposition and recombination as a result of temperature, together with loss of volatile substances.

In order to obtain more representative extracts, methods have grown more complex:

-- performing successive extractions at varying pH;

-- using a variety of solvents;

-- using several types of resin or activated carbon;

-- applying several different extraction methods in succession.

At the same time, sample volume has continually increased, to improve sensitivity.

The IRCHA "total" method is an example of this tendency:

--   extraction with chloroform at pH 7;

--   extraction with chloroform at pH 2;

--   extraction with chloroform at pH 10, after entrainment over aluminum hydroxide;

--   neutralization, adsorption on resin, elution with butanol;

--   adsorption on resin, elution with a mixture of ethanol and methylene chloride;

--   adsorption on carbon, washing with ethanol, extraction with chloroform.

In order to avoid the disadvantages of solvent techniques, another method has been explored, concentration by freezing the water sample at -15°C. This method also avoids loss of volatile substances by evaporation, as well as thermal decomposition of others.

However, a certain number of substances may be adsorbed by precipitated mineral salts. Likewise, the increase in the salinity of the concentrate may seriously interfere with toxicity testing.

Attempts to reduce inorganic salt concentration by a preliminary dialysis treatment have not succeeded in producing a satisfactory solution. An effort has been made to improve this procedure, by carrying out a solvent extraction after concentration by freezing. However, the results obtained have always been inferior to those obtained by direct concentration.

Thus, the freezing concentration method appears to be of interest only for waters with low mineral contents.

2.3 Comparison of methods used in France. The characteristics of the methods most widely used in France are given in Appendix 1. Depending on the objective desired, one or another of the methods will be chosen:

--   if a simple method is sought, that will be rapid in use with a small sample volume, then direct chloroform extraction is indicated;

-- if volatile substances are under study, then cyclohexane extraction or inert gas entrainment will be chosen;

-- if toxicity tests are planned on a water of low mineral content, then the freezing method may be of interest;

-- if the micropollutant concentration is very low, large volumes of water will be required, and the resin or activated carbon column adsorption method will be advisable;

-- if priority is given to the representative value of the sample, then a more complex method, such as the IRCHA method, will be selected.

3 Evaluation of total micropollutant load

3.1 *Methods used*. The most widely used method consists of weighing the dry residue obtained by the extraction/concentration procedure. Thus, the representativeness of the measurement depends on the technique used in concentrating the micropollutants.

Another technique involves determining the organic carbon content. As a rule, this measurement requires no concentration, and is therefore very fast and easy. This makes it suitable for examining changes in water quality, either over time or along a river's course. This is an example of a direct method.

3.2 *Comparison of CES and IRCHA method results*. It is of interest to compare these two methods, since they are respectively the simplest and the most complex of the extraction/concentration methods. Comparative measurements have been made with the CES and the IRCHA methods, at the inlet and outlet of a treatment plant.

Before treatment, the pollution level in the water tested was very high, from a variety of origins: industrial, domestic, agricultural, and natural. The treatment applied in this installation is quite sophisticated, and includes activated carbon and ozone treatment, in addition to settling/filtration and chlorination.

Figure N° 1 points up the selectivity of the CES method, since the value it gives corresponds roughly to

30 % of the IRCHA measurement. However, it may be suggested that the pollutant load determined by the CES method gives good comparative indications as to water quality, since the dispersion of the CES:IRCHA ratio is low.

On account of its simplicity, therefore, the CES method remains attractive in the initial analysis of a problem.

3.3 Comparison of activated carbon adsorption and IRCHA method results. The few comparative measurements that have been performed show a factor of 0.6 between the pollutant load determined by adsorption and that given by the IRCHA method. However, this factor is only indicative, and is no doubt strongly dependent on the nature of the water.

3.4 Variation of pollutant load with treatment. CES measurements have been made, over a period of more than a year, in various installations in the Paris region, and at various stages of treatment.

On an average, the reduction in micropollutants, as a result of the various treatments, has been on the order of the following:
-- 40 to 50 % for settling and powdered activated carbon treatment
-- 10 to 25 % for filtration
-- 15 to 20 % for ozone treatment.

These percentages are with respect to the concentration in the raw water.

4 Evaluation of toxicity

The presence of micropollutants in water poses a certain number of threats to human health. These concern the long-term risks, which appear to be the most serious and the most difficult to detect.

Toxicity tests have been developed to determine the following:

-- overall risks: chronic toxicity;

-- specific risks: carcinogens, mutagens, etc.

4.1 _Chronic toxicity evaluation._ These tests make it possible to determine a relation between a given micropollutant concentration and its effects on a certain number of living organisms. As a rule, the result is expressed as a DI 50 value, that is, as the dosage required to change the behavior of 50 % of the organisms tested. The test principle is based on numeration of organisms affected by progressively increasing levels of toxic agents.

Depending on the laboratory involved, use is made of the following:

-- unicellular organisms: bacteria
　　　　　　　　　　　　　 algae
　　　　　　　　　　　　　 isolated human cells;

-- multicellular organisms: daphnia
　　　　　　　　　　　　　　 fish (trout)
　　　　　　　　　　　　　　 rats and mice.

　　The criterion selected may be:
　　-- growth of colonies (bacteria)
　　-- mobility (daphnia, algae)
　　-- cytotoxicity (human cells)
　　-- death of specimens.

The response period covered by these various techniques is in the range of 48 hours to one week, depending on the nature of the organisms in question.

At the present time, the sensitivity of these methods is too low to show the toxicity of the dosages of micropollutants ordinarily present in water. As a result, it is necessary to carry out a preliminary concentration procedure, by the methods described in chapter 2 above. Research is currently under way to develop tests directly applicable to water samples. Some encouraging results have already been obtained in this line, with the use of human cells.

4.2 _Evaluation of specific risks._ The presence of toxic agents can have very diverse consequences, such as: -- cancer formation
　　　　-- mutations, chromosome anomalies
　　　　-- blood and enzyme changes.
The most widely studied specific risks are those concerning carcinogenic and mutagenic effects.

Carcinogenic effects have been studied by the following methods:
- inclusion of micropollutants in the food ration of rats and mice
- painting, in the dorsal area of mice
- sub-cutaneous injection of rats.

However, it requires from six months to two years to obtain statistically valid results with these tests. The surface painting method can give an indication after a matter of weeks.

Hence, to overcome this disadvantage, numerous laboratories have studied mutagenic effects. This is a useful approach because of the high correlation: 90 % of carcinogens are also mutagens.

A number of mutagen tests have been developed, dealing with:
- bacteria
- yeasts
- animal cells
- human cells.

These tests are much more rapid: significant results can be obtained in 48 or 72 hours.

In studying either carcinogenic or mutagenic effects, it is necessary to make use of concentrated extracts prepared by the methods described in chapter 2. Nonetheless, in certain cases, the relationship between dosage and effect is difficult to specify.

4.3 Qualities of a toxicity test. The chief qualities that a toxicity test should have are as follows:
- rapid response
- convenience in use
- low dispersion of results
- sensitivity
- reaction specificity
- direct applicability to water sample.

The tests available at present do not fully meet these requirements. For this reason, numerous studies are under way:
- to devise more suitable tests
- to develop higher-performance batteries of tests.

4.4 Correlation of results from various tests. The correlation among the results obtained with these various tests is extremely low. Thus, for instance, inclusion of concentrated extracts in the food ration of rats and mice has not been able to demonstrate a carcinogenic effect.

Furthermore, though there is good correlation between carcinogenic and mutagenic effects, it is difficult to cross check the results given by the various mutagen tests in use. In particular, human cells, in contrast to bacteria, appear to be sensitive to the nature of the substances present, and provide more specific responses. This leads to difficulty in interpreting results.

In addition, it must be noted that the large disparities in concentrations used further complicate the problem of interpretation: this emphasizes the desirability of finding a method directly applicable to water as sampled.

4.5 Results obtained. Toxicity tests carried out on extracts prepared from water samples taken at treatment plant inlet and outlet points appear to show more important carcinogenic effects after treatment.

However, account must be taken of the dosages in fact contained in water. In addition, the manner in which micropollutants are ingested may have an influence: inclusion of extracts in the feed of mice and rats has not been able to show a carcinogenic effect.

5 Conclusion

Measurement of total micropollutant load by concentration has led to interesting results in the fields of micropollution level surveillance and treatment procedure effectiveness evaluation.

In the near future, it should become possible to evaluate this load rapidly, without prior concentration. This parameter could then be measured systematically in regular operation.

However, the lack of correlation among toxicity tests makes them difficult to interpret. The efforts that have been made to develop more significant tests should therefore be continued. In particular, we must hope that the encouraging results obtained with human cells, using unconcentrated samples, may achieve confirmation.

APPENDIX 1

Characteristics of principal concentration methods used in France

1   CES Method (Chloroform-Extractible Substances)

principle: liquid phase extraction with chloroform at pH 7; evaporation at atmospheric pressure

sample volume: 10 to 20 l

duration of analysis: 48 hr

application: very convenient: the solvent is non-flammable

changes in pollutant character: since the sample is heated to 60°C, there is a significant risk of thermal decomposition or recombination

representativeness: the method is very selective, as a single solvent is used; in addition, volatile substances are eliminated during the evaporation step

toxicity testing: this is easy to perform with the extract obtained

use: periodic testing in plant operation

2   Cyclohexane Extraction Method

This method was developed, and is currently being used, by SETUDE of Paris, France.

principle: extraction with cyclohexane; concentration by freezing; vacuum sublimation

sample volume: 100 to 200 l

duration of analysis: 48 hr

application: relatively convenient: a single solvent is used; the absence of heating cuts down the risk from the flammability of the solvent

changes in pollutant character: very restricted

representativeness: the method is very selective, on account of the use of a single solvent; volatile substances are taken into account

toxicity testing: this is very difficult, as the solvent is itself toxic; the extract requires further preparation for this purpose

use: for measurements on volatile substances

3   Carbon or Resin Adsorption Method

SETUDE has developed a method on this basis; the U.S. Standard Methods include a similar technique.

principle: adsorption on carbon or resin; elution with several solvents; concentration

Sample volume: several m^3; as the adsorption column is installed in the plant, sampling is convenient

application: relatively complex, on account of the use of several solvents; special precautions may be necessary if certain solvents are flammable

changes in pollutant character: the risk of such changes is strongly dependent on the concentration method, and particularly on sample heating

representativeness: the use of multiple solvents takes into account a large number of substances; however, some selectivity may result from the choice of carbon or resin type

toxicity testing: convenience is dependent on solvent choice

use: for measurement of very low micropollutant concentrations

4   **Inert-Gas Entrainment**

<u>principle</u>: entrainment by inert gas; adsorption on carbon column; solvent elution; concentration

<u>sample volume</u>: very large

<u>duration of analysis</u>: long

<u>application</u>: convenience depends on solvent

<u>changes in pollutant character</u>: very restricted

<u>representativeness</u>: takes into account only volatile substances

<u>toxicity testing</u>: convenience is dependent on solvent choice

<u>use</u>: measurement of volatile substances: must be considered a supplementary technique to correct the selectivity of certain other methods

5   **IRCHA Total Method**

<u>principle</u>: chloroform extraction at pH 7; chloroform extraction at pH 2; chloroform extraction at pH 10, after entrainment over aluminum hydroxide; neutralization, resin adsorption, butanol extraction; resin adsorption, elution with mixture of ethanol and methylene chloride; carbon adsorption, washing with ethanol, chloroform extraction

<u>sample volume</u>: several m^3

<u>duration of analysis</u>: long

<u>application</u>: complicated

<u>changes in pollutant character</u>: restricted

<u>representativeness</u>: very good, as several solvents and extraction techniques are used

<u>toxicity testing</u>: relatively convenient

<u>use</u>: in research measurements, or when representativeness is the most important criterion

6   Freezing Method

Technique developed by SLEE, LePecq, France

sample volume: small, 5 to 10 l

duration of analysis: concentrate obtained in 20 hr

application: very convenient

changes in pollutant character: very restricted

representativeness: there is a risk of adsorption on precipitated inorganic substances

toxicity testing: the inorganic substance concentration can interfere with toxicity tests

use: with waters of low mineral content

APPENDIX 2

Laboratories that have participated in total micropollutant load evaluation

CGE	Compagnie Générale des Eaux, Laboratoire Central 52 rue d'Anjou 75008 Paris
IRCHA	Institute National de la Recherche Appliqué B.P. N° 1 91170 Vert le Petit
LHVP	Laboratoire d'Hygiène de la Ville de Paris 1bis rue des Hospitalières St Gervais 75004 Paris
SETUDE	Société d'Etudes pour le Traitement et l'Utilisation des Eaux 27 boulevard des Italiens 75002 Paris
SLEE	Société Lyonnaise des Eaux et de l'Eclairage, Laboratoire 46 rue du 11-novembre 78230 Le Pecq

Laboratories that have participated in toxicity testing

CEA	Commissariat à l'Energie Atomique B.P. N° 6 92260 Fontenay aux Roses
CNRS	Centre National de la Recherche Scientifique, Laboratoire de Médecine Expérimentale 16bis avenue P.V.Couturier 94800 Villejuif
INSERM	Laboratoire de Toxicologie Alimentaire 44 chemin de Ronde 78110 Le Vésinet

IPL	Institut Pasteur de Lyon 77 rue Pasteur 69006 Lyon
LTFP	Laboratoire de Toxicologie de la Faculté de Pharmacie 4 avenue de l'Observatoire 75006 Paris
IRCHA	Institut National de la Recherche Appliqué B.P. N° 1 91170 Vert le Petit
LHVP	Laboratoire d'Hygiène de la Ville de Paris 1bis rue des Hospitalières St Gervais 75004 Paris

fig. 1

COMPARISON BETWEEN RESULTS OBTAINED EITHER BY:
— CES METHOD
— OR IRCHA METHOD

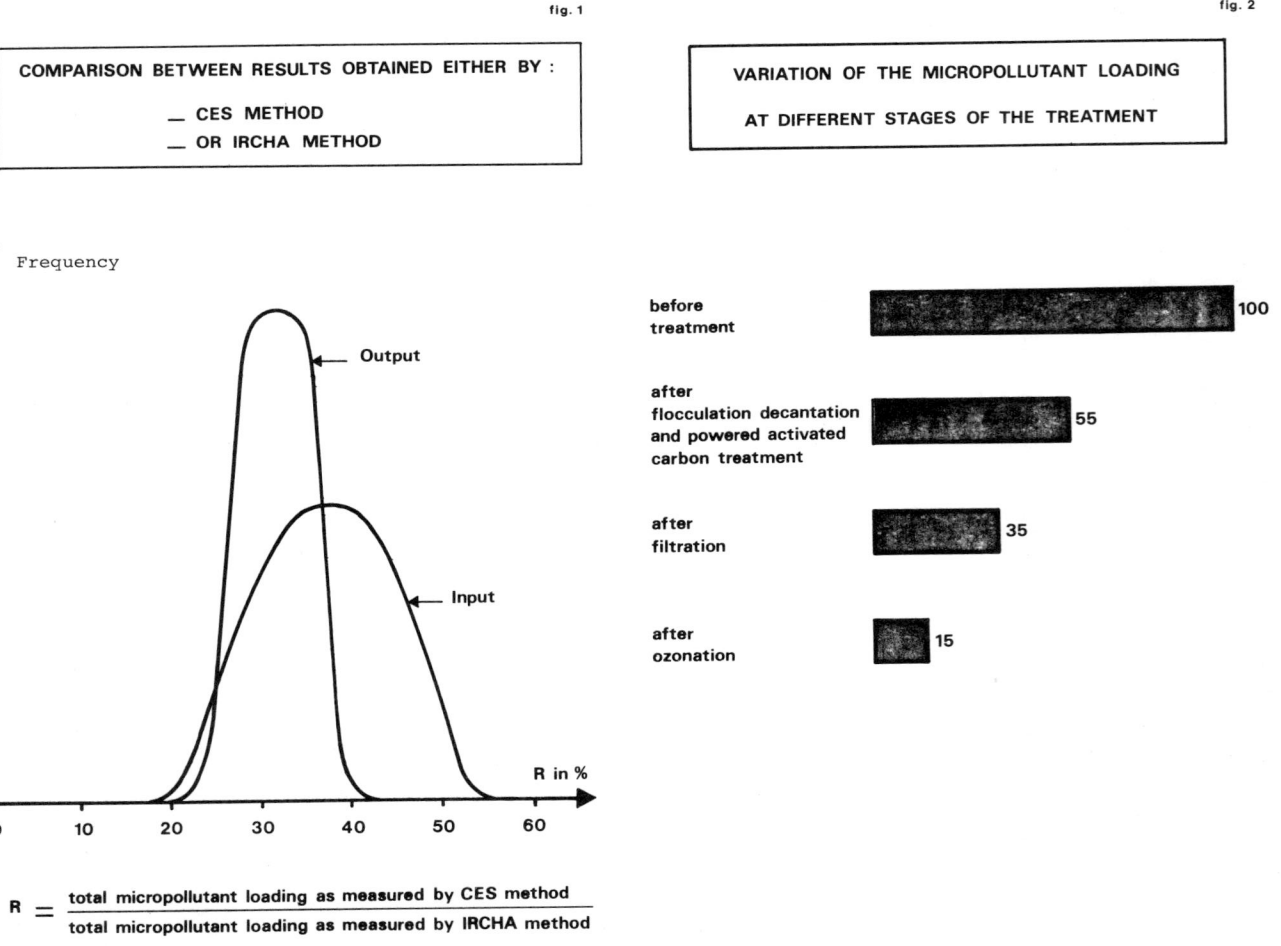

Output
Input

R in %

Frequency

$R = \dfrac{\text{total micropollutant loading as measured by CES method}}{\text{total micropollutant loading as measured by IRCHA method}}$

fig. 2

VARIATION OF THE MICROPOLLUTANT LOADING AT DIFFERENT STAGES OF THE TREATMENT

before treatment — 100
after flocculation decantation and powered activated carbon treatment — 55
after filtration — 35
after ozonation — 15

IDENTIFICATION OF END PRODUCTS RESULTING FROM OZONATION OF COMPOUNDS COMMONLY FOUND IN WATER

P. P. K. Kuo, E. S. K. Chian* and B. J. Chang

Environmental Engineering
Department of Civil Engineering
University of Illinois
Urbana, Illinois 61801
U.S.A.

ABSTRACT

Ozonation of 2-propanol, acetic acid and oxalic acid was conducted with and without UV irradiation. It was found that removal of organics was greatly improved under UV irradiation. Upon ozonation, 2-propanol was converted to acetone, which in turn was oxidized to acetic and oxalic acids. Trace amounts of formaldehyde and formic acid were also detected in the ozonated acetone solution. Further ozonation of acetic acid resulted in the formation of glyoxylic acid, which was oxidized readily to oxalic acid. The latter was oxidized to carbon dioxide rather slowly. Methylene chloride, chloroform, carbon tetrachloride, bromodichloromethane, chlorodibromomethane and bromoform were observed in the reaction mixtures of chlorinated ultrafiltration (UF) and reverse osmosis (RO) retentates of secondary effluent. On the other hand, ozonation of UF and RO retentates did not result in the formation of these volatile halogenated organics.

---
*to whom correspondence should be addressed

Introduction

Since the detection of halogenated organics in drinking water (1,2), much research effort has been directed toward finding water treatment processes to remove precursors of halogenated organics and toward finding disinfectants other than chlorine. Great interest has been focused upon ozonation because both disinfection and organic removal can be accomplished with this process.

To gain insight into the kinetics of the ozonation process, it is necessary to have knowledge of the intermediate and end products formed during ozonation. This information can be helpful in operating ozonation treatment systems more effectively and economically. It is also of great interest when the end products are potentially toxic or are precursors of other toxic compounds which would result if the ozonated effluent is subsequently chlorinated to provide residuals for disinfection. Such ozonated end products would also accumulate in wastewater after repeated cycles of reuse. Determination of ozonation by-products using model compounds commonly found in water is therefore of great practical importance.

Fractionation and characterization of secondary effluents has shown that they consist mostly of humic acid-like material in the ultrafiltration (UF) retentate and fulvic acid-like material in the reverse osmosis (RO) retentate (3). Chlorination of these acids during chlorine disinfection processes could result in formation of potentially toxic halogenated organics. Rook (4) has indeed shown that fulvic acids are precursors of haloforms.

In this study 2-propanol, acetic acid and oxalic acid were subjected to ozonation and UV-ozonation. Their ozonation pathways, the organic removal efficiencies, and the effects of UV irradiation were studied. Also, UF and RO retentates of secondary effluent samples were subjected to both chlorination and ozonation to examine the formation of volatile halogenated organics.

Experimental

Solutions of the model compounds were prepared from analytical grade reagents and stripped distilled water. The ozonation studies were performed with a

4-liter New Brunswick fermentor (New Brunswick Scientific, New Brunswick, N.J.) operated at a constant pH and a constant temperature of 25°C. The speed of the stirring impellers was set at 695 RPM. Ozone was generated from dried oxygen by a Welsbach Model T-408 ozone generator (Philadelphia, PA). The feed gas to the fermentor had an ozone concentration of 25 mg/l and a flow rate of 4 liter/min. The fermentor was also equipped with a 15-W, low-pressure mercury germicidal lamp (General Electric, Schenectady, NY) for the UV-ozonation study.

Every 15 or 30 minutes during the course of ozonation or UV-ozonation, 20 ml of solution was drawn and stored in a refrigerator at 4°C for later chemical analysis. In addition, another 5 ml of solution was drawn and immediately analyzed for dissolved ozone using Schechter's iodometrical method (5). The dissolved ozone concentration was determined to be 3 ppm or higher in the reaction mixtures, showing that none of the ozonation runs was conducted under mass-transfer-limited conditions.

A Beckman Model 915 TOC Analyzer (Fullerton, CA) was used for total organic carbon (TOC) determinations. Gas chromatographic (GC) analyses were performed on a HP 5750B gas chromatograph (Hewlett Packard, Palo Alto, CA) equipped with dual flame ionization detectors. A 6-ft x 1/4-in-O.D. glass column packed with 0.2% Carbowax 1500 on 80/100 Carbopack C (Supelco, Bellefonte, PA) was used for determining alcohols and ketones. The derivative-GC procedures of Bethge and Lindstrom's method (6) were followed to determine formic, acetic and propionic acids. The benzyl esters of acids were separated and determined on a 6 ft x 1/4-in-O.D. glass column packed with 3% butane-1,4-diol succinate polyester on 100/120 AW chromosorb W. Oxalic acid was determined on the latter GC column after the compound was converted to dimethyl ester using diazomethane (7). Glyoxylic acid was determined by a colorimetric method described by Kramer et al. (8). The chromotropic acid method was used to determine formaldehyde (9).

Wastewater was collected from a Champaign-Urbana sewage main and treated, producing a secondary effluent which was then fractionated into UF retentate, UF permeate, RO retentate and RO permeate. (The detailed experimental procedures have been described elsewhere (10)). The UF and RO retentates were subjected to ozonation and chlorination after their concentrations were adjusted to 5 ppm of TOC.

The retentates were ozonated with the unit described previously. A constant temperature of 25°C was maintained, and the stirring speed was set at 355 RPM. The pH of the solution was not controlled. The ozonated mixture was sampled after 2, 10 and 20 minutes of ozonation. The dissolved ozone concentration was found to increase from 0 to 5 ppm as ozonation time increased from 0 to 20 min. Sodium thiosulfate was added to the ozonated mixtures immediately after sample collection to quench dissolved ozone. Control mixtures were obtained by the same procedures but without the ozonation step. Both the ozonated and the control mixtures were analyzed for volatile halogenated organics by a stripping-GC and GC/mass spectrometric (MS) method. 125 ml of solution was stripped at 60°C and the stripped volatile organics were adsorbed in a Tenax-GC column. The adsorbed organics were thermally released and injected into the GC for quantitation and into the GC/MS for identification. A detailed description of the procedures has been given elsewhere (11,12). For GC determinations a 6-ft x 1/8-in-O.D. stainless-steel column packed with 0.4% Carbowax 1500 on 80/100 Carbopack A was used. The column was conditioned at 200°C. A 12-ft x 1/4-in-O.D. glass column was used for GC/MS determinations, which were performed on a Varian-MAT 311 A (Palo Alto, CA) combined GC/MS with a Varian-Aerograph 2700 GC using a two-stage Watson-Biemann sample enricher.

The UF and RO retentates were dosed with sodium hypochlorite and mixed for 30 min. The reaction mixtures were then transferred into 125 ml glass bottles leaving no headspace volume, topped with Teflon-lined rubber septa, sealed with crimped-on aluminum caps, and stored in a dark place. Three bottles were removed for chemical analyses after storing for 1 day, three more were removed after 3 days, and 3 after 7 days. Sodium thiosulfate was added to two of the samples to remove residual chlorine before the stripping-GC and GC/MS determinations. At the same time, one bottle was used for determining residual chlorine by the DPD method (13). Control samples (without chlorine dosage) were treated and analyzed similarly.

Results and Discussion

Ozonation and UV-Ozonation of 2-Propanol, Acetic Acid and Oxalic Acid. 2-Propanol, acetic acid and oxalic acid were selected as model compounds for the ozonation study because they are low molecular weight polar organics and are poorly removed by the two advanced wastewater treatment processes: reverse osmosis and activated carbon treatment (14,15). Moreover, other investigators (11,16-19) have found acetic and oxalic acids to be present in most of the ozonation products. In addition, excretion of acetate by microorganisms contributes further to the acetic acid concentration in water (20). Oxalic acid is also rated as toxic at high concentrations (21).

Figures 1-6 show the ozonation pathways for these model compounds as well as the removals of organic carbon (TOC) and specific organic compounds. 2-Propanol was oxidized to acetone, which in turn was oxidized to acetic and oxalic acids (Figure 1, 2 and 4). Trace amounts of formaldehyde and formic acid were also detected in the ozonated acetone mixtures. Ozonation of acetic acid resulted in the formation of glyoxylic and oxalic acids (Figure 4). Glyoxylic acid was, however, present in trace amounts (< 1 ppm). The organic carbon (TOC) can be fully accounted for by acetic and oxalic acids in the reaction mixtures resulting from the UV-ozonation of acetic acid (Figure 4). No other organics except oxalic acid itself were detected in reaction mixtures resulting from the ozonation of oxalic acid.

Glyoxylic, oxalic, formic and acetic acids were found to be the common ozonation end products prior to their complete oxidation to CO_2. These compounds were also determined in the reaction mixtures which resulted from the ozonation of phenol (16), malonic acid (17), chlorophenol (18), maleic acid (18), humic acids (19), and many aliphatic and aromatic compounds (11). The oxidation of oxalic and formic acids to carbon dioxide is partially responsible for the removal of organic carbon by ozonation. Acetic acid was oxidized through ω-oxidation to yield glyoxylic acid. Gilbert (18) has proposed that further ozonation of glyoxylic acid causes either direct oxidation to carbon dioxide or the formation of oxalic acid. Glyoxylic acid is an intermediate which appears to have a very rapid rate of oxidation to oxalic acid under the experimental conditions because only a very small amount of it could be detected during the ozonation of the acetic acid solution.

The effect of UV irradiation in enhancing the removal of organics during ozonation is apparent from the data in Table 1 and Figures 1-6. The percentage of organic removal at the end of a two-hour ozonation period was greatly improved with UV irradiation. The organic carbon removal and reactant removal followed either zeroth or first order reaction kinetics. The rate constant for organic carbon removal was improved under UV irradiation by a factor of approximately 8 and 2, respectively, for 2-propanol and oxalic acid. For acetic acid, the removal rate was improved by a factor of 6. Oxalic acid is the ozonated end product resulting from the ozonation of many organic compounds. Its removal, however, is relatively slow, as shown in this study and by Dorfman (22). Therefore, attention should be given to the removal of oxalic acid, particularly in view of its toxicity (21).

<u>Chlorination and Ozonation of UF and RO Retentates of Secondary Effluent</u>. It has been found that UF and RO retentates contain primarily humic acid-like and fulvic acid-like materials, respectively (3). The molecular weight distribution of soluble organics in these retentates was determined by fractionation on Sephadex gel permeation columns (10). The UF retentate contained two major molecular weight peaks, i.e., > 50 K and 2-5 K; the RO retentate had only one major peak at an apparent molecular weight of approximately 250. The formation of volatile halogenated organics as a result of ozonation and chlorination was tested on these retentates.

Table 2 shows that the concentrations of free chlorine and combined chlorine decrease with storage time, indicating that chlorination still proceeded one week after chlorine dosage. It was also observed that chlorine was consumed more rapidly in the RO retentate than in the UF retentate (Table 2). The rapid rate of chlorine uptake in the RO retentate can be attributed to the higher density of organic functional groups present in this fraction than in the UF retentate, as reported by DeWalle and Chian (3). Since RO itself also concentrates inorganics, higher concentrations of chlorine oxidizable salts, such as bromide, in the RO retentate may also contribute to the higher chlorine demand (23). The latter possibility is supported by the finding of an appreciable amount of brominated compounds in the chlorinated RO retentate. These compounds are not found in the chlorinated UF retentate, especially after seven days of storage (Table 3).

Using the stripping technique, volatile halogenated organics were detected in the chlorinated UF and RO retentates and their concentrations were determined. The results are summarized in Table 3. These compounds were identified by GC/MS and quantitated by GC. The data shown in Table 3 have been corrected for blank concentrations, which were obtained from control determinations and were generally less than 2 ppb. The concentrations of most of the compounds in the samples increased upon storage. Chloroform was present in both retentates at concentrations higher than those of other halogenated compounds. Brominated organics were found in appreciable amounts only in the RO retentate, possibly because bromide is concentrated by the RO process. Bromide can be oxidized by hypochlorite to form hypobromite, a compound with which bromination can take place to produce brominated compounds (24). A material balance for chlorine has shown that approximately 5 to 6% and 3 to 4% of the chlorine consumed ended up in the volatile chlorinated compounds in the UF and the RO retentates, respectively. Approximately 1 to 2% of the chlorine ended up in the volatile brominated compounds present in the RO retentate.

Ozonation, on the other hand, did not result in the formation of any of these volatile halogenated compounds. The concentration of these compounds actually decreased during ozonation, possibly resulting from both stripping and oxidation of these compounds by the ozone. This finding is, however, somewhat speculative because the initial concentrations of these compounds were near the detection limit using the stripping-GC method with a flame ionization detector.

Conclusion

The use of chlorine for disinfection in water treatment processes could result in the formation of volatile halogenated organics, as was observed in the present study as well as by other investigators (4,24,25). The precursors of these halogenated compounds may be either humic acid-like or fulvic acid-like materials. Rook (4) has found that m-dihydroxy aromatic compounds and methyl ketones were precursors of haloforms. Advanced water treatment process should therefore be employed to remove these precursors prior to chlorine disinfection to minimize the formation of such potentially toxic compounds. The ozonation process appears to be a good candidate, as it provides disinfection and prevents the formation of halogenated compounds. To protect against post-contamination in the distribution system, chlorine can be applied after ozonation treatment at much lower dosages than would be required in a conventional treatment system. Cost, however, appears to be the major obstacle in the wide application of ozonation. Future studies should therefore be directed toward minimizing costs by optimizing conditions for ozonation.

Acknowledgements

This work was supported by the U.S. Army Medical Bioengineering Research and Development Laboratory under Contract No. DAMD 17-75-C-5006.

Literature Cited

(1) B. J. Dowty, D. R. Carlisle and J. L. Laseter. Environ. Sci. Technol. 9:762 (1975).

(2) J. M. Symons et al. J. Amer. Water Works Assoc. 67:634 (1975).

(3) F. B. DeWalle and E. S. K. Chian. J. Environ. Engr. Div., ASCE, Vol. 100, No. EE5, Proc. Paper 10867, Oct. (1974).

(4) J. J. Rook. J. Amer. Water Works Assoc. 5:168 (1976).

(5) H. Schechter. Water Res. 7:729 (1973).

(6) P. O. Bethge and K. Lindstrom. Analyst 99:137 (1974).

(7) R. G. Webb, A. W. Garrison, L. H. Keith and
     J. J. McGuire. U.S. EPA Report No.
     EPA-R2-73-277 (1973).

(8) D. N. Kramer, N. Klein and R. A. Baselice.
     Anal. Chem. 31:250 (1959).

(9) M. J. Houle, D. E. Long and D. Smette. Anal.
     Lett. 3:401 (1970).

(10) E. S. K. Chian, S. S. Cheng, F. B. DeWalle and
     P. P. K. Kuo. Prog. Wat. Tech. 9:761 (1977).

(11) E. S. K. Chian and P. P. K. Kuo, "Fundamental
     Study on the Post Treatment of RO Permeates
     from a MUST Hospital Wastewater," Second Annual
     Report to U.S. Army Medical Bioengineering R&D
     Command (1976).

(12) E. S. K. Chian et al., "Monitoring to Detect
     Previously Unrecognized Pollutants," Progress
     Report No. 3, EPA Contract 68-01-3234 (1976).

(13) "Standard Methods for the Examination of Water
     and Wastewater," 13th Ed., American Public
     Health Assoc., Washington, D.C. (1971).

(14) H. H. P. Fang and E. S. K. Chian. Environ.
     Sci. Technol. 10:364 (1976).

(15) D. M. Guisti, R. A. Conway and C. T. Lawson.
     J. Water Poll. Control Fed. 46:947 (1974).

(16) J. P. Gould and W. Weber Jr. J. Water Poll.
     Control Fed. 48:47 (1976).

(17) F. Dobinson. Chem. and Ind. 853 (1959).

(18) E. Gilbert, "Ozonolysis of Chlorophenols and
     Maleic Acid in Aqueous Solution," presented at
     2nd International Ozone Symposium, Montreal,
     Canada (May 11-14, 1975).

(19) M. D. Ahmed and C. R. Kinney. J. Amer. Chem.
     Soc. 72:559 (1950).

(20) R. I. Matels and E. S. K. Chian. Environ. Sci.
     Technol. 3:569 (1969).

(21) M. N. Gleason, R. E. Gosselin, H. C. Hodge and R. P. Smith, "Clinical Toxicity of Commercial Products," 3rd Ed., Williams and Wilkins Co., Baltimore (1969).

(22) L. M. Dorfman, NSRDS-NBS 46, US Dept. of Commerce, Natl. Bu. Standards, Washington, D.C.

(23) H. H. P. Fang and E. S. K. Chian. J. Appl. Polymer Sci. 19:2889 (1875).

(24) W. W. Bunn, B. B. Haas, E. R. Deane and R. D. Kloepfler. Environ. Lett. 10:205 (1975).

(25) F. C. Kopfler, R. G. Melton, R. D. Lingg and W. E. Coleman, Chapter 6, "GC/MS Determination of Volatiles for the National Organics Reconnaissance Survey (NORS) Drinking Water," in *Identification and Analysis of Organic Pollutants in Water*, ed. by L. H. Keith, Ann Arbor Science, Ann Arbor, MI (1976).

DISCUSSION

Michael Kavanaugh, University of California/Berkeley: Do you have any speculations as to what is going on in your UV/ozone system? Do you have any concepts as to the process flow diagrams that you are talking about, as far as applying the combination of UV and ozone, or, with respect to the second half of your talk, which was the use of ozone to treat permeates from reverse osmosis processes?

My first question is, do you have any speculations as to the reaction mechanisms? What's going on with UV and ozone together that doesn't happen when ozone is used alone?

Kuo: We believe that the UV effect is to activate the ozone molecule and also to activate the substrate.

Kavanaugh: So you see this as a process that you could use with ozone and UV. How do you envision the process mechanically? I am expressing my bias as an engineer; I just want to know what kind of a process you are going to use.

Ed Chian, University of Illinois: Are you asking what process diagrams we will propose as a result of our studies?

Kavanaugh: Yes.

Chian: For those of you who are pro-chlorination, we will say that if activated carbon (which has been found to be very effective in removal of humic and fulvic acid materials) can be placed in front of chlorination, then you would eliminate or reduce the chances to form potentially toxic chlorinated compounds. This has been shown in our work. We actually chlorinate only those humic substance fractions in the UF retentate. We are pretty much talking about the humic acid fractions, these are high molecular weight fractions which will precipitate at pH 1.

The lower molecular weight fraction is fulvic acids, which has a much higher functional density of organic compounds, such as carboxylic groups or phenolic groups. If you can remove those compounds, especially those compounds with the methoxyl groups, then you at least reduce the chances of formation of such halogenated compounds. Now this is for those who are pro-chlorination.

For those who are pro-ozone we have another approach. It depends upon how you look at it. If your purpose is disinfection, I would say, regardless of what organics end up in the final effluent, we would say that ozone oxidation would be advantageous to reduce the possible formation of halogenated compounds if chlorination were used later on to provide residuals in the distribution system.

As I presented in my talk at the Second International Symposium on Ozone Technology in Montreal, ozonation also tends to remove all those organics present in reverse osmosis and ultrafiltration retentates very effectively; it will take on the order of at least 15 to 30 minutes of ozonation before those compounds are completely removed.

But only 2-5 minutes of contact time are required for good disinfection, such as in the paper we presented at the IOI Forum on Disinfection in Chicago this past June. So if you combine ozone in front, ozone tends to reduce some of those compounds that may give rise to the halogenated compounds when you chlorinate it to provide residuals. So ozonation and chlorination would make a good pair.

To what extent should you ozonate? What amount of chlorine residual you should give is a matter of interest for a more extensive study. How far should you ozonate it, then wait for how long until the ozone all decomposes? Then you should use chlorine, just enough to maintain residuals --- enough to keep the microorganisms from regrowing, because you realize that after ozonation you tend to make the organics more biodegradable, and by ozonizing you convert more of those organics to become available food for the microorganisms. You _may_ have to add a little bit more chlorine than normally used for chlorine residual.

Table 1. Organic Removal by Ozonation and UV-Ozonation

	2-Propanol		Acetic Acid		Oxalic Acid	
pH	7		7[1]		9	
UV	no	yes	no	yes	no	yes
Initial TOC, mg/l	114	116	84	83	46	41
Ozonation Time, hr	2		2		2	
% TOC Removal	17	82	14	92	36	64
TOC Removal						
Order	1st	1st	1st	0th	1st	1st
Period, min	0-120	30-105	0-120	0-60	0-75	0-60
Rate Constant	1.66×10^{-3} min^{-1}	1.29×10^{-2} min^{-1}	1.52×10^{-3} min^{-1}	0.892 mg/l min	6.24×10^{-3} min^{-1}	1.18×10^{-2} min^{-1}
Reactant Removal						
Order	1st	0th	1st	0th		
Period, min	0-90	0-15	0-120	0-60		
Rate Constant	1.35×10^{-2} min^{-1}	5.00 mg/l min	3.23×10^{-3} min^{-1}	1.17 mg/l min		

[1] Initial pH.

Table 2. Chlorination of UF and RO Retentates Upon Storage

			Concentration of Chlorine[*], mg/l	
Retentate	Time, Day	pH	Free Chlorine	Combined Chlorine
UF	0	7.6	1.30	0.10
	1	7.4	0.60	0.10
	3	7.5	0.28	0.08
	7	7.5	0.09	0.08
RO	0	8.2	1.32	0.24
	1	8.1	0.32	0.23
	3	8.1	0.00	0.18
	7	8.0	0.00	0.09

[*]Chlorine Determination by DPD Method (13).

Table 3. Concentrations of Volatile Halogenated Organics in Chlorinated UF and RO Retentates, ppb

Retentate	Day	CH_2Cl_2	$CHCl_3$	CCl_4	$CHCl_2Br$	$CHClBr_2$	$CHBr_3$
UF	1	15	35		2		
	3	18	42				
	7	23	61			2	
RO	1	4	28		2	12	7
	3		43		2	17	16
	7		50	5	26	25	21

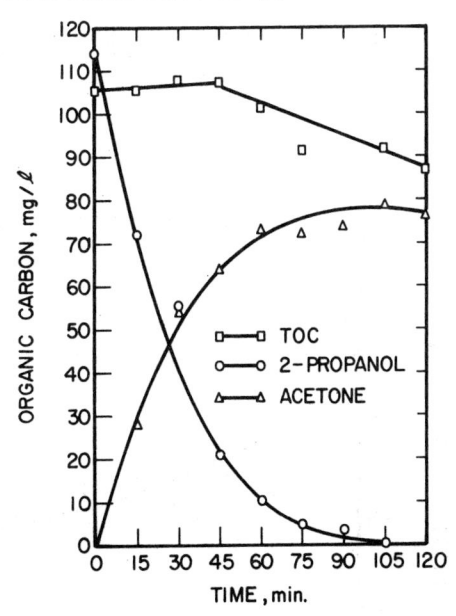

Figure 1   Ozonation of 2-Propanol

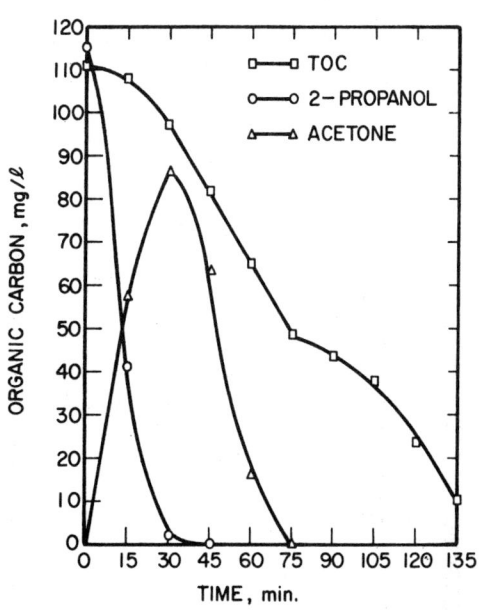

Figure 2   UV-Ozonation of 2-Propanol

-165-

Figure 3 Ozonation of Acetic Acid

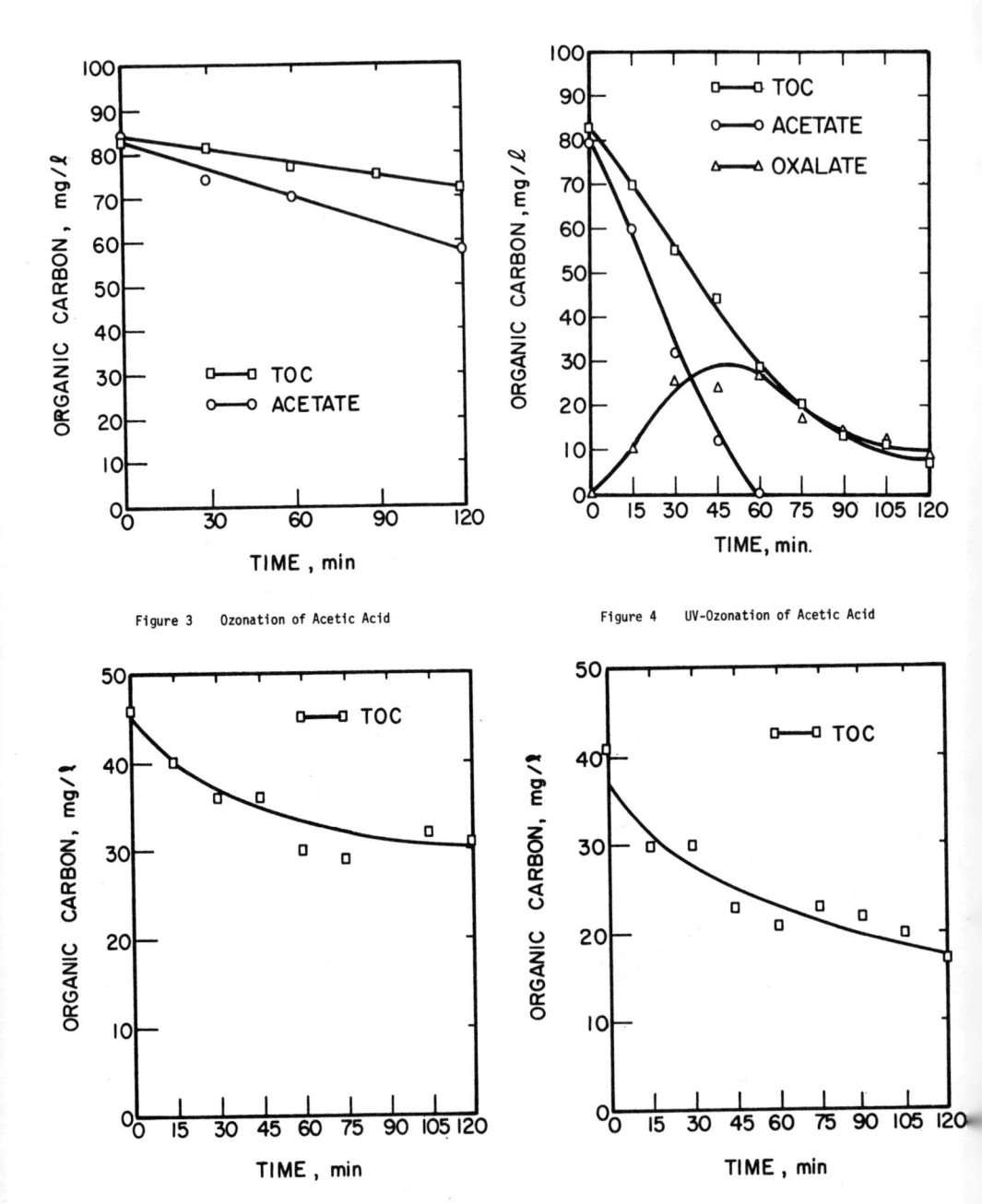

Figure 4 UV-Ozonation of Acetic Acid

Figure 5 Ozonation of Oxalic Acid

Figure 6 UV-Ozonation of Oxalic Acid

COMMENTS ON FIRST DAY SESSION

Harvey Rosen

Harvey Rosen: Before we begin the session this morning, I would like to make a few comments about some of the topics presented yesterday, particularly those dealing with toxicological data of ozonated organic compounds. The papers presented by scientists at Stanford Research Institute reported data on mutagenic products from the reactions of ozone with certain organic compounds, and these results can be interpreted in number of different ways.

It is important to point out that there were 29 compounds that were ozonized and tested, and only three showed this activity. Of those that did show this activity, only ethanol and NTA (nitrilotriacetic acid) showed this activity. These compounds, specifically NTA and ethanol, happen to be highly biodegradable, so there is a question as to the likelihood of those materials being in a real secondary effluent or in natural waters to any major degree relative to other organic compounds that are certainly a lot less biodegradable.

Another point is that those compounds, particularly in relative terms, are somewhat refractory to ozonation, and these compounds are not going to be in wastewater or natural waters alone. There are going to be a lot of other materials that are going to react with ozone before ethanol and NTA react.

Another thing to remember is that ten thousand parts per million, one weight percent, of these materials were ozonated in the laboratory studies, as opposed to perhaps at the most a few parts per million that might be found in wastewater or natural water. These relatively concentrated organics were hit very hard with ozone in order to produce these compounds that in further study showed mutagenic activity. So, I look at it in a positive way. I was also told that if these same compounds had undergone the same treatment with chlorine, a lot more of them would have shown mutagenic activity. So, just keep all that in mind when considering what you heard.

The reason I have said all this is that I was confused, and I asked some questions. I have talked to others at this meeting who were not familiar with this kind of work and they were also confused, so I wanted to put that into perspective before we start. If anybody has an argument with what I've said, you can have a minute or two here. No? OK, then let's get started with today's program.

ORGANIC MATERIALS PRODUCED
UPON OZONIZATION OF WATER

Y. RICHARD / L. BRENER

Physico-chemistry Research Department, Société
DEGREMONT, 183 avenue du 18 juin 1940,
92500 RUEIL MALMAISON (France)

Introduction

During the last few years, the increased chemical pollution of waters has led to the diversification of polishing techniques designed to satisfy the main objective of maintaining all of a water's "drinking water" qualities while reducing chemical pollution to a minimum. Various polishing treatments (ozonization and the use of powdered or granular activated carbon) have been added to the classical treatment scheme comprising clarification, prechlorination, coagulation, flocculation, sedimentation, and filtration on sand.
Concurrently, the development of analytical techniques, and especially of gas chromatography, has permitted detection on a µg/l level of organic derivatives ves and laboratory study of the action of ozone and chlorine on organic matter on a comparative basis. Thus, studies effected conjointly in laboratories and plants permit definition of a new field of application for chemical oxidants in water treatment.
In France, over 15,000 tons of phytosanitary products are presently sold each year; although water pollution resulting from the use of such products is low due to various administrative measures restricting their use, there still remains a risk of accidental pollution similar to the recent contamination of the Rhine (by endosulfan).

Research on the oxidation of organophosphorous pesticides by ozone

Our initial work on the oxidation of organo-phosphorous pesticides (1) began in 1970 and was carried out in collaboration with the Laboratory A - Chemistry of Waters and the Environment, Ecole Nationale Supérieure de Chimie in Rennes - Director : Prof. MARTIN. Mr. LAPLANCHE, assistant at this same laboratory, participated in this research.

Table 1 contains the results of ozonization tests carried out on parathion, malathion, methylparathion, fenitrothion and ronnel. Rapid degradation of these compounds is observed when the ozone treatment rate increases from 1 to 5 mg/l of O_3.

As the result of this study, parathion was selected for investigation on the various stages of ozonization of an organo-phosphorous compound (2).

Laboratory study and search for metabolites.

Procedure. To detect the presence of metabolites that might result from the action of ozone on an organo-phosphorus compound, ozone was bubbled in a pure solution of this insecticide.

In the case of parathion, as revealed by figure 1, disappearance of the parathion was followed by gas chromatography on a column QF_1 15% + DC 200 10% on Gas Chrom Q coupled with a flame ionization detector. Analysis reveals that disappearance of the parathion is accompanied by the appearance of a new compound whose retention time, in comparison with the parathion, is 1,14.

Figure 2 shows that thin layer analysis of an ozonized parathion extract on a Merck fluorescent silica plate, eluted with ethyl acetate, reveals the presence of 5 metabolites that were identified by combining elementary analysis with analysis by nuclear magnetic resonance and infrared light (figure 3).

Ozonization of parathion. In this way, for the ozonization of parathion, we identified paraoxon, 2,4-dinitrophenol, picric acid, sulfuric and phosphoric acid.

In order to become more familiar with the reaction, we followed transformation of the parathion into paraoxon as a function of the ozonization time. The parathion and paraoxon are followed by gas chromatography.

Figure 4 reveals rapid attack of the parathion, with the obtention of paraoxon as the main product, and slower attack of the paraoxon.

Figure 5 illustrates the oxidation mechanism for the parathion and the metabolites that are formed.

Ozonization of malathion. For malathion (figure 6) we isolated malaoxon, then observed the cracking of the malaoxon molecule followed by the formation, in a last stage, of phosphoric acid. Certain metabolites have not yet been identified.

Ozonization of phosalone. In the case of phosalone (figure 7), we were never able to prove the presence of the oxon derivative of phosalone since this compound must be highly unstable. In contrast, we isolated two metabolites :

ether bis (chloro-6-oxo-2-benzoxazolyl-3-methyl)

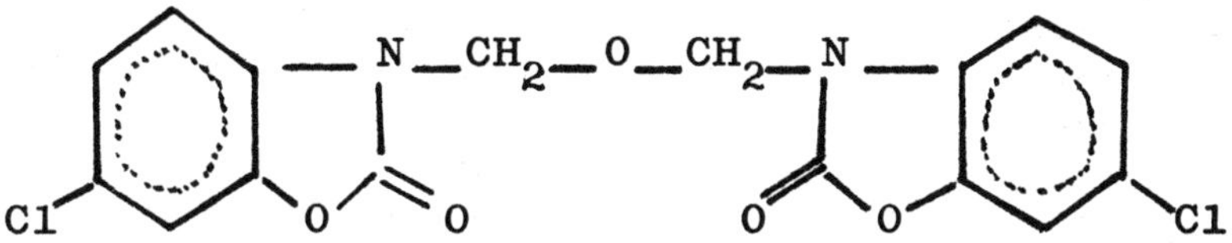

hydroxymethyl-3-chlorobenzoxazolone

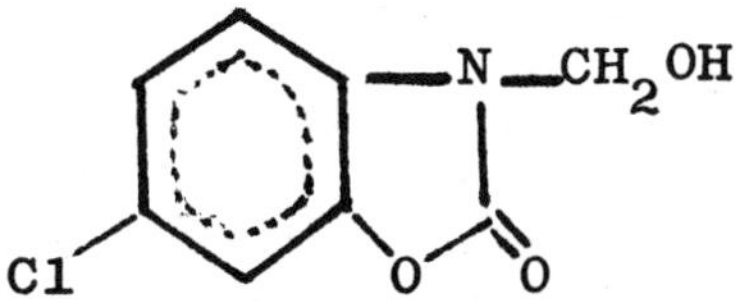

Action of ozone on organo-phosphorous derivatives : proposed mechanism. Study of the ozonation of these three organo-phosphorous insecticides (figure 8) reveals that the pesticide breaks down at the level of the bond : $P\begin{bmatrix}O\\S\end{bmatrix}$ - R' This can be represented by the reactions :

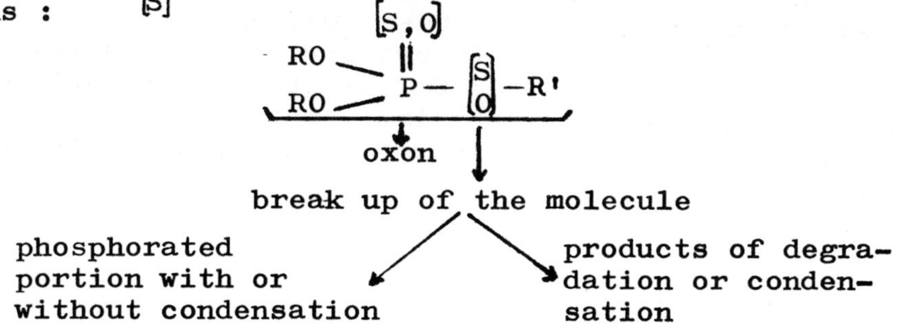

$$RO\diagdown\atop{RO\diagup}P\overset{[S,O]}{\overset{\|}{-}}\begin{bmatrix}S\\O\end{bmatrix}-R'$$

↓ oxon

↓ break up of the molecule

phosphorated portion with or without condensation ← → products of degradation or condensation

Study using a pilot oxidation tower.
Laboratory study was completed by tests on a pilot ozonation tower (figure 9) fed by a flow of up to 1 m/hr.

The inflow of water contained very low parathion concentrations of less than 0,1 mg/l. During this study, we followed the evolution of the parathion and of the most toxic and abundant metabolite (paraoxon), and varied the following parameters :
- contact time in the column : 6 and 12 minutes
- ozone treatment rate : 1, 2, 3, 4 and 5 mg/l of O_3 were employed
- pH : two pH zones were studied (6,7 - 6,8 and 8,2 - 8,3).

The experimental test conditions are listed in table 2 and the experimental results are summarized in table 3. The percentages of parathion and paraoxon are expressed as percentages of the initial mass of parathion.

Points A, B, C and D represent sampling points in the column. For each of these points we can define an apparent contact time corresponding to one quarter, one half, three-quarters and the entirety of the contact time studied.

Figure 10 illustrates the general shape of all of these curves, characterized by disappearance of the parathion, and of the paraoxon bell curves, which confirm the following mechanism :

PARATHION ⟶ PARAOXON ⟶ VARIOUS OXIDATION PRODUCTS

These tests allowed us to conclude that:
- the ozone treatment rate (figure 11) is the most important parameter since it directly determines speed of reaction. At least 3 g/m_3 of ozone are required for parathion removal under the test conditions, and at least 5 g/m_3 of ozone are needed for paraoxon elimination.
- as shown by figure 12, disappearance of the parathion depends on the pH; this occurs more easily in an acid medium while paraoxon degradation, in contrast, is more rapid in a basic medium.

These tests in the laboratory and on ozonization columns reveal that a careful kinetic study must be made of oxidation of the primary pesticide to avoid delivering a water that is more toxic after ozonization.

Study of the interference between chlorine and ozone

Concurrently with these studies on the oxidation of organo-phosphoros derivatives by ozone, we continued our research on the simultaneous use of chlorine and ozone for water treatment. Already in 1975, at the 2nd Symposium on Ozone held in Montreal (3), we reported on the interferences between chlorine and ozone at both the level of the dosing of these 2 oxidants and during treatment itself. This report revealed that it is indispensable to accurately check a water's chlorine demand after ozonization.

Recently, we were invited to intervene at a water treatment plant located on the Loire river. This plant treats 150,000 m^3 of water per day and supplies a population of 500,000 inhabitants. The treatment scheme at this plant can be summarized as follows:
- prechlorination using 16 g/m^3 of chlorine
- coagulation, flocculation, clarification and filtration on sand with a free chlorine residual of 0.3 g/m^3
- ozonization using 1.6 g/m^3 of O_3
- post-chlorination using 0.6 g/m^3 of Cl_2

As revealed by table 4, the bacteriological analyses for this plant never revealed the presence of any germs in the water delivered by the installation. In contrast, these same analyses reveal considerable bacterial development in the reservoir located 7.5 km from the plant. At the same time, the water in this reservoir does not contain any trace of free chlorine.

When the operator decided to treat the reservoir locally with chlorine, the germs disappeared.

In an effort to define the nature of the chlorine demand, the latter was measured 24 hours for the water after ozonization as well as at each point in the treatment line. For this purpose, 3.5 g of chlorine were added to each of the waters and evolution of the free chlorine was studied from 0 to 24 hours after injection.

Figure 13 reveals that the ozonized water's chlorine demand : 2.08 mg/l, is of the same order of magnitude as that of water filtered on sand (1.84 mg/l). In contrast, filtration of ozonized water on granular activated carbon reduces the chlorine demand after 24 hrs to 0.t mg/l.

The cause of chlorine interference in the presence of ozone must therefore be sought among a number of pollutants that incomplete treatment involving coagulation, flocculation, clarification and sand filtration does not totally eliminate.

Humic acids, substances of vegetal origin that are abundant in all waters, appeared to us as one of the probable causes of a water's chlorine demand.

The humic acids were estimated by the colorimetric method described on page 346 in the 1971 edition of Standard Methods. Since we did not dispose of a reference source of humic acids from Loire river water taken locally, the humic acid content was compared with a reference volume of Riedel Hann sodium huminate.

Table 5 gives parallel indications of the humic acid indices and the corresponding chlorine demand after 24 hours, from which it can be seen that :
- treatment by coagulation, flocculation, clarification and sand filtration permits, at best, elimination of only 83% of the humic acid in water. The humic acid residual provokes a high chlorine demand after 24 hours (1,84 mg/l of chlorine).
- ozonation leads to a very slight increase in the humic acid index (2.20 mg/l) due to new formation of hydroxylated derivatives and to an increased chlorine demand (2.08 mg/l of chlorine).
- filtration of the ozonized water on activated carbon eliminates 77% of the humic acids in the water; this reduces the ozonized water's chlorine demand by 76%.

Table 5. Influence of humic acids on the chlorine consumption of the water :

Treatment	chlorine consumption mg/l	Humic acid concentration mg/l
Raw water	16	11
Settled sand and sand filtered water	1.84	1.90
Ozonated water	2.08	2.20
Filtered water on activated carbon after ozonation	0.5	0.5

This study (figure 14) revealed that a direct relationship exists between the humic acid content and the chlorine demand of a water. This relationship can be expressed as follows : $D = KH^a$, where
- D = chlorine demand after a contact time T
- H = the water's humic acid index
- K and a = experimental conditions depending on the experiment.

As a result, to obtain a water that is as stable as possible in the network, it is indispensable that humic acid elimination be as complete as possible. On this matter, it should be noted that treatment using activated carbon, either as a powder in the clarifier, or in granular form in the second filtration state, or even as a combination of the two during periods of maximum pollution, constitutes one of the best solutions.

After having discovered the role of humic acids as one of the agents responsible for a water's chlorine demand, we studied the interaction of chlorine on humic acids. In agreement with the work carried out by Rook (4) and T.A. Bellar and J.J. Lichtenberg of the EPA (5), we showed that the presence of humic acids entrains the formation of halogenous derivatives. Figure 15 indicates the main halogenous derivatives that we encountered in water from the Loire and Seine rivers.

Measurement protocol for halogenous derivatives. The halogenous derivatives are extracted from the water using Merch Uvasol pentane (pentane extract must never be concentrated) then analyzed by gas chromatography on Porasil C - Carbowax 400 columns and on Tenax GC columns, coupled with an Ni63 electron capture detector.

Results. Figure 16 shows the typical chromatogram of an extract of halogenous derivatives found in Loire river water after chlorination. The following substances were identified:
- carbon tetrachloride in the trace state
- chloroform in large amounts
- trichloroethylene, perchloroethylene and ethane tetrachloride in the trace state
- dichlorobromomethane in large amounts
- dibromochloromethane in the trace state.

Table 6 summarizes the chloroform and dichlorobromethane concentrations found at the different stages of treatment at the plant:
- prechlorination of the raw water using 16 g/m^3, followed by a contact period of 2½ hours in the clarifier, led to the appearance of 50 µg/l of chloroform and 16 µg/l of dichlorobromomethane in the water filtered on sand.
- ozonization reduced the dichlorobromomethane content from 50 to 35 µg/l; this reduction of the 2 halogenous derivatives is due to the stripping process that occurs in the ozonisation tower.

- post-chlorination of the ozonized water, followed by a 5 to 8 hour contact period, led to a new increase in the halogenous derivatives. The chloroform increased from 35 to 47.5 µg/l while the dichlorobromomethane rose from 14.5 µg/l to 25 µg/l. We find this quite normal since the humic acid index found in the ozonized water was 2.20 mg/l. Our measurements confirm the work carried out in the US which shows that the concentration of halogenous derivatives is proportional to the humic acid content as well as to the contact period between the chlorine and the humic acids.

- filtration of the ozonized water on new activated carbon gave contents of 3 µg/l of chloroform and 0.5 µg/l of dichlorobromomethane, i.e. 95% reduction of these two main halogenous derivatives.

Table 6. Dose of chloroform and bromodichloromethane.

Treatment	chloroform µg/l	bromodichloromethane µg/l
Raw water	0	0
Prechlorinated raw water	30	9.3
Settled and sand filtered water	50	16
Ozonated water	35	14.5
Finished water after postchloration	47	25
Filtered water on activated carbon after ozonation	3	0.5

Conclusions

Both chloroform and carbon tetrachloride have been recognized as being carcinogens and mutagens, and it is therefore obviously desirable to reduce the content of or completely eliminate such halogenous material which results from the action of chlorine on humic substances in water.

The easiest solution would be to eliminate the chlorine and replace it by ozone, since ozone does not give rise to halogenous derivatives. However, we feel that ozone's low remanence will entrain risks of bacterial pollution in distribution networks (this point was discussed at Montreal) and that it will always remain prudent to treat discharged water with chlorine or chlorine dioxide. Study of the ozonization of parathion has revealed that under-treatment with ozone can lead to the appearance of metabolites that are just as toxic as halogenous derivatives.

The most promising solution will be continued research into the various means available for removing humic substances from water. Investigation should cover coagulation, flocculation, clarification and filtration, with or without activated carbon in powder or granular form, and modification of the chlorine or chlorine dioxide injection points.

To the end, and in collaboration with the Ministère de la Qualité de la Vie et de l'Environnement, we have initiated a wide scale study program including both laboratory work and tests on industrial plants.

Figure 17 shows the various treatment line possibilities offered by the Morsang-sur-Seine plant :
- the first line comprises : coagulation, flocculation, clarification with the addition of powdered activated carbon, filtration on sand and ozonization.
- the second line includes : coagulation, flocculation, clarification, sand filtration, ozonization and filtration on granular activated carbon as a second stage.
- the third line will compare the preceding treatment with filtration on granular activated carbon followed by a pilot ozonization unit.

All of these waters will be treated with chlorine.
Investigation will include not only study of the evolution of the organic and mineral substances at various points in the plant, but also toxicity tests effected in parallel.

Bibliography

(1) A. LAPLANCHE, G. MARTIN, Y. RICHARD, "Contribution à l'étude de la dégradation par l'ozone de quelques insecticides du groupe des organo-phosphorés". Techniques et Sciences Municipales 5 (1972), p. 271-273.

(2) A. LAPLANCHE, G. MARTIN, Y. RICHARD, "Ozonation d'une eau polluée par le parathion". Techniques et Sciences Municipales 7 (1974), p. 407-413.

(3) Y. RICHARD, "Interference chlore-ozone en traitement d'eau potable". Second symposium international IOI Montréal 1975.

(4) J.J. ROOK, "Formation of Haloforms during chlorination of natural water". Water Treatment and Examination 23 (1974), p. 234-243.

(5) T.A. BELLAR, J.J. LICHTENBERG, "The occurence of organohalides in chlorinated drinking waters". J.A.W.W.A. 66:12 (1974), p. 703-706.

TABLE 1

OZONATION OF ORGANOPHOSPHORUS PESTICIDES

PESTICIDE / OZONE DOSE (mg/l)	0	1	2	3	4	5
MALATHION (μg/l)	100	65	61	45	37	12
PARATHION (μg/l)	87	67	58	41	20	15
METHYL-PARATHION (μg/l)	125	98	83	40	28	20
FENITROTHION (μg/l)	120	72	33	16	13	5
RONNEL (μg/l)	26	23	21	15	11	8

TABLE 2
OZONATION OF PARATHION -- EXPERIMENTAL CONDITIONS

EXPERIENCE NUMBER	Co OF PARATHION g/l	pH	CONTACT TIME mn	OZONE DOSE mg/l O_3	AIR FLOW l/h	DOSE OF OZONE IN AIR mg/l
I	80	8,2-8,3	6	1	225	3,34
				2		6,7
				3		10
				4		13,3
				5		16,7
II	65	8,2-8,3	12	1	225	1,66
				2		3,34
				3		5
				4		6,7
				5		8,35
III	75	6,7-6,8	6	1	225	3,34
				2		6,7
				3		10
				4		13,3
				5		16,7
IV	82	6,7-6,8	12	1		1,66
				2		3,34
				3		5
				4		6,7
				5		8,35
V	80	8,2-8,3	6	1	62,5	12
				2	125	
				3	187,5	
				4	250	
				5	312,5	
VI	70	8,2-8,3	12	1	31,2	12
				2	62,5	
				3	94	
				4	125	
				5	156	
VII	75	6,7-6,8	6	1	62,5	12
				2	125	
				3	187,5	
				4	250	
				5	312,5	
VIII	70	6,7-6,8	12	1	31,2	12
				2	62,5	
				3	94	
				4	125	
				5	156	

TABLE 3
OZONATION OF PARATHION -- EXPERIMENTAL RESULTS

EXPERIENCE		1 mg/l A	B	C	D	2 mg/l A	B	C	D	3 mg/l A	B	C	D	4 mg/l A	B	C	D	5 mg/l A	B	C	D
I	% parathion	89	72	36	33	78	51,5	27	13	70	47	22	4	66	22	3	0	62	11	0	0
	% paraoxon	10	12	20	18	10	21	21	20	17	26	31	21	20	27	23	18,5	20	30	14	9
II	% parathion	81	39	11	8,5	45	13	2	0	41	10	0	0	24	0	0	0	6	0	0	0
	% paraoxon	17	13	12	12	20	14	7,5	6	16	13	6,5	0	16	7,5	0	0	9	3	0	0
III	% parathion	82	55	24	21	78	39	32	10	66	17	0	0	61	11	0	0	52	4	0	0
	% paraoxon	18	32	39	32	22	35,5	35,5	27	23,5	41	27,5	22	34	42	32	22	23,5	35	22	18
IV	% paration	92	62	20	20	55	13	3	0	19,5	0	0	0	11	0	0	0	9	0	0	0
	% paraoxon	8	20	56	35	27	35	39	36	40	42	32	21	35	38	26	13	57	48	19,5	10
V	% parathion	96	90	51	36	92	75	8	6	80	50	3	0	63	20	0	0	42,5	8	0	0
	% paraoxon	0	0	14,5	17	0	10	20	18	14	20	14	10	13	18	12	10	16,5	20	7,5	0
VI	% parathion	91	78,5	21	13,5	83,5	62	6	0	70	36,5	0	0	51	12	0	0	28	0	0	0
	% paraoxon	1	6,5	18	18	5,5	13	19,5	13	9,5	17	17	8	12,5	19	8	2	19	10	0	0
VII	% parathion	90	95	40	30	90,5	72,5	40	10	77	39	3	0	59	11,5	1	0	28,5	8,5	0	0
	% paraoxon	0	1	30,5	35	1	9,5	18	29	21	31,5	23,5	22,5	28	40	20	20	21	30	12	10
VIII	% parathion	97	80	13	10	83	55	3,5	3,5	71	13	2	1	40	3	0	0	15	2,5	0	0
	% paraoxon	3	11	39	41	6,5	21	28	31	17,5	46	22,5	21	29	39,5	15	12	39,5	26,5	10	6,5

TABLE 4 - BACTERIOLOGICAL ANALYSES OF PLANT AND RESERVOIR WATERS

	Finished Water	Non-Chlorinated Water of the Tank	Locally Chlorinated Water of the Tank
Total Bacterial Counts (per ml) - at 37° C. after 24 hrs. - at 20-22° C. after 72 hrs.	0	1,280	0
Number of coliforms/100 ml in membrane filter (37° C.)	0	3,500	16
Number of E.Coli/100 ml in membrane filter (44° C.)	0	0	0
Number of Fecal Streptococcae /100 ml in membrane filter	0	0	0
Clostridium (Sulfite Reductor)	10	10	10
Dose of Chlorine (g/m$_3$)	0.6	0	0.7

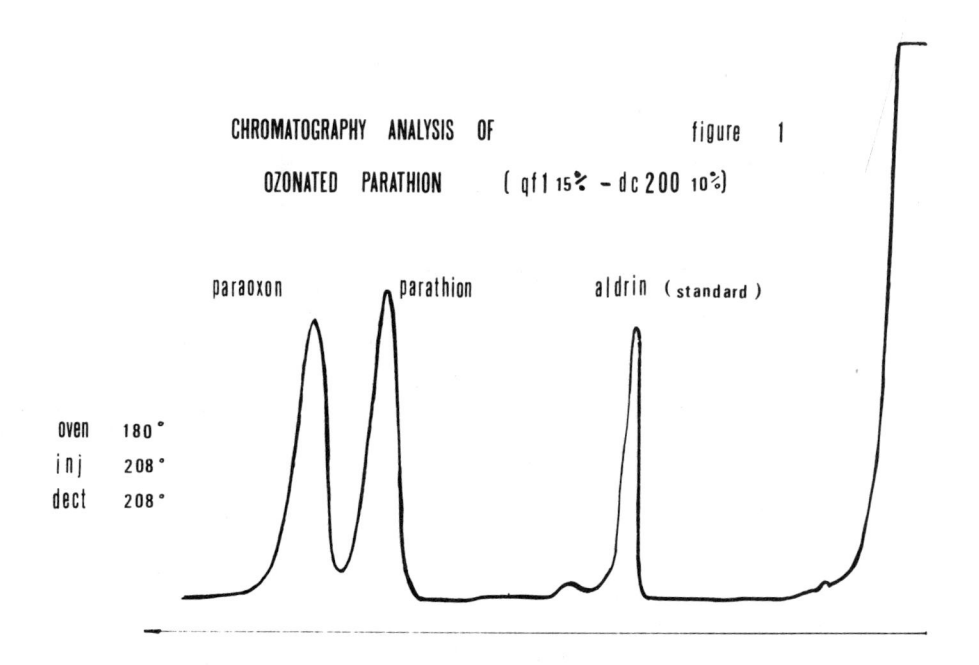

CHROMATOGRAPHY ANALYSIS OF OZONATED PARATHION (qf1 15% - dc 200 10%)

figure 1

paraoxon   parathion   aldrin (standard)

oven 180°
inj 208°
dect 208°

THIN LAYER CHROMATOGRAPHY OF OZONATED PARATHION   figure 2

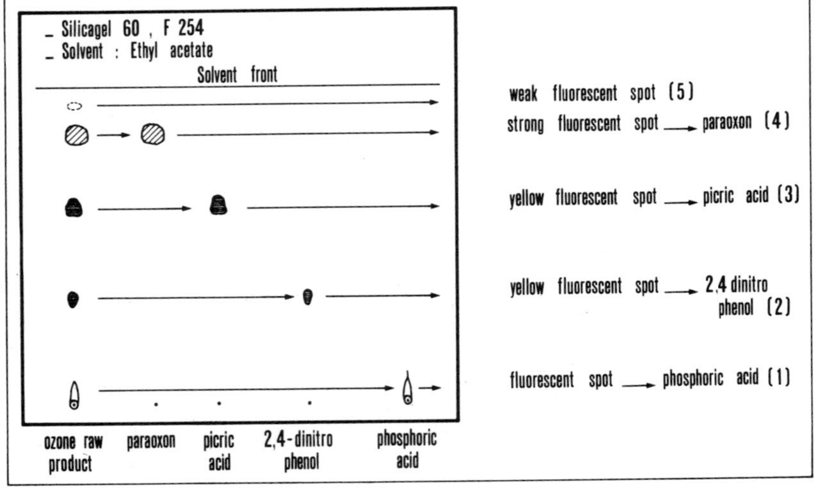

- Silicagel 60, F 254
- Solvent: Ethyl acetate

weak fluorescent spot (5)
strong fluorescent spot ⎯⎯ paraoxon (4)

yellow fluorescent spot ⎯⎯ picric acid (3)

yellow fluorescent spot ⎯⎯ 2,4 dinitro phenol (2)

fluorescent spot ⎯⎯ phosphoric acid (1)

ozone raw product | paraoxon | picric acid | 2,4-dinitro phenol | phosphoric acid

RMN ANALYSIS OF OZONATED PARATHION   figure 3

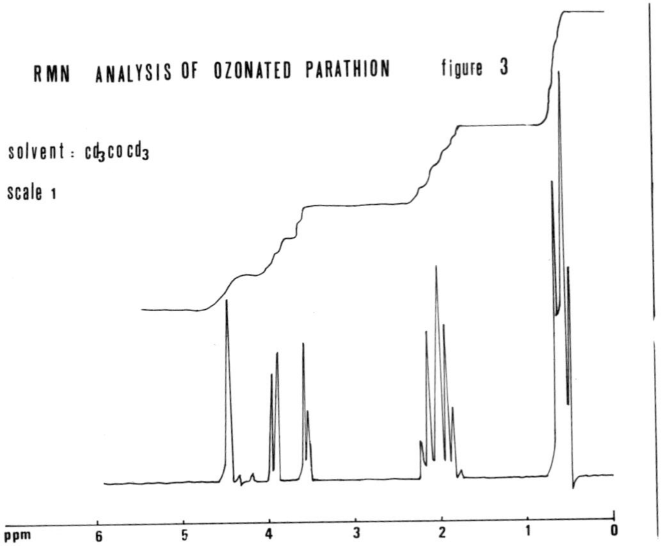

solvent: cd_3cocd_3

scale 1

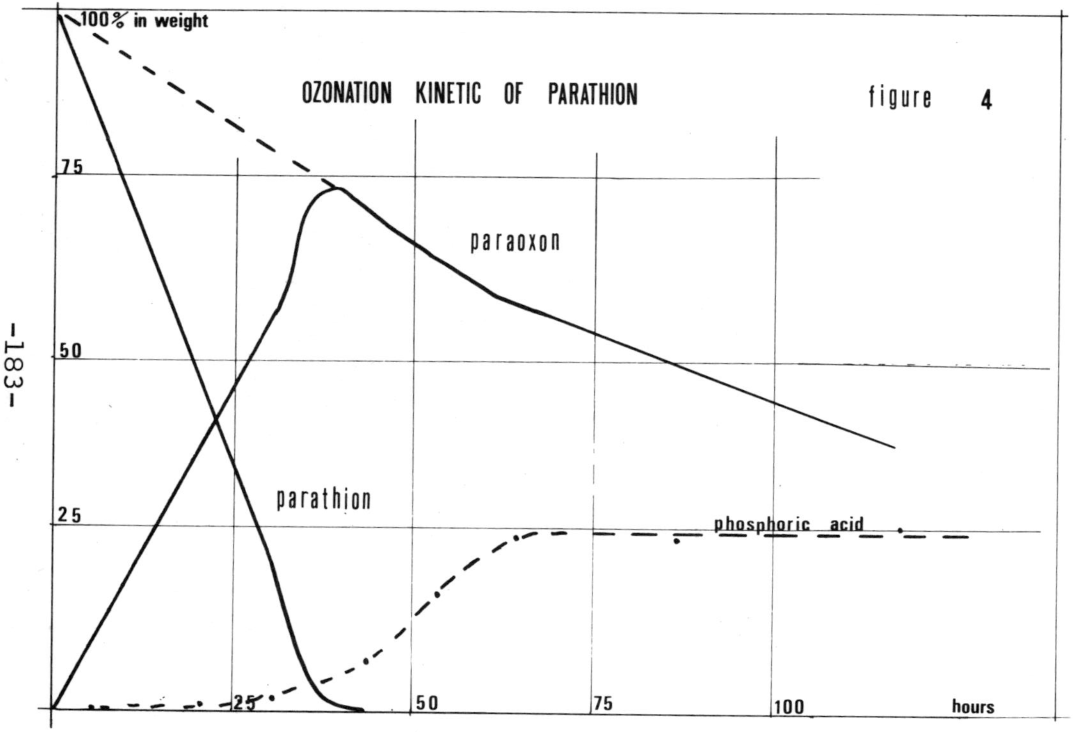

OZONATION KINETIC OF PARATHION figure 4

OZONATION OF PARATHION figure 5

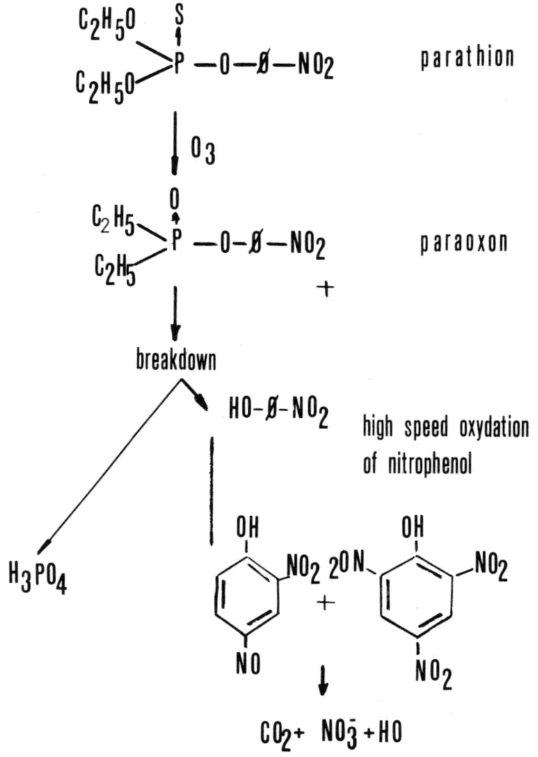

OZONATION OF MALATHION figure 6

Malathion → (O₃) → Malaoxon → (O₃) → H₃PO₄

OZONATION OF PHOSALONE figure 7

o,o-diethyldithiophosphorylmethyl-3-chloro-6-oxo-2-benzazolone

→ (O₃) → hydroxymethyl-3-chloro-6-oxo-2-benzoxazolone

→ (O₃) → bis(chloro-6-oxo-2 benzoxazolylméthyl-3) éther

OZONE OXYDATION OF figure 8
ORGANOPHOSPHORUS INSECTICIDES

$$RO(OS)\\ \diagdown\|\\ P-R'\\ \diagup\\ RO$$

↓

OXON

BREAKDOWN OF THE MOLECULE

- phosphorus parts with or without condensation
- degradation or condensation of R' radical

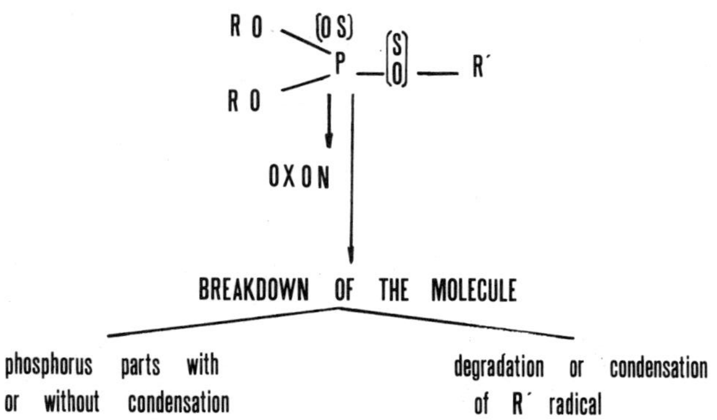

figure 10

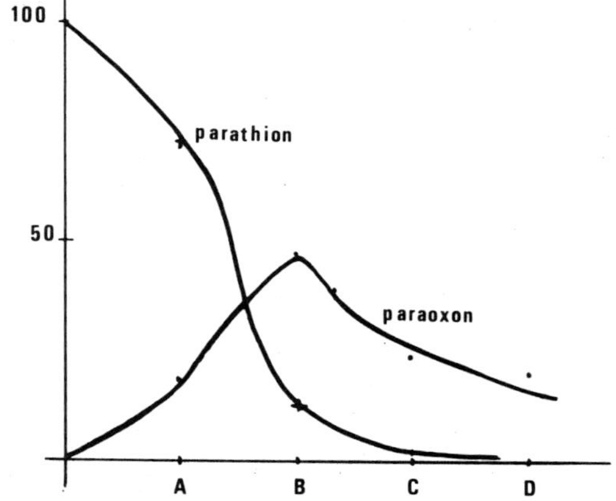

parathion / paraoxon

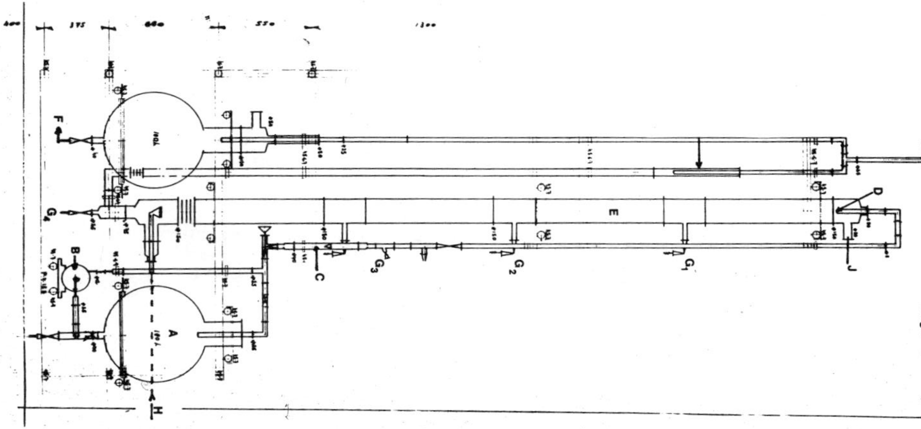

OZONATION COLUMN figure 9

figure 11
INFLUENCE OF OZONE DOSE ON THE ELIMINATION OF PARATHION AND PARAOXON

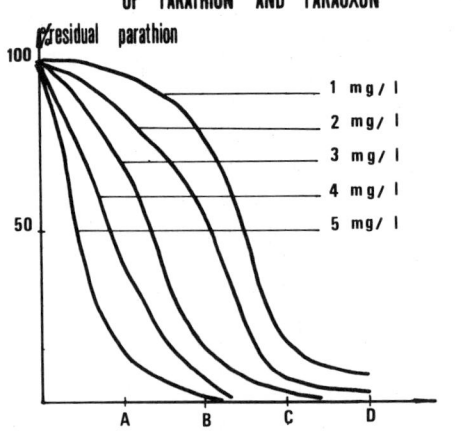

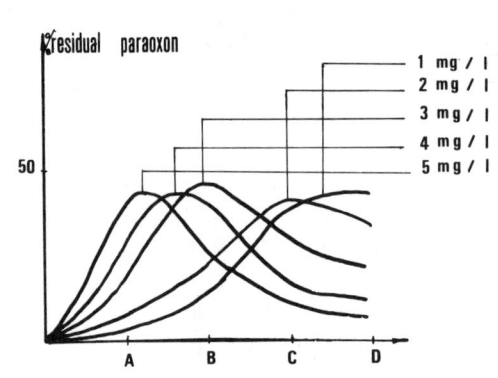

OZONATION OF PARATHION AND PARAOXON IN COLUMN figure 12

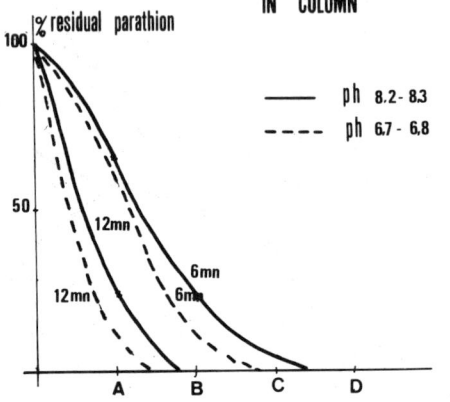

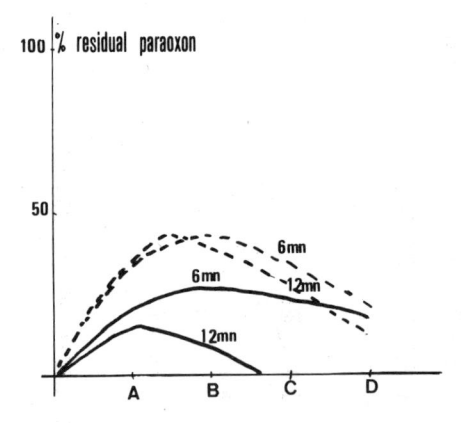

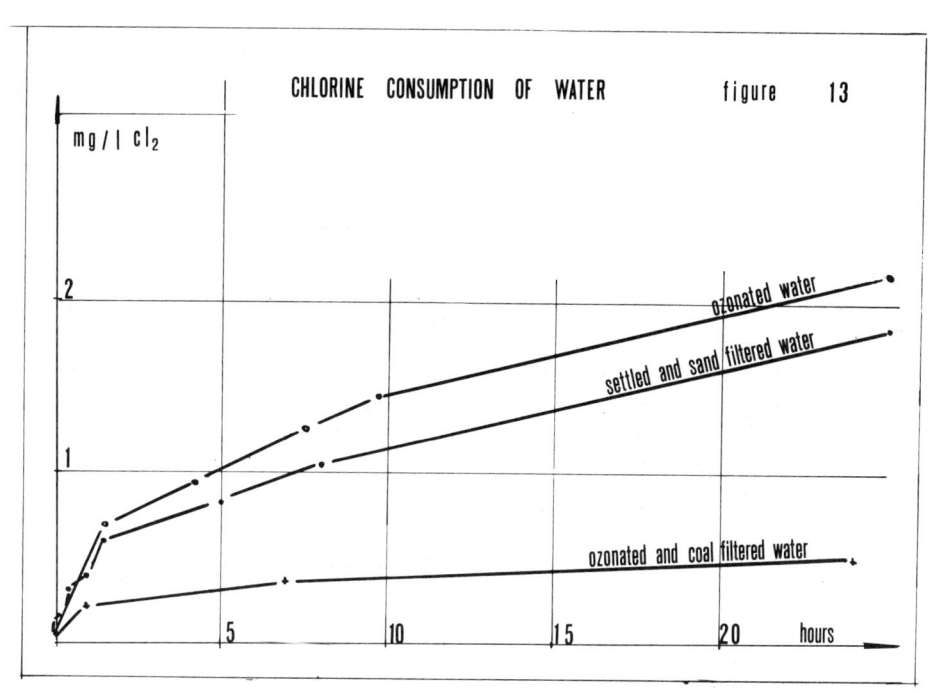

CHLORINE CONSUMPTION OF WATER   figure 13

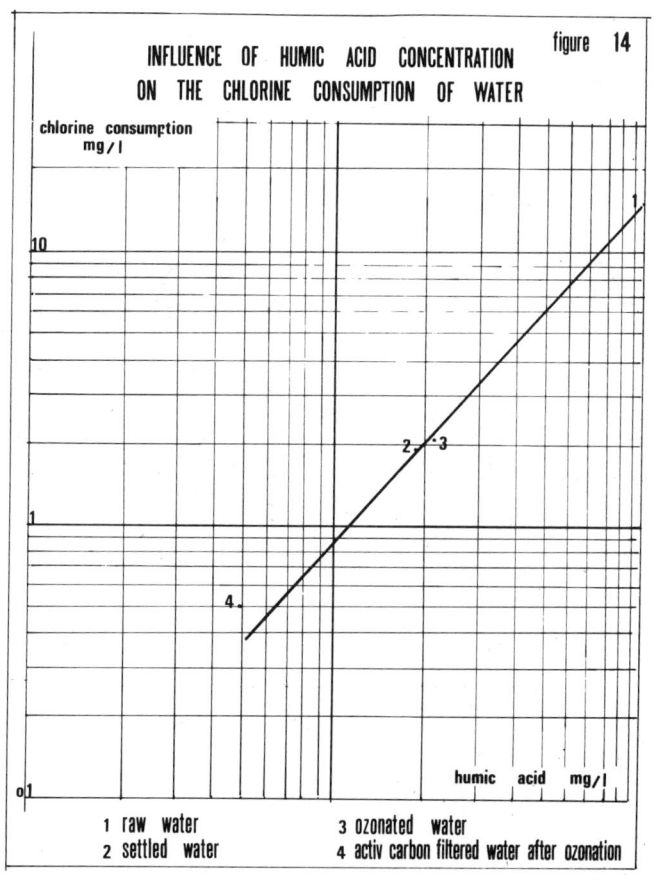

INFLUENCE OF HUMIC ACID CONCENTRATION ON THE CHLORINE CONSUMPTION OF WATER   figure 14

1 raw water
2 settled water
3 ozonated water
4 activ carbon filtered water after ozonation

-188-

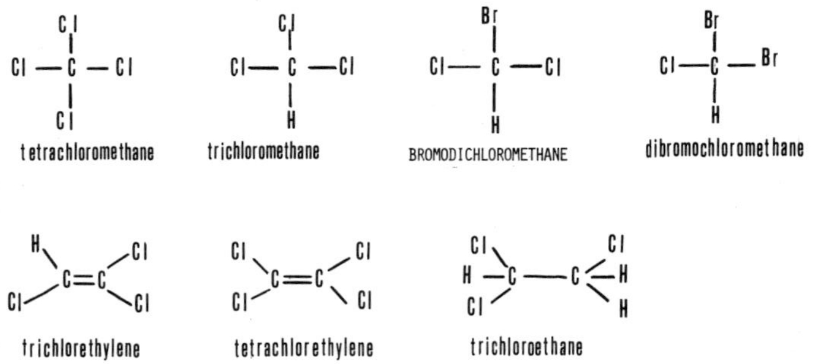

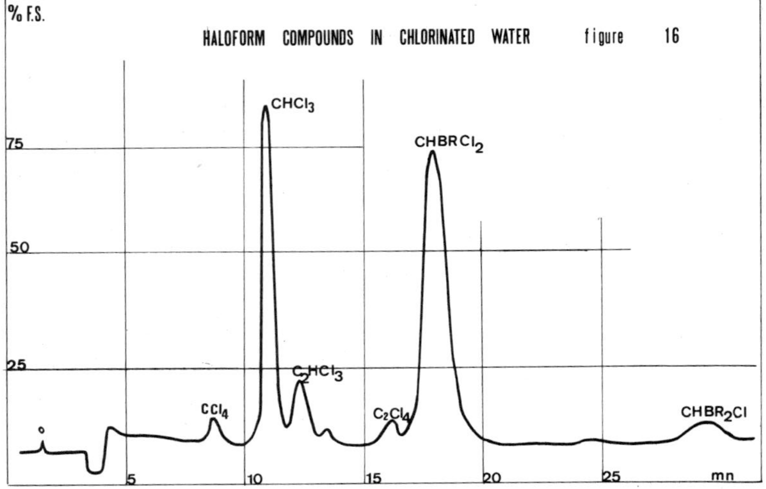

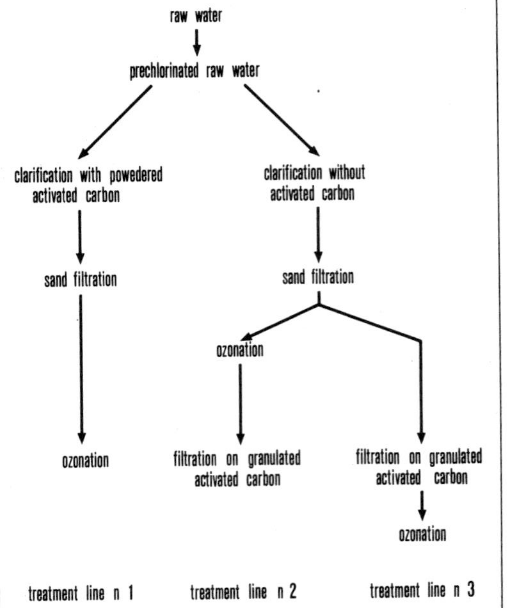

THE DEGRADATION OF HUMIC SUBSTANCES IN WATER BY VARIOUS OXIDATION AGENTS (OZONE, CHLORINE, CHLORINE DIOXIDE)

J. MALLEVIALLE, Y. LAVAL, M. LEFEBVRE, C. ROUSSEAU.

SOCIETE LYONNAISE DES EAUX ET DE L'ECLAIRAGE
27, rue de la Liberté - 78230 LE PECQ- France

In previous publications we pointed out the problems created by the presence of humic substances (HS) in water (1) (2). Until a few years ago this problem seemed to exist only in those countries whose surface water contained large amounts of these substances. As those responsible for the treatment of water for human consumption are increasingly concerned with the compounds which may be formed during oxidation treatments (chloration, ozonization), the identification of of humic-type compounds becomes particularly important, even where no problem is apparent (3) (4). In fact, a certain fraction of HS remains after most of the conventional treatments, and traces almost invariably appear in the distributed water. For this reason we must examine in greater detail the action of different oxidation agents on the humic and fulvic acids (HA) & (FA) in water.

1 - <u>Equipment</u>

 - WELSBACH ozonator, supplying 8 g/hour of ozone when fed with oxygen.

 - Two pyrex reactors used for the ozonization; one measuring 2.5 liters described in a previous publication (2), the other 50 liters.

 - METTLER automatic titrator with TACUSSEL Cu^{++} selective solid electrode.

 - VARIAN 1440 and 2440 gas chromatographs with Ni electron capture detectors and flame ionization.

- TOCSIN TOC-analyzer which measures the carbon as methane
- IONICS 225 TOD-analyzer

2 - Chemical characteristics of the humic substances studied

We worked with natural water containing large amounts of humic matter, and mainly with the water of a depression in the forest of Fontainebleau (near Paris) whose average characteristics are given in Table 1. The values in the first column correspond to the experiments described in paragraphs 3 and 4, those in the second column to those described in paragraphs 5 and 6.

We separated the HS by ultrafiltration through PM10 and UM2 DIAFLO membranes (5). In theory, these membranes retain substances whose molecular weights are greater than 10,000 and 2,000, respectively. We designated as humic acids (HA) the compounds retained by the PM10 membrane, and as fulvic acids (FA) those retained by the UM2 but not the PM10. In this case, the HA were 70 % of the organic carbon, the FA, 20 %. The average molecular weight measured by this technique seemed higher than 10,000, but J. BUFFLE found an average of 2,500 with the use of electrochemical techniques (6).

The HA and FA contain hydroxyl aromatic functional groups which react with the tungstophosphoric and molybdophosphoric acids, producing a blue coloration which may be used to estimate their concentration (7).

The chemical characteristics of the HS which we isolated are given in Table 2. We can see that the total acidity and the number of carboxyl groups are much higher in the case of the FA. According to these results, the HS contain many electron-donating functional groups which can thus form complexes and chelates with metallic ions. These groups, moreover, will enable us to make hydrogen bonds with organics molecules such as pesticides (2) (8). It is therefore useful to determine the complexation ability of the HS, and for this we chose the method based on the use of a Cu^{++} selective solid electrode.

According to Nernst law, the potential measured by this electrode, in the case of a solution which does not contain a complexing agent is :

$$E = E_o + p \log [Me^{2+}]$$

where $[Me^{2+}]$ is the Cu^{++} concentration and E_o the potential of the electrode

and in the presence of such an agent :

$$E' = E_o + p \log [Me_t] - p \log \alpha$$

where $[Me_t]$ is the total copper concentration and

$$\alpha = \frac{[Me_t]}{[Me^{2+}]}$$

Increasing quantities of copper are added to a buffering solution (0.1 M KNO_3), as well as to the same buffering solution containing a given amount of a complexing agent. We plot the curves of the potentials measured as a function of $\log [Me_t]$ and can thus calculate α [Fig 1(6)].

3 - The action of ozone on the humic substances

We carried out several ozonization series by varying the flow of gas as well as the concentrations of ozone and HS. We shall only give a few curves here, which illustrate the changes in the different parameters during the ozonization.

Figure 2 shows these changes as a function of time, color, total acidity, and the concentration of polyhydroxyaromatic derivatives. The color decreases rapidly; in ten minutes more than 90% disappeared. We also observe that the point of maximum acidity corresponds to the almost total disappearance of color, and we can detect polyhydroxyaromatic derivatives for 90 to 120 minutes.

Figure 3 compares the percentages of color eliminated, the TOC, the COD, TOD and BOD_5 obtained as a function of the ozonization time. The TOD decreases much more rapidly than the TOC, which shows that there is a gradual oxidation of the organic compounds present, which do not, however, reach total mineralization. We also observe that during the depolymerization phase, the BOD5 increases gradually, which indicates the formation of more biodegradable substances.

It was also important to examine the average values of ozone consumption in relation to the variations in some of the parameters used (Fig. 4). We see that for 525 mg of HS, 100 mg of ozone are needed to eliminate 95% of the color, 320 mg to eliminate 95% of the polyhydroxyaromatic derivatives while 380 mg and 500 mg are required, respectively, to decrease by 75% the COD and TOC values.

4 - The action of ozone on HS-cation and HS-pesticide complexes

Fig. (5) shows the diagrammatic representation of HA's structure (9).

During ozonization of water containing large amounts of humic substances, we often observe a turbidity and residual color due to the liberation and oxidation of iron into ferric hydroxide. In the water which we studied, we observed a violet tint, due to the oxidation of liberated manganese, which gave 0.5 mg of Mn and 0.9 mg of Fe with 150 mg of HS. These examples demonstrate clearly that the complexed metals are liberated through depolymerization by ozone.

We tried to determine, as a function of the ozonization time, the complexation ability of the water as defined in paragraph 2. The results obtained are a bit surprising (Fig. 1) for this ability increases as the color disappears. This can probably be explained by the fact that the ozone immediately breaks the side chains linking the different cyclic compounds of the core, thus liberating a greater number of functional groups (COOH, C = O, ...) which can increase the complexation ability of the water. This type of test is quite interesting, but requires a more systematic study for a better interpretation of results.

As we have already mentioned, the HS are able to bind organic compounds such as pesticides. We therefore prepared solutions of various HS pesticide complexes, and then placed HS filtered samples in an ultrafiltration apparatus (UM2 membrane) where they were washed with distilled water until no further traces of pesticides remained in the filtrate. This technique thus enabled us to purify the complexes formed without having to dry them out. Figure 6 shows the infrared spectra of a herbicide - 2,4-dichlorophen-oxyacetic acid - in the form of its triethanolamine salt, of the SH, and finally of the purified HS-herbicide complex. We were thus concerned with the evolution of these HS-bound organic micropollutants during ozonization. We worked on two pesticides, lindane and aldrin. The ozonization was carried out on ten liters of water, from which samples were extracted at one minute intervals, using a mixture of sulfuric ether and petroleum ether (50/50), in order to show the presence of lindane or aldrin. We were never able to demonstrate the liberation of aldrin during the ozonization, probably because aldrin is rather easily degraded by ozone. However, Figure 7 shows that lindane is released very rapidly into the solution and that the highest concentrations obtained correspond to the total removal of color from water. In one liter of water, originally containing 25 mg of TOC, we were thus able to detect 0.45 µg of lindane.

This series of experiments, which we intend to continue, furnishes very interesting data, for it demonstrates that the HS can not only hinder the analysis of micropollutants such as pesticides, but can also protect them in certain types of treatment.

5 - The combined action of chlorine and ozone

In recent years attention has been drawn to the haloform compounds which may be formed during chlorination treatments. We therefore tested the combined action of chlorine and ozone on the HS, without, however, making a complete study of the problem because we are just beginning this study. The tests were made in a 50 liter reactor, on water which contained a much smaller amount of HS than that used in the preceding experiments (Table 1, column 2).

5-1 Ozonization followed by chlorination

Three samples of the water of Fontainebleau were taken, and two of them ozonized, so that we had:

(A) = a sample of the raw water

(B) = a sample of the ozonized water with 50% of the color removed.

(C) = a sample of the ozonized water with 75% of the color removed.

These three samples were then chlorinated with 40 ppm of chlorine, a dose which corresponds to the breakpoint of the raw water. Extractions were made with pentane and the extracts obtained, before and after chlorination, were analyzed by electron capture gas chromatography (at 70°C with 4 m-Carbowax 20 M column). The results are summarized in Table 3.

We cannot observe significant differences between sample water B and C. The amount of chloroform detected after chlorination seems less significant when the chlorination is performed on ozonized water, although the total amount of halogenated compounds of low boiling point is greater. It can be explained by the degradation of the precursor functional groups. These by-products are not halomethanes, but we have not yet identified them.

5-2 - Chlorination followed by ozonization

In this case we had three water samples; two samples were chlorinated with 40 ppm of chlorine and one of them was then ozonized up to 75% disappearance of color. The third one was ozonized during the same time and then chlorinated with 40 ppm of chlorine.

The results are given in Table 4.

Here again we have the formation of chloroform which seems to dissipate during the ozonization, probably due to stripping. The total amount of halogenated by-products seems greater when the ozonization is performed on chlorinated water.

6 - Comparison of the action of chlorine and chlorine dioxide on the HS

We used three different amounts of chlorine and chlorine dioxide; 2, 20 and 40 ppm for three periods of contact: 1, 4 and 22 hours. The parameters measured were color, TOC, fluorescence and concentration of chlorinated compounds.

From the overall results obtained, we can conclude that there is not a big difference between chlorine and chlorine dioxide as far as color removal, fluorescence or TOC are concerned but chlorine dioxide does not induce the formation of even a slight amount of halogenated compounds of low boiling point under these present experimental conditions (Fig. 8).

Conclusions

The presence of humic-type substances in most water treated for human consumption is now an accepted fact, and the role of these compounds in the formation of by-products during oxidation reactions cannot be ignored. Ozonization leads to the formation of compounds which ought to be removed by filtration through activated charcoal. Whereas, on the other hand, chlorine oxidation produces trihalomethanes which are considered hazardous to human health, their production seems considerably smaller in the case of chlorine dioxide, which could thus be used particularly in the final stage of treatment. The ideal solution perhaps might be the combined use of these three oxidation agents in the same treatment chain.

REFERENCES

(1) - J. Mallevialle: "Les agents complexants naturels des eaux - Etude des propriétés physico-chimiques des matières humiques et de leurs transformations par ozonation" - Thèse (1974) - University Paul Sabatier - Toulouse (France).

(2) - J. Ph. Buffle, J. Mallevialle: "Le rôle des matières humiques et envisagées comme agent d'accumulation et véhicule des substances" - T.S.M. 6.74

(3) - A.J. Drapeau, M. Trudeau - "Du chloroforme dans votre eau potable" - Eau du Quebec, 8(2):19-23 (1975)

(4) - J.J. Rook - "Formation of haloforms during chlorination of natural waters" - J. Water Treatment Exam., 23:234 (1974)

(5) - E.T. Gzessing - "Physical and chemical characteristics of Aquatic humus" Ann Arbor Science Pub.

(6) - J. Buffle and coll. - "Study of complexation of heavy metal ions in natural waters using ion selective electrodes and voltammetric methods" (to be printed in Anal. Chem.).

(7) - Standard Methods (13e ed. p. 346).

(8) - M. Schitzer, S.U. Khan "Humic substances in the environment" - Marcel Dekker Inc. - New York, 1972.

(9) - R. D. Haworth, Soil Sci. - 111:71 (1971).

TABLE 1 - CHEMICAL CHARACTERISTICS OF THE FONTAINEBLEAU FOREST'S WATER

Resistivity @ 20° C.	10,615	5,000
pH @ 20° C.	6.2	6.5 - 7.5
T.C.C. (mg/L)	80	25 - 30
Ca^{++} (mg/L)	14	28
Cl^- (mg/L)	7	10
SO_4^{--} (mg/L)	0	0
NO_3^- (mg/L)	0	3
NH_4^+ (mg/L)	1	1.5
Fe (mg/L)	1.2	1.8

TABLE 2
FUNCTIONAL GROUP ANALYSIS

Functional Groups	Fulvic Acid (meg/g)	Humic Acid (meg/g)
Total Acidity	9.2	5.4
Carboxyls	8.9	2.3
Carbonyls	4	9.5
Quinones	3	2.7
Phenolic Hydroxyls	5.6	5.7
Total Hydroxyls	6.3	3.1

TABLE 3

	INITIAL WATER (A)	OZONIZED WATER (50% of the color removed) (B)	OZONIZED WATER (75% of the color removed) (C)	A + Cl$_2$ [40 ppm 2 h]	B + Cl$_2$ [40 ppm 2 h]	C + Cl$_2$ [40 ppm 2 h]
COLOR O.D. 420nm d = 1 cm	0.098	0.047	0.026	0.070	0.035	0.020
T.O.C. (mg/l)	25	24	19	24	23	15
CHCl$_3$ (µg/l)	--	--	--	10	6	6
HALOGENATED BY PRODUCTS (µg/l)	--	--	--	60	120	125

Halogenated by-products are calculated in µg/l of water by reference with the chloroform.

TABLE 4

	COLOR [O.D. 420 nm d = 1 cm]	T.O.C. (mg/l)	HALOGENATED BY-PRODUCTS (µg/l)
INITIAL WATER (I)	0.1	25.5	--
I + Cl$_2$ [40 ppm 2 h]	0.072	24.5	90
I + O$_3$ (75% of the color removed) tmn	0.027	21.0	--
I + Cl$_2$ [40 ppm 2 h] + O$_3$ (tmn)	0.020	20.0	235
I + O$_3$ 75% of the color removed + Cl$_2$ [40 ppm 2 h]	0.018	17.5	160

Halogenated by-products are calculated in µg/l of water by reference with the chloroform

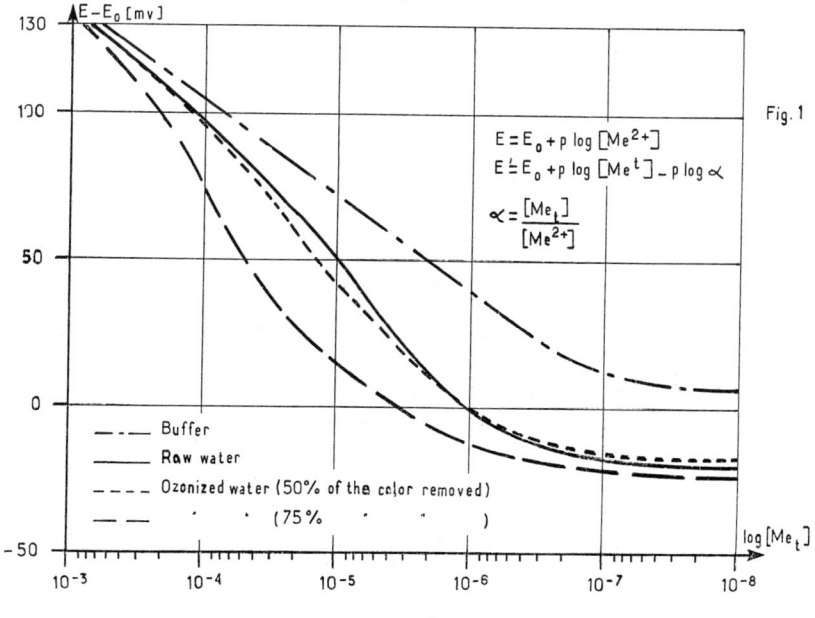

Fig. 1

$$E = E_0 + p \log [Me^{2+}]$$
$$E = E_0 + p \log [Me_t] - p \log \alpha$$
$$\alpha = \frac{[Me_t]}{[Me^{2+}]}$$

— · — Buffer
——— Raw water
- - - - Ozonized water (50% of the color removed)
— — " (75% " ")

-197-

Fig. 2

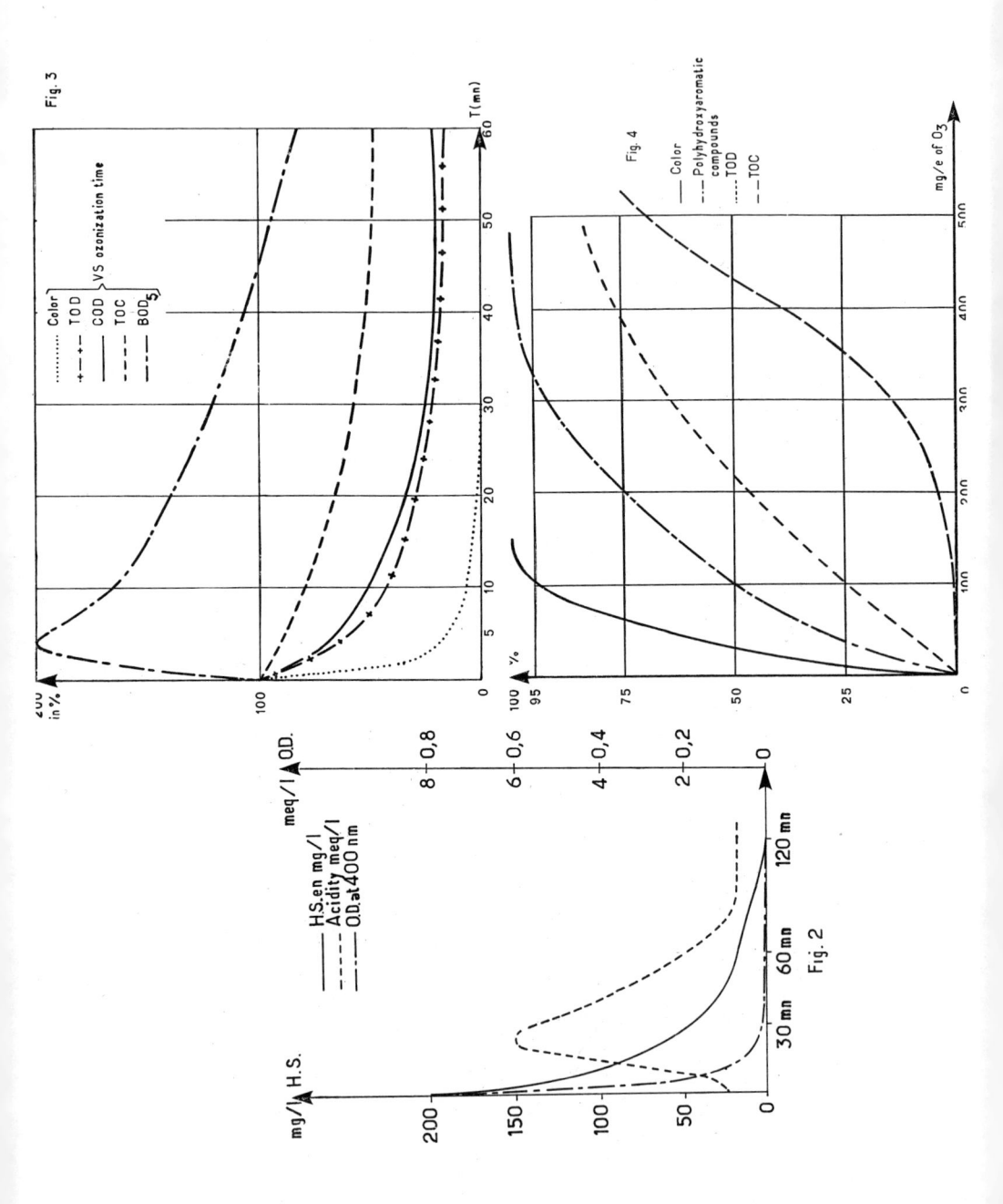

Fig. 3

Fig. 4

-198-

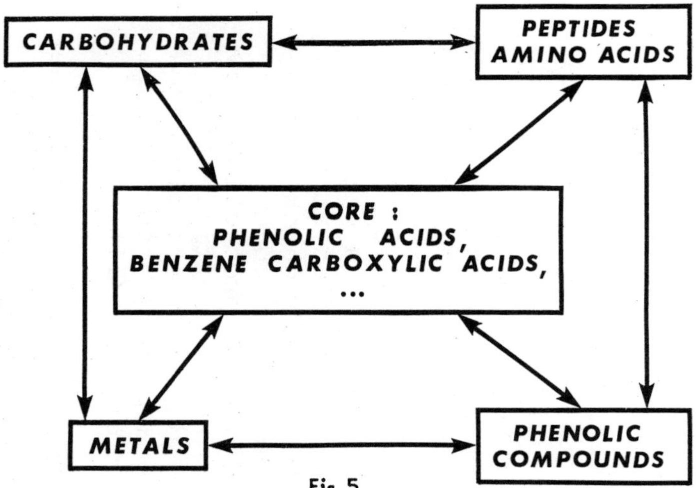

Fig.: 7 Released lindane concentration vs ozonization time

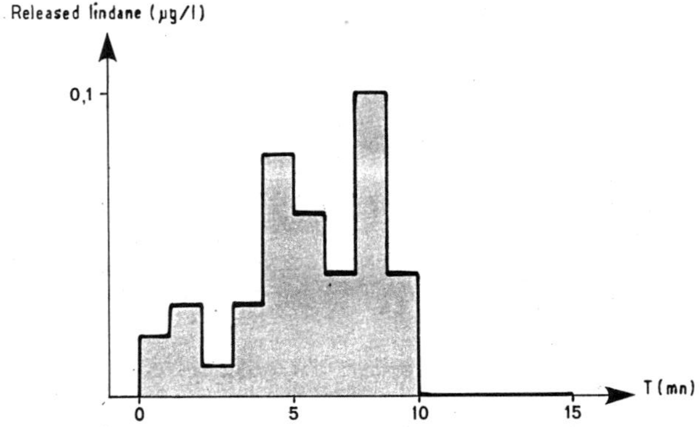

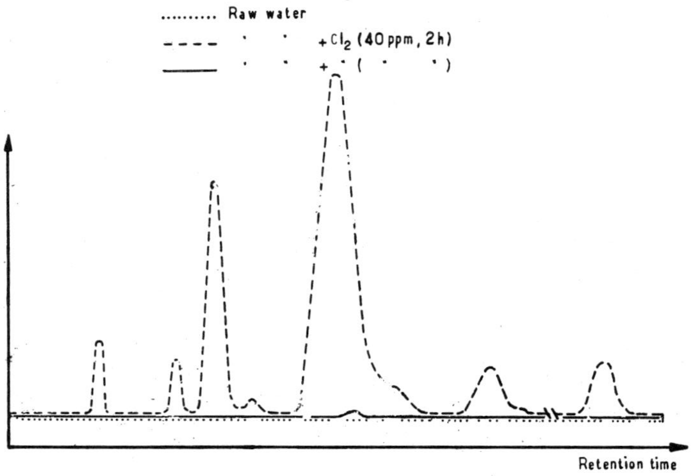

Fig. 5

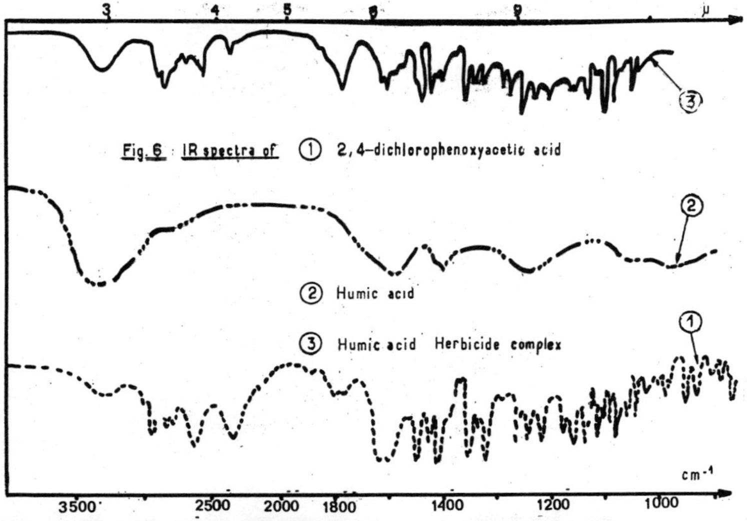

Fig. 6 IR spectra of ① 2,4-dichlorophenoxyacetic acid
② Humic acid
③ Humic acid Herbicide complex

Fig. 8 — ORGANIC PROFILE (E.C.D., 4m. CARBOWAX 20M column, 70°C)

-199-

THE OXIDATION OF HALOFORMS AND HALOFORM PRECURSORS UTILIZING OZONE

Stephen A. Hubbs, Research Engineer

Louisville Water Company
435 South Third Street
Louisville, Kentucky 40202
United States of America

The Louisville Water Company in conjunction with the University of Louisville is currently conducting a series of investigations into various techniques of reducing haloforms in drinking water supplies. One portion of the study involves investigations into the use of ozone both as a primary disinfectant in lieu of chlorine and as an oxidizing agent for the removal of haloforms and precursors. This presentation is a progress report on work analyzing the oxidizing capacity of ozone for chloroform and its precursors in drinking water.

Introduction

In anticipation of a Federally imposed standard on trace organic contaminants in drinking water, the Louisville Water Company is considering several process changes designed to lower the concentration of trihalomethanes-particularly chloroform-in the product water. Previous investigations (1) have indicated that product water concentrations of chloroform can be reduced by introducing chlorine to the highest quality water possible in the treatment chain of a particular system. This phenomenon has been verified at the Louisville Water Company by experiments in which a change in chlorine feed point from raw to settled water corresponded to a decrease in chloroform production. However, the benefit realized may not be sufficient to insure that the product water will consistently have a chloroform concentration below the suggested EPA Maximum Contaminant Level of 100 micrograms/liter.

The introduction of chlorine at any point further downstream in the Louisville Water Company treatment chain would result in a significant increase in the

[OCl⁻]/[HOCl] ratio in the water and a corresponding decrease in bactericidal potential due to elevated pH values in the softening process. Thus, to insure adequate disinfection, a higher concentration of chlorine would be required. When coupled with the enhanced formation of haloforms at elevated pH values, the practice of adding chlorine at or beyond the coagulation process point to reduce chloroform in the product water could be self-defeating.

A treatment chain indicating promise towards the goal of chloroform reduction involves the use of ozone as a primary disinfectant, introduced to either raw, coagulated, or softened process waters, followed by filtration and minimal post-chlorination to insure against recontamination of the distribution system and to meet Federal guidelines. The disinfective capabilities of ozone have been documented in previous works (2); however, literature on the specific oxidation of chloroform and precursors by ozone is limited (1)(3). To determine at what process point the oxidative capacity of ozone could be best exploited, a series of batch tests was designed to allow a focusing of future work on particularly promising alternatives.

Experimental Work

The Louisville Water Company employs pre-sedimentation, coagulation (with organic polymer and alum), softening (lime-soda ash), stabilization (CO_2), filtration, and disinfection in its treatment system. Chlorine is currently applied after pre-sedimentation (at the influent of the coagulation process) and in the clearwell, immediately after filtration. Chemicals used and typical feed rates are shown in Table #1.

Experiments were designed to mimic actual treatment plant conditions. The water used in the precursor-chlorine-ozone reactions was either process water from the plant or raw water treated with chemicals from the plant when the exclusion of prechlorination was desired. The reaction time for ozone oxidation of chloroform and precursors was 20 minutes. The reaction time of chlorine with precursors to form chloroform was 48 hours, the reaction being stopped with sodium thiosulfate. The reaction volume was 20 liters. All reactions (except for initial contact) proceeded in dark to avoid loss of chlorine by sunlight.

Five basic experiments were conducted, simulating the addition of ozone to pre-settled, coagulated, softened, coagulated and softened, and finished water. Reaction conditions for these experiments are listed in Table #2.

The analytical technique used for the detection of chloroform and associated trihalomethanes in the microgram per liter range was the modified Bellar (4) technique, utilizing a 25 ml sample purged for 10 minutes with nitrogen. The sample was trapped on a Tekmar LSC Concentrator and desorbed for 12 minutes at 180°C. The gas chromatograph was programmed at 80°C for 4 minutes, 8°C per minute for 15 minutes, and 200°C for 4 minutes. The G.C. was equipped with a six-foot column containing Chromosorb 101 and a flame ionization detector.

The reproducibility of the analytical technique described above in the range of 25 micrograms per liter of chloroform is defined with a coefficient of variation of 4.5%, a standard deviation of 1.1, in a range of 22.8 to 25.7. The mean value for the five identical samples run to define the reproducibility was 24.52 micrograms per liter. The samples were tap water, individually capped, with sodium thiosulfate added to stop the chloroform reaction. Analytical reproducibility for chloroform in the 100 microgram per liter range was similar.

For all experiments performed, the analyses for any batch were run within a 10 hour time span to eliminate any fluctuations in daily operation of the gas chromatograph. The reaction water used for any one experiment was taken from one batch. However, the water used from experiment to experiment varied with the quality of the river.

To determine the removal of the organic precursor responsible for chloroform formation, the potential of a treated water to form chloroform was measured by chlorinating the water with 10 mg/l of chlorine and allowing the reaction to proceed at room temperature in the dark for 48 hours. This sample was then analyzed for chloroform, the decrease in chloroform detected corresponding to a decrease in the precursor organic. In this investigation, no effort was made to adjust pH in determining chloroform potential; rather, the reaction was allowed to proceed at the naturally occurring pH. Thus, no comparison of analytical results of chloroform potential outside of experimental batches is possible.

Discussion of Results

Ozone As An Oxidant of Chloroform.

The effectiveness of ozone in oxidizing chloroform in Louisville Water Company finished water is illustrated in Figure #1. The graph indicates that ozone in doses up to 8.0 mg/l has no significant effect on the concentration of chloroform. This phenomenon may be explained by the relatively high organic content in the product water (1 to 3 mg/l NVTOC). It is suspected that the ozone is spent on the more readily oxidized portion of the total organic loading, and that until that portion is removed, no effect is exerted on the chloroform. Further experiments are underway to firmly establish this concept.

Ozone As An Oxidant of Chlorofrom Precursor Organics.

The ability of ozone to remove the organic precursor to chloroform in various Louisville Water Company process waters was investigated in four separate experiments. In each experiment, raw Ohio River water which had settled for 24 hours in the system's pre-sedimentation basins was reacted with ozone after some degree of pre-treatment, then chlorinated and analyzed for chloroform (see Table #2).

Figure #2 shows the results of applying ozone in doses up to 8 mg/l to pre-settled water. A definite decrease in the potential of the water to form chloroform corresponds to increasing ozone dosages. An ozone dosage of 8 mg/l roughly corresponds to a 40% reduction in chloroform produced upon chlorination. Unfortunately, the zero dosage point on this curve was lost during analysis; however, the trend of decreasing chloroform potential with increasing ozone dosage is clearly established.

To evaluate the effect of adding ozone to a higher quality water, pre-settled water was coagulated and settled, then ozonated up to 4.8 mg/l. The results of this experiment shown in Figure #3 indicate that no trend in chloroform production versus ozone concentration was established. The absence of a definite trend may be tentatively explained by two factors: the removal of the easily oxidizable portion of the precursor by coagulation, and the addition of an ozone demand in the form of the organic polymer coagulant aide. Under the first factor, that portion of the precursor easily oxidized by ozone may be also easily removable by coagulation. If this is the case, then no actual benefit is realized by ozonation if chlorination is delayed until coagulation is complete (in the range of ozone dosages considered). The second factor of increased ozone demand created by the addition of the organic polymer coagulant aide would exhibit similar experimental traits as did the first factor, due to the preferred oxidation of the organic polymer over the chloroform precursor. However, due to the low dosage of polymer added (0.2 ppm) and the removal of a great deal of this polymer in the coagulation/settling process, it is expected that the second factor exerts a minimal ozone demand in the coagulated water. Experiments are being conducted to verify this assertion, utilizing chemicals other than organic polymers to accomplish coagulation.

The results of Experiment #4 illustrate the effect of ozonating pre-settled water that has been softened via a lime-soda ash precipitative process (Figure #4). After an apparent increase in chloroform potential upon initial ozonation, a general trend of decreasing chloroform potential with increasing ozone dosage is established. The relatively higher chloroform concentrations formed upon chlorination in this experiment represent the effect of pH on the chloroform reaction, and not the effect of ozonation at this higher pH. The initial increase in chloroform potential upon ozonation may indicate the formation of more sites for the chloroform reaction on the humic acid precursor; however, the absence of this trend in previous experiments discounts the possibility of this occurrence.

The results of an investigation of a treatment chain simulating the current processes of pre-sedimentation, coagulation, and softening utilized at the Louisville Water Company with the exclusion of pre-chlorination and the inclusion of ozonation after softening is presented in Figure #5. Again, no definite trend of chloroform potential and ozone dosage exists, for reasons which may be similar to those pertaining to Experiment #3. The effect of pH is again illustrated in the higher concentrations of chloroform produced at pH 10.3, and is not an effect of ozonation at this high pH.

Conclusions

Conclusions based on these preliminary investigations should be considered in the context of one water supply and one particular treatment system. These investigations indicate that ozonation offers little relief from the problem of chloroform formation via the route of precursor or chloroform oxidation within a reasonable range of ozone dosages. The extent of organic precursor removal realized by ozonation of pre-settled water is also realized by coagulation. Thus, ozonation offers no enhancement of precursor removal within the Louisville Water Company treatment system. The oxidation of chloroform is likewise negligible under the described conditions of ozonation of finished water. This is presumably caused by ozone demand exerted by more readily oxidizable organic compounds in the water.

Although these preliminary experiments indicate a limited applicability of ozone as an oxidative remedy to the chloroform problem in public water supplies, the author acknowledges the potential of ozone as a substitute disinfectant for chlorine, and as such an agent for the avoidance of chloroform. Further experiments are planned to support conclusions drawn from these preliminary investigations under conditions of varying raw water quality.

Acknowledgements

The author acknowledges the analytical support received from the University of Louisville Environmental Engineering Staff.

References

(1) J.M. Symons, et.al., "Interim treatment guide for the control of chloroform and other trihalomethanes", USEPA, Water Supply Research Division, Municipal Environmental Research Laboratory, Office of Research and Development, U. S. Environmental Protection Agency, Cincinnati, Ohio 45268 (1976).

(2) R.G. Rice, M.E. Browning, Editors, "Proc. First Intl. Symposium on Ozone for Water and Wastewater Treatment", Intl. Ozone Inst., Cleveland, Ohio (1975).

(3) J.J. Rook, "Haloforms in drinking water", J.Am. Water Works Assn., 68:168-172 (1976).

(4) T.A. Beller, J.J. Lichtenberg, "Determining volatile organics at microgram-per-liter levels by gas chromatography", J.Am. Water Works Assn., 66:739-744 (1974).

DISCUSSION

Maggiolo: When you softened the water were there any surfactants? You can have non-ionic surfactants, which were mentioned the other day, and these have ethoxy groups in the chains. What are the chances of your affecting the chloroform in the presence of these non-ionic surfactants that you wouldn't take out with an ionic exchange water softener? You actually should have produced an alcohol.

Hubbs: No, this is precipitant softening with lime-soda ash.

Maggiolo: I don't think you take out the non-ionic softeners.

Hubbs: I'm not sure.

Maggiolo: I don't think you do, chemically, because they are not ionic. So I don't think you remove the ethoxy surfactants, what we call non-ionic surfactants. If they are available in any kind of quantity at all, your chances of affecting the chloroform are just about nil.

Hubbs: Until they are totally oxidized, right? Is that what your point is?

Maggiolo: In other words, you're going to ethoxylate the non-ionic surfactant.

Hubbs: First.

Maggiolo: Much more, you won't go near the chloroform with ozone. So you might reevaluate that area.

Harvey Rosen: You used organic coagulants, is that correct, polymers?

Hubbs: Right.

Rosen: I think those are going to create an ozone demand as well, and that's why maybe you didn't see any reductions after coagulation, whereas you did before. We have done some experiments on water supplies where we've had the ozone react with the coagulant, adding ozone post-coagulation.

Hubbs: Another point, too, is that in our best systems we were using identical water to the process, and we did see a reduction in the chloroform concentration, and with the use of ozone and with the use of just coagulation. I see your point though.

Rosen: You are spending your ozone on the coagulant when you add it after, that's why you saw a bigger effect than when you added it before.

Ed Chian: I understand that you used Bellar's method--- you used this 5-10 cc sample?

Hubbs: Twenty-five cc.

Chian: Because of this background contamination, especially in the apparatus and on the strip, I think it could introduce a tremendous amount of error in quantitation, especially on chloroform. Now Bellar's method is good to detect those compounds, but it was of questionable value to quantify them. You really have to use a modified method. The method we are using in our lab uses 150 cc, a much larger volume. Then we take all the precautionary steps, such as keeping it under a nitrogen atmosphere prior to any analysis so you can eliminate the problem of chloroform variations.

Hubbs: We did check the standard deviation of our technique in the range of 70 micrograms per liter, and we were working in this experiment in the range of 40 to 50. We got a coefficient of variation of 6%, which means that with this experiment, if it is extrapolatable down to those varied values, a difference of about 4 to 5 units. This is the efficiency that we were able to get out of our particular techniques.

Table #1: TYPICAL FEED RATES OF CHEMICALS USED AT THE LOUISVILLE WATER COMPANY

Process	Chemicals Used	Feed Rate
COAGULATION	alum organic polymer	100#/MG 0.2 ppm
SOFTENED	lime soda ash CO_2	250#/MG 90#/MG 50#/MG
DISINFECTION	pre-chlorine post-chlorine	33#/MG 10#/MG
OTHER	sodium silicofluoride	12#/MG

Table #2: REACTION CONDITIONS FOR EXPERIMENTS #1 THROUGH #5

Experiment	Degree of Pretreatment	pH
1	finished (tap) water	8.5
2	pre-settled water	7.1
3	pre-settled coagulated water	7.1
4	pre-settled, softened water	9.8
5	pre-settled, coagulated, softened water	10.3

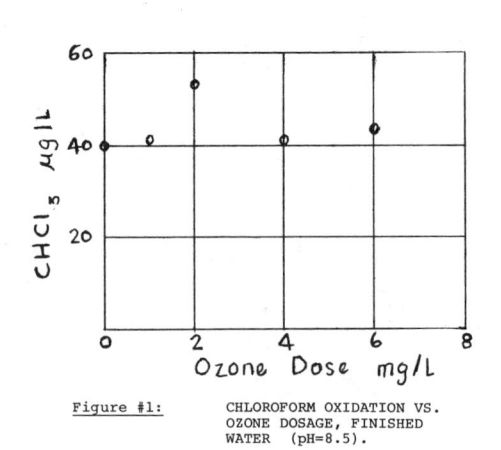

Figure #1: CHLOROFORM OXIDATION VS. OZONE DOSAGE, FINISHED WATER (pH=8.5).

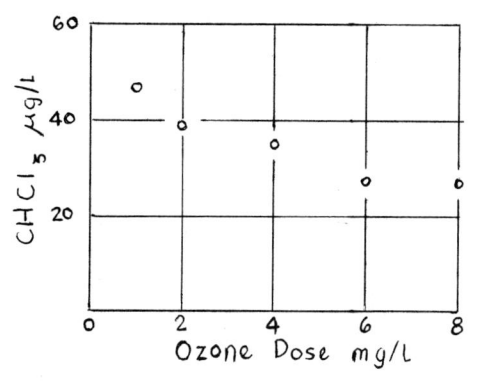

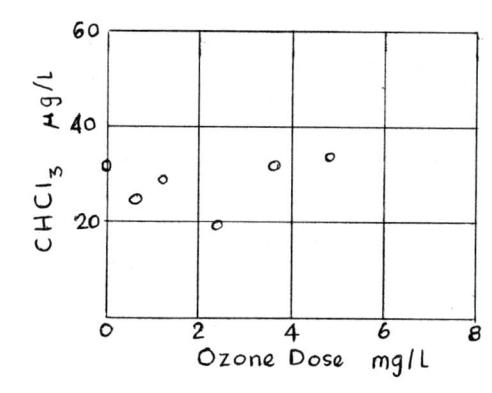

Figure #2: CHLOROFORM PRECURSOR OXIDATION VS. OZONE DOSAGE, PRE-SETTLED WATER (pH=7.1).

Figure #3: CHLOROFORM PRECURSOR OXIDATION VS. OZONE DOSAGE, COAGULATED WATER (PH=7.1).

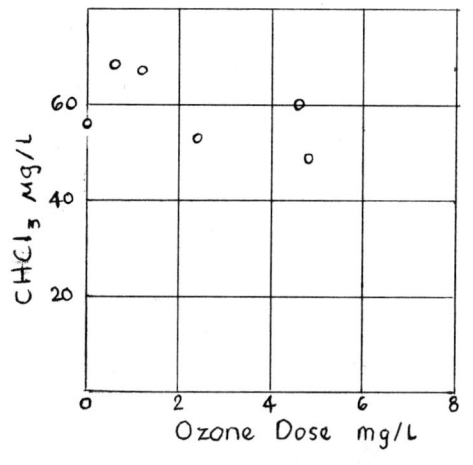

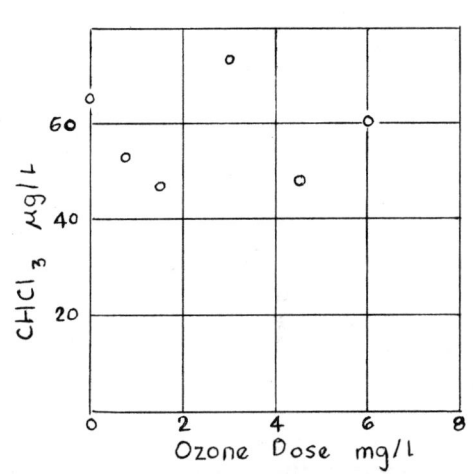

Figure #4: CHLOROFORM PRECURSOR OXIDATION VS. OZONE DOSAGE, SOFTENED WATER (PH=9.8).

Figure #5: CHLOROFORM PRECURSOR OXIDATION VS. OZONE DOSAGE, COAGULATED AND SOFTENED WATER (pH=10.3).

OZONATION OF HAZARDOUS AND TOXIC ORGANIC COMPOUNDS IN AQUEOUS SOLUTION

Kozo Ishizaki,[a] Richard A. Dobbs[b] and Jesse M. Cohen[c]

(a) Visiting Researcher at EPA from Government Industrial Development Laboratory, Sapporo, Japan

(b) Research Chemist, and (c) Chief, Physical-Chemical Treatment Section, Wastewater Research Division, Municipal Environmental Research Laboratory, U.S. Environmental Protection Agency, Cincinnati, Ohio, USA 45268.

Oxidation of toxic organic compounds by ozone has been investigated. Comparative reaction rates and oxidation efficiencies have been measured in aqueous solution for eleven selected compounds over a wide pH range.

Introduction

Numerous incidents of environmental pollution by toxic compounds such as methyl mercury, polychlorinated biphenyls, kepone, pesticides and other synthetic organics have been discussed in the literature. Discharge of toxic organics to the environment at apparently harmless levels may result in magnification to toxic levels by accumulation in the food chain. In order to avoid adverse effects from synthetic organic compounds in the environment, it may be necessary to apply advanced treatment processes.

The high oxidation potential of ozone and the wide range of organic groupings which are susceptible to oxidation suggest application of this oxidant for removal of objectionable compounds from water and wastewater. On this basis, a number of studies have been conducted on ozone treatment of municipal wastewaters (1-5). Results have shown substantial removal of Chemical Oxygen Demand (COD) and Biological Oxygen Demand (BOD) by ozone treatment. Recent evidence (6,7) that chlorination produces toxic halogenated organic compounds has increased interest in ozone as a possible alternative for disinfection of wastewater (8).

Before ozone can be used on a practical scale for municipal or industrial wastewater treatment, it is necessary to have some understanding of the chemistry involved. Few detailed studies of the chemical reactions of ozone with organics in water have been made (9) and much more data are needed to evaluate the capability of ozone to oxidize toxic compounds. Phenols have been studied in great detail (10,11,12,13). Other compounds which have been investigated include: alkylbenzene sulfonate (1,14), amino acids and proteins (15), lactic acid (16), alcohols (18), pesticides (1), polyethylene glycol (19), and organics in photographic processing wastes (20).

In view of the potential of ozone for oxidizing organics, the present study was undertaken to investigate the chemistry of ozonation in aqueous solution. Eleven organic compounds were selected for experimental study. They were: diethylene glycol, acetone cyanohydrin, triethanolamine, phenol, nitrobenzene, dimethyl phthalate, α-naphthol, benzidine, N-nitrosodiphenylamine, benzothiazole and lindane. Most of the compounds selected appear on various EPA Lists of Toxic Pollutants published in the Federal Register and elsewhere. In addition, the selected compounds included a wide variety of chemical groups for reaction with ozone.

A major objective of this study was to determine the capability of ozone to remove toxic and hazardous organic compounds from water and wastewater and to gain some insight into the chemistry involved.

Experimental Apparatus

The experimental apparatus used for this study is shown in Figure 1. A cylindrical pyrex glass reactor (11.4 cm in diameter x 21.6 cm high) containing a gas inlet tube, a gas diffuser, a gas outlet adapter, a liquid sampling tube, and a tube for acid or base addition was used for ozonation. A combination electrode was inserted in the reactor for pH control. The gas diffuser, a medium porosity sintered glass disc was placed near the bottom of the reactor. Improved transfer of the ozone to the solution was provided by a magnetic stirrer.

The ozone generator (Welsbach T-408 Laboratory Ozonator) was fed with dry zero-grade oxygen at a rate of 0.75 l/min. The ozone-oxygen mixture from the ozonation was introduced into the reactor at a constant rate. The exit gas from the reactor passed through a 500-ml gas washing bottle containing 5 percent KI solution to determine unreacted ozone and through a wet test meter to determine the gas volume fed into the reactor. Unreacted ozone was determined by titration with standardized sodium thiosulfate using starch as indicator (21). In order to determine the ozone concentration of the inlet gas into the reactor, a portion of the gas flow from the ozonator was passed through a second KI trap and wet test meter. All gas lines were constructed of glass, aluminum or teflon.

The pH of the reaction solutions was controlled continuously during ozonation by addition of NaOH and H_2SO_4 using an automatic pH controller (Radiometer Co., SBR-2C) and liquid pumps.

Experimental Procedure

Ozonation of each of the eleven organic compounds was carried out at pH 4, 7, 9, and 11. An initial volume of 1.5 l and an initial concentration of 50 mg/l was used for all test compounds except N-nitrosodiphenylamine and lindane. Starting concentrations for the latter two compounds were 35 mg/l and 2.7 mg/l, respectively, due to solubility limitations.

Ozone-oxygen gas mixture was fed into the reactor at 745 $\pm$ 20 ml/min. The average ozone concentration of 30.0 $\pm$ 2.0 mg/l in the gas mixture gave an ozone dosage rate of 22.4 $\pm$ 1.5 mg/min. for all compounds unless otherwise noted. Reaction times varied from 60 to 120 minutes.

Six to twelve samples for analysis were withdrawn from the reactor during the course of each run. The volume of each sample was approximately 50 ml and was collected after discarding the 10 ml volume in the sampling tube. Removal of samples for analysis reduced the starting volume of 1.5 l to a final volume of 1.1-0.8 l during the course of a run. As a result, the amount of ozone applied per unit weight of organic compound increased during the run. Corrections for reduced volume were made only for mass balance calculations. All analytical samples were vigorously purged with helium for 2 minutes immediately after sampling to remove residual ozone.

Reaction solutions were maintained at the desired pH using the automatic controller. At pH 4, 9, and 11, values were maintained within $\pm$ 0.1 unit while at pH 7 control was $\pm$ 0.3 unit. In general, ozonation caused a decrease in pH and NaOH was used for control. In the late stages of some runs, the pH tended to increase slightly and H_2SO_4 was added using the liquid pump with manual control to maintain the pH constant. The total volume of acid or base added to the reactor was less than 25 ml in all cases and analytical values were not corrected for the small amount of dilution which occurred

Ozonation was carried out at room temperature (24°C ± 2°C). Blank tests, using distilled water only, confirmed that no significant ozone demand was produced from the reactor and all components in the experimental setup.

Chemicals

Lindane was supplied by the Water and Hazardous Material Laboratory of this Research Center. Other organic compounds were obtained from the following commercial sources: phenol and nitrobenzene (MC/B Manufacturing Chemists); dimethyl phthalate (Aldrich Chemical Co., Inc.); α-naphthol (Fisher Scientific Co.); all others (Eastman Kodak Co.).

Analytical Methods

Analytical parameters used in this study were chemical oxygen demand (COD), total organic carbon (TOC), total Kjeldahl nitrogen (TKN), NH_3-N, NO_2^--N, NO_3^--N, and ultraviolet absorbance. Gas chromatographic analysis was employed for lindane. COD, TKN, NH_3-N, NO_2^--N, and NO_3^--N were determined according to Standard Methods (21). TOC was measured using a Beckman Model 915 Carbonaceous Analyzer. A Beckman Model 25 Spectrophotometer was used for ultraviolet absorbance. Ozone was determined by the iodometric method (21). Lindane was extracted from aqueous solution by hexane. After drying the extract with anhydrous sodium sulfate, the lindane was analyzed by injection into a gas chromatograph (Perkin-Elmer Model 3920) equipped with a hydrogen-flame detector. A 183 cm x 0.32 cm OV-1 (3%) on 80/90 mesh Chromosorb W column was used at 170°C with nitrogen as the carrier gas. Concentrations in the samples were determined from peak height after calibration of the gas chromatograph with standards.

In order to determine the effect of stripping by the gas flow and reaction with oxygen on the concentration of test compounds, tests were conducted with the ozone generator power off. After 90 minutes of stripping with an oxygen flow of 750 ml/min., the decreases in concentrations measured were 20, 18, and 8 percent for acetone cyanohydrin, lindane, and nitrobenzene, respectively. Less than 3 percent decrease in concentration was observed for all the other test compounds. Observed changes were attributed primarily to stripping since no changes in ultraviolet spectral patterns or nitrogen forms were produced by oxygenation. No correction was made for the stripping effects encountered.

Results and Discussion

Reduction of COD by ozonation for selected compounds is shown in Figures 2 - 4 in the conventional semilog plot. The compounds were selected to illustrate various aspects of the chemistry involved in the oxidation reactions. In the case of diethylene glycol (Figure 2), a large pH dependence on the reaction rate was obtained. In contrast, ozonation of benzidine (Figure 3) showed little or no pH dependence. The curves for triethanolamine (Figure 4) were typical except for the increase in COD in the early stages of ozonation. This effect was observed with some of the other compounds studied.

The COD removal data conform to a pseudo-first order reaction mode with ozone dosage rate constant. Removal of COD occurred in two phases for many of the compounds. In the initial phase, oxidation of the compound was rapid and removal proceeded at this rate until some percentage of the total COD was removed. In the second phase, the oxidation rate slowed drastically and removal of COD became slow or even zero in some cases. The transition from the fast to the slow oxidation rate will be called "rate-change" in COD reduction. It is expressed as the percent COD removed at the rate change.

Apparent first order rate constants and percent COD reduction at the "rate-change" are summarized in Table 1. The rate-change occurred at a given percent removal of the total COD. The first order rate constants ranged from 0 to 6.0×10^{-2} min.$^{-1}$. Reaction rates were generally higher at higher pH values although for most compounds the rates at pH 9 and pH 11 were quite similar.

The rate-change values shown in Table 1 are useful in evaluating the treatability of a given compound by ozone. Rate-change occurred at 50 to 95 percent COD reduction. If removals greater than the rate-change values are desired, ozone requirements and reaction times will be greatly increased.

Rate-change values and ozone utilization at pH 9 are presented in Table 2. Published studies (4,5,14,16) have shown drastic changes in reaction rates during the course of ozonation. This is readily explained on the basis of conversion of the original compound to an oxygen-rich oxidation product which is resistant to further oxidation. Qualitative conclusions can be drawn from the ozone utilization values. Low values probably indicate participation of the whole ozone molecule and/or decompositon products (i.e., $O_3 + OH^- \rightarrow O_3^- + HO_2$; $O_3 \rightarrow O_2 + O$. etc.) in the oxidation reaction.

In fact, in the ozonation of phenol the mechanism in which whole ozone molecules participate has been demonstrated (11,12,13). The ozone utilization value of 1.6 for phenol in the present study is consistent with such a mechanism. A high ozone utilization value, as in the case of triethanolamine, indicates inefficient use of the total oxidant and can be explained on the basis that only one species in the oxidation system is effective for oxidation. For example, if $O\cdot$ or HO_2 were the only species involved in the oxidation mechanism, the ozone utilization value would be increased, as is observed for triethanolamine. Possible oxidant forms in the ozone system are numerous and attempts to elucidate mechanisms require consideration of factors other than ozone consumption.

The higher oxidation rates in basic solution are in agreement with previous results (4,5,14). Hoigné and Bader (22) have shown that hydroxyl radical, formed by the hydroxide ion catalyzed decomposition of ozone, plays an important role in ozone reactions in water. The pH effect observed in the present study is consistent with such a mechanism. In addition, the ionic form of an organic acid or base is an important factor in the ozonation of these compounds. For example, the dissociated form of a compound seems more reactive toward ozone, as indicated by results of the present study and by others (13).

Oxidation Efficiencies

Organic oxidation efficiencies were calculated on the basis of mass balance and change in COD. Calculations were based on the assumption that one oxygen atom from each ozone molecule was available for oxidation. The organic oxidation efficiency (O.O.E.) was defined according to Roan et al. (5) as follows:

$$O.O.E. = \frac{COD \text{ removed (mg)}}{O_3 \text{ reacted (mg)} \times 1/3} \times 100$$

In the case of nitrogen containing compounds, a term for the nitrogen oxidation demand (N.O.D.) was added. The total oxidation efficiency (T.O.E) can be defined as follows:

$$T.O.E. = \frac{COD \text{ removed} + N.O.D. \text{ removed (mg)}}{O_3 \text{ reacted (mg)} \times 1/3}$$

As mentioned by Roan et al. (5), only portions of the organic nitrogen may be oxidized in the COD analysis. Therefore, T.O.E. values are approximate. Since organic nitrogen was oxidized to nitrate by ozone, the N.O.D. values were calculated on the basis of four oxygen atoms per nitrogen atom to convert to nitrate. Mass balance calculations were corrected for sample volumes withdrawn during the course of the run. Oxidation efficiencies for selected compounds are shown in Figure 5 as a function of COD reduction. Oxidation efficiencies usually decrease with increasing COD reduction. In the curves for nitrobenzene, the top curve represents total oxidation efficiency. Since the nitrogen atoms were readily converted to nitrate ions by ozonation, the difference between O.O.E. and T.O.E. was large. On the other hand, the differences were small in the case of benzidine and benzothiazole where oxidation of the organic nitrogen was slow. Values for O.O.E. and T.O.E. at an arbitrarily selected 70 percent COD reduction are summarized in Table 1. Most of the O.O.E. values are between 100 and 200 percent. Lower values at pH 4 may be attributed to the low oxidation rates; while lower values at pH 11 may be due to rapid decomposition which is not effective in oxidation of the organic substrate. Oxidation efficiencies greater than 100 percent have been reported in previous studies (1,5).

TOC Reduction

In the determination of total organic carbon, the sample is acidified and purged to remove inorganic carbonates or carbon dioxide. During purging significant amounts of nitrobenzene and acetone cyanohydrin were stripped from the samples. As a result TOC data were not reported for these two compounds. Some losses of phenol and α-naphthol were also observed during the purging step. Therefore, only a rough evaluation of the TOC data can be made. TOC removal by ozonation for six of the compounds is shown in Figures 6 and 7. Results show that ozonation produced significant removal of TOC although reduction rates are slower than those for COD removal.

Ultraviolet Absorption

The characteristic absorption peaks of the aromatic compounds in the UV region of the spectrum rapidly disappeared after ozonation. Typical results are shown in Figure 8 where ultraviolet scans are plotted at various reaction times for benzidine. It should be noted that in the initial stages of ozonation, colored solutions were produced as indicated by the increase in absorbance at higher wavelengths. Subsequent ozonation resulted in removal of color.

The disappearance of the absorbance at the peak wavelength did not follow the first-order reaction equation. The measured rate was between zero and first order. The rapid decrease in UV absorbance is considered to indicate rapid destruction of the aromatic ring by ozone.

Nitrogen Oxidation

Six of the eleven compounds used in the present study contained one or more nitrogen atoms in various functional groups. In order to determine the effect of ozone on the nitrogen content of the compounds, separate runs were made for nitrogen analyses under the same experimental conditions previously described for the other parameters. Samples were withdrawn after 0, 30, 60 and 90 minutes of reaction time with ozone. Results are summarized in Table 3. Organic nitrogen was determined as the difference between total Kjeldahl nitrogen and ammonia nitrogen. Nitrite nitrogen was omitted because values were 0.1 mg/l or less in all samples.

It was not surprising that little nitrite was detected in the samples since nitrite is oxidized to nitrate very rapidly with ozone (23).

Nitrogen in acetone cyanohydrin was only slightly oxidized at pH 4. At the other pH values, organic nitrogen readily decreased and ammonia and nitrate were produced.

Organic nitrogen was reduced to nearly zero in 60 minutes of ozonation in the case of triethanolamine. Ammonia production was small, and nitrate was the major product.

Organic nitrogen in nitrobenzene was rapidly oxidized, and nitrate was produced nearly stoichiometrically.

Approximately 50 percent of the organic nitrogen of benzidine was oxidized readily. The residual 50 percent was not decomposed by further ozonation. Ammonia was a major product and was subsequently oxidized to nitrate at pH 7, 9, and 11.

In the case of N-nitrosodiphenylamine, 80 percent of the organic nitrogen was oxidized. Nitrate was the major product.

Oxidation of the organic nitrogen in benzothiazole was slow. Apparently nitrogen in a ring structure is more difficult to oxidize.

With the exception of acetone cyanohydrin the effect of pH on the oxidation of organic nitrogen was small. However, the variations of ammonia and nitrate were very strongly affected by pH. Previous studies (24,25) have reported that oxidation of ammonia to nitrate by ozone is slow at pH 7 but increases with increasing pH.

Lindane

Because experimental conditions for the ozonation of lindane were different, this compound will be discussed separately.

An aqueous solution of lindane was prepared which contained 2.5 to 3.0 mg/l. Reported solubility of lindane in water is 5-10 mg/l (26,27). Ozonation of lindane was carried out at a smaller ozone dosage rate of

3.7 mg/min. The decreased dosage rate was based on the smaller initial concentration of lindane in solution and the rapid reaction rates observed in basic solution in preliminary runs. Results of lindane oxidation are shown in Figure 9. The oxidation rate at pH 4 was nearly the same as the rate obtained by oxygen alone. Therefore, lindane does not appear to be oxidized by ozone at pH 4. Rapid oxidation was achieved at pH 9 and 11, while the rate at pH 7 was significantly slower. Previous studies (1,28) have also shown that the concentration of lindane in aqueous solution can be reduced by ozonation. However, no pH effects were investigated in the previous work.

Conclusions

All eleven toxic and hazardous organic compounds were oxidized by ozone at pH 7, 9, and 11. Acetone cyanohydrin, diethylene glycol and lindane were not effectively oxidized at pH 4.

The COD's were reduced rapidly in a first-order reaction mode until the removal reached approximately 70 percent. COD removal rates increased with increasing pH. Beyond the rate change, COD removal was very slow.

Oxidation efficiencies decreased gradually with increasing removal of COD. Oxidation efficiencies greater than 100 percent were calculated which means that whole ozone molecules and/or decomposition products participate in the reaction mechanism.

LITERATURE CITED

(1) D.K. Gardner and H.A.C. Montgomery, "The treatment of sewage effluents with ozone." Water and Waste Treatment, 12:92 (1968).

(2) J.W. Chen, "Catalytic oxidation in advanced waste treatment." AIChE Symposium Series 69: No.129, 61 (1972).

(3) C.G. Hewes and R.R. Davison, "Renovation of wastewater by ozone." ibid 69: No. 129, 71 (1972).

(4) C.S. Wynn, B.S. Kirk and R. McNabney, "Pilot plant for tertiary treatment of wastewater with ozone." EPA-R2-73-146, U.S. EPA, Washington, DC (Jan. 1973).

(5) S.G. Roan, D.F. Bishop and T.A. Pressley, "Laboratory ozonation of municipal wastewaters." EPA-670/2-73-075, U.S.EPA, Washington, DC (Sept. 1973).

(6) T.A.Bellar, J.J. Lichtenberg and R.C. Kroner, "The occurrence of organohalides in chlorinated drinking waters." Jour.AWWA 66: 703 (1974).

(7) J.J. Rook, "Formation of haloforms during chlorination of natural water." Water Treatment and Examination 23: Part 2, 234 (1974).

(8) C. Nebel, R.D. Gottschling, P.C. Unangst, H.J. O'Neil and G.V. Zintel, "Ozone provides alternative for secondary effluent disinfection." Part 1. Water and Sewage Works 123:76 (1976).

(9) P.S. Bailey, "Organic groupings reactive toward ozone: Mechanisms in aqueous media." in "Ozone in Water and Wastewater Treatment (Edited by F.L.Evans. Ann Arbor Science Publishers,Inc., Ann Arbor, Michigan (1972).

(10) S.J.Niegowski, "Destruction of phenols by oxidation with ozone." Ind. Eng. Chem. 45:632(1953).

(11) H.R.Eisenhauer, "The ozonization of phenolic wastes." Jour.WPCF 40:1887 (1968).

(12) Y. Skarlatos, R.C. Barker, G.L.Haller and A.Yalom, "Ozonation of phenol in water studied by electron tunneling." Jour.Phys.Chem 79:2587 (1975).

(13) J.P.Gould and W.J. Weber, "Oxidation of phenols by ozone." Jour.WPCF 48: 47 (1976).

(14) F.L. Evans and D.W. Ryckman, "Ozonated treatment of wastes containing ABS." Proc. 18th Ind. Waste Conf., Purdue Univ., p.141 (1963).

(15) J.B. Mudd, R. Leavitt, A. Ongun and T.T.McManus, "Reaction of ozone with amino acids and proteins." Atmospheric Environment 3: 669 (1969).

(16) R.H. Walter and R.M. Sherman, "Ozonation of lactic acid fermentation effluent." Jour.WPCF 46: 1800 (1974).

(17) B.P. Krasnov, D.L. Pakul and R.V. Kirillova,"Use of ozone for the treatment of industrial waste waters." International Chem.Eng. 14: 747 (1974).

(18) W.L. Waters, A.J. Rollin, C.M. Bardwell, J.A. Schneider and T.W. Aanerud, "Oxidation of secondary alcohols with ozone." Jour.Org. Chem. 41: 889 (1976).

(19) J. Suzuki, "Study on ozone treatment of water-soluble polymers: Ozone degradation of polyethylene glycol in water." Jour.Applied Polymer Sci 20: 93 (1976).

(20) T.W. Bober and T.J. Dagon, "Ozonation of photographic processing wastes." Jour.WPCF 47: 2114 (1975).

(21) "Standard Methods for the Examination of Water and Wastewater." 14th Ed., American Public Health Association, Washington, DC (1975).

(22) J. Hoigné and H. Bader, "The role of hydroxyl radical reactions in ozonation processes in aqueous solutions." Water Research 10:377 (1976).

(23) S.A. Penkett, "Oxidation of SO_2 and other atmospheric gases by ozone in aqueous solution." Nature Physical Science 240: 105 (1972).

(24) A. Ikehata and T. Shimizu, "Removal of ammonia nitrogen in drinking water by ozone." (Text in Japanese). Kogyo Yosui 164: 13 (1972).

(25) P.C.Singer and W.B. Zilli, "Ozonation of ammonia in wastewater." Water Research 9:127 (1975).

(26) R.T.Skrinde, J.W. Caskey and C.K. Gillespie, "Detection and quantitative estimation of synthetic organic pesticides by chromatography." Jour.AWWA 54: 1407 (1962).

(27) C.A. Buescher, J.H. Dougherty and R.T. Skrinde, "Chemical oxidation of selected organic pesticides." Jour. WPCF 36:1005 (1964).

(28) G.G. Robeck, K.A. Dostal, J.M. Cohen, and J.F. Kreissl, "Effectiveness of water treatment processes in pesticide removal." Jour. AWWA 57:181 (1965).

DISCUSSION

Singer: Am I correct in saying that the ammonia nitrogen persisted when you had low pH, but we only see oxidation of the ammonia nitrogen at the high pH?

Ishizaki: Yes.

Singer: Nitrate was released at low pH from the organic compound if it was a nitro compound, but if it was an amino or a reduced ammonia compound, the ammonia released at low pH persisted.

Ishizaki: Probably nitrate was produced by both reaction processes and by subsequent oxidation of ammonia. And probably some portion of the nitrate was produced more directly from the organic nitrogen.

TABLE 2 - OZONE UTILIZATION TO RATE CHANGE AT pH 9

Compound	% COD Reduction at the Rate Change	O_3 Utilized to Rate Change mg O_3/mg COD
Acetone Cyanohydrin	50	3.8
Diethylene glycol	88	3.1
Triethanolamine	90	5.3
Phenol	92	1.6
Nitrobenzene	87	2.2
Dimethyl phthalate	78	2.7
α-Naphthol	89	1.7
Benzidine	77	2.6
N-Nitroso-diphenylamine	87	2.2
Benzothiazole	95	1.8

TABLE 1 - COD REDUCTION AND OXIDATION EFFICIENCIES
FOR OZONATION OF ORGANIC COMPOUNDS

Compound	pH	Apparent 1st-order rate constant x 10^2 (min^{-1})	% COD Reduction at the Rate Change	O.O.E. % at 70% COD Reduction	T.O.E % at 70% COD Reduction
Acetone cyanohydrin	4	--	--	--	--
	7	0.9	50	142 [a]	195 [a]
	9	1.0	50	80	120
	11	1.6	60	73	95
Diethylene glycol	4	--	--	--	
	7	1.3	>85	162	
	9	3.5	88	115	
	11	4.2	89	140	
Triethanolamine	4	1.8	>70	56	80
	7	2.2	>80	60	90
	9	4.2	90	70	95
	11	4.0	90	78	105
Phenol	4	3.8	87	140	
	7	5.7	91	178	
	9	6.7	92	210	
	11	6.9	93	202	
Nitrobenzene	4	2.9	80	155	195
	7	4.1	87	164	200
	9	5.2	87	176	220
	11	5.0	88	148	185

(a) O.O.E. and T.O.E. are calculated at 50% COD reduction for acetone cyanohydrin.

Compound	pH	Apparent 1st-order rate constant x 10^2 (min^{-1})	% COD Reduction at the Rate Change	O.O.E. % at 70% COD Reduction	T.O.E. % at 70% COD Reduction
Dimethylphthalate	4	2.0	50	125	
	7	5.0	75	185	
	9	4.8	78	145	
	11	6.0	83	150	
α-Naphthol	4	3.3	86	210	
	7	3.7	85	205	
	9	4.5	89	220	
	11	4.5	89	153	
Benzidine	4	3.2	77	126	130
	7	3.5	77	129	140
	9	3.3	77	120	125
	11	3.7	78	120	125
N-Nitrosodiphenylamine	4	3.7	77	140	170
	7	4.5	76	162	195
	9	5.0	87	150	180
	11	5.5	85	137	170
Benzothiazole	4	3.0	70	188	195
	7	4.3	90	230	230
	9	4.8	95	245	245
	11	4.5	95	173	173

TABLE 3 - OXIDATION OF NITROGEN CONTENT OF ORGANIC COMPOUNDS

Compound	pH	Org-N (mg/l) Reaction Time (min)				NH$_3$-N (mg/l) Reaction Time (min)			NO$_3^-$-N (mg/l) Reaction Time (min)		
		0	30	60	90	30	60	90	30	60	90
Acetone	4	6.9	6.5	5.6	5.3	0.1	0.1	0.2	<0.1	<0.1	<0.1
cyano-	7		0.5	0.2	<0.1	3.6	3.7	3.4	2.7	2.9	3.0
hydrin	9		<0.1	<0.1	<0.1	4.3	3.1	1.4	3.0	4.5	6.2
	11		<0.1	0.5	0.4	5.9	4.7	3.1	1.5	3.3	3.4
Trietha-	4	4.7	1.2	0.2	<0.1	0.6	0.7	2.2	2.0	3.0	3.5
nolamine	7		2.0	0.9	<0.1	0.5	1.2	<0.1	2.2	3.6	4.4
	9		1.4	0.2	0.2	0.3	0.4	0.1	3.2	4.2	4.4
	11		1.5	0.1	<0.1	0.3	0.2	0.3	3.7	4.0	4.3
Nitro-	4	5.5	<0.1	0.3	0.2	0.2	0.1	0.1	5.1	5.6	5.6
benzene	7		0.1	<0.1	<0.1	0.2	0.2	0.1	4.7	5.2	5.3
	9		0.1	0.3	0.3	0.1	0.1	0.1	4.9	5.5	5.5
	11		0.4	<0.1	<0.1	0.1	0.2	0.1	4.6	5.4	5.3
Benzidine	4	7.1	3.2	3.0	3.1	3.1	3.2	3.2	0.6	0.7	0.8
	7		3.7	3.7	4.1	2.6	2.5	2.0	0.9	1.6	2.0
	9		3.8	3.5	3.0	2.2	1.5	0.9	1.1	2.3	3.2
	11		2.8	2.5	2.8	2.3	1.8	1.3	1.5	2.2	2.8
N-Nitroso-	4	4.7	0.9	0.7	0.8	1.1	1.2	1.1	2.5	2.8	2.9
diphenyl-	7		1.1	1.1	1.5	0.6	0.6	0.4	2.9	2.9	2.9
amine	9		0.7	0.9	0.9	0.7	0.3	0.1	3.2	3.6	3.7
	11		1.2	1.2	1.1	0.9	0.6	0.6	2.6	2.9	3.0
Benzo-	4	5.1	4.0	3.6	3.2	0.8	1.1	1.1	0.1	0.3	0.5
thiazole	7		3.4	3.4	3.1	1.1	1.3	1.1	<0.1	0.6	0.9
	0		2.9	2.7	3.6	1.8	1.6	0.8	<0.1	0.8	1.6
	11		2.3	1.1	2.3	2.3	2.1	1.6	0.3	0.6	1.1

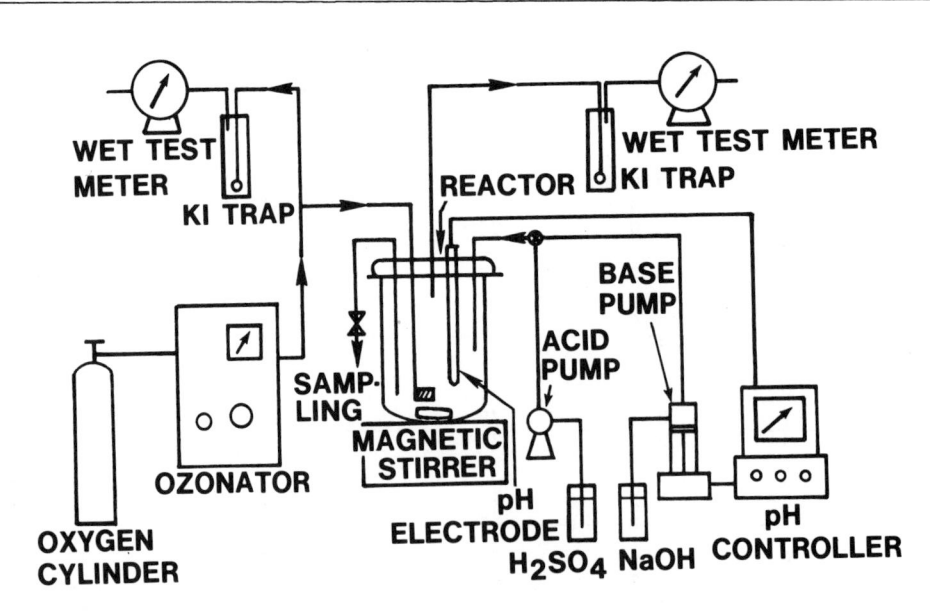

FIGURE 1. EXPERIMENTAL APPARATUS.

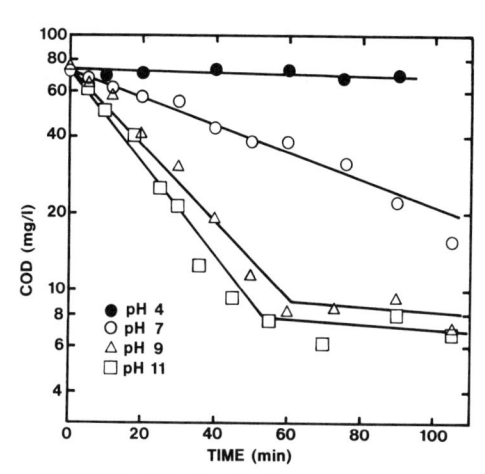

FIGURE 2. COD REDUCTION BY OZONATION OF DIETHYLENE GLYCOL.

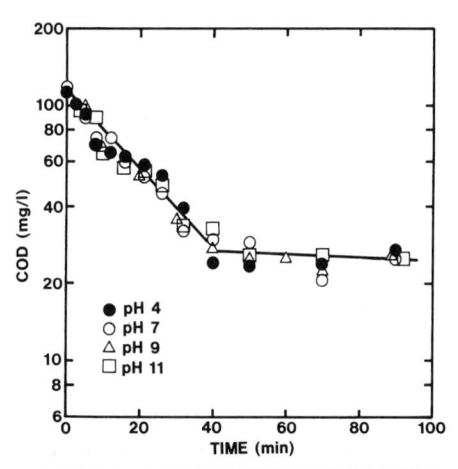

FIGURE 3. COD REDUCTION BY OZONATION OF BENZIDINE.

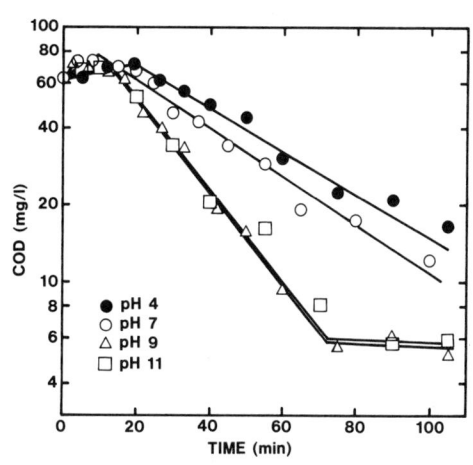

FIGURE 4. COD REDUCTION BY OZONATION OF TRIETHANOLAMINE.

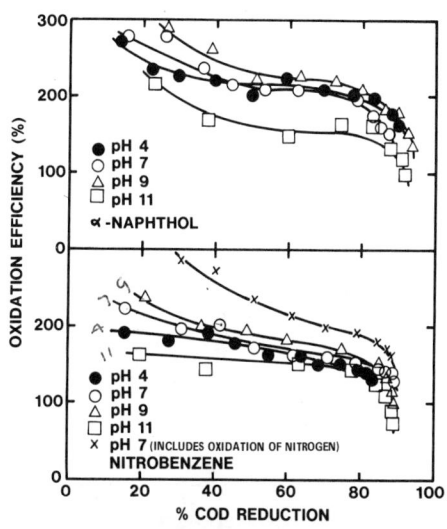

FIGURE 5. OXIDATION EFFICIENCY OF OZONATION OF α-NAPHTHOL AND NITROBENZENE.

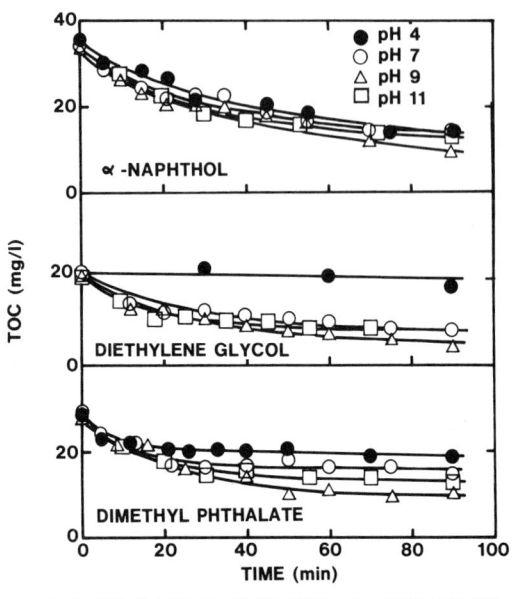

FIGURE 6. TOC REDUCTION BY OZONATION.

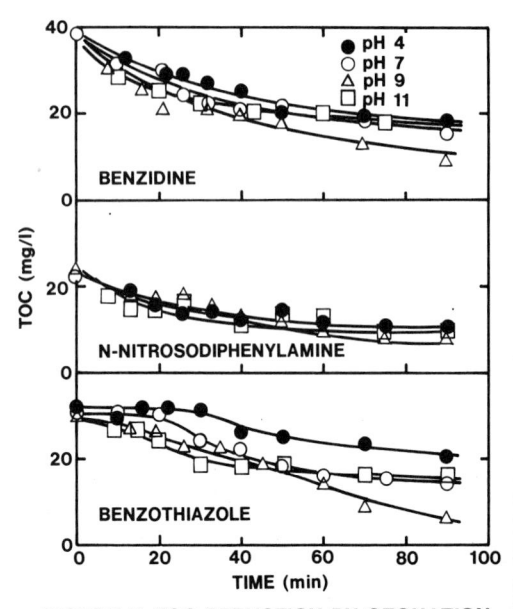

FIGURE 7. TOC REDUCTION BY OZONATION.

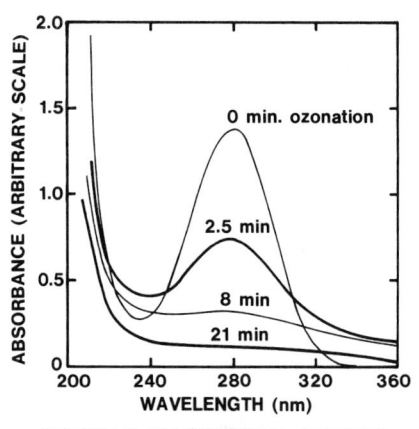

FIGURE 8. UV-SPECTRAL CHANGE BY OZONATION OF BENZIDINE AT pH 9.

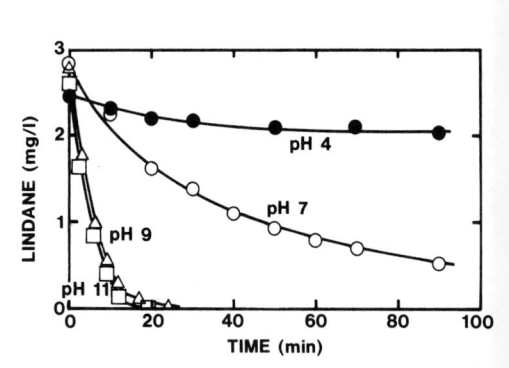

FIGURE 9. REDUCTION OF LINDANE BY OZONATION.

REACTIONS OF OZONE WITH ORGANIC COMPOUNDS IN DILUTE AQUEOUS SOLUTION: IDENTIFICATION OF THEIR OXIDATION PRODUCTS

Ernst Gilbert

Kernforchungszentrum Karlsruhe
Institut für Radiochemie
Bereich Wasserchemie
75 Karlsruhe, Postfach 3640
Federal Republic of Germany

ABSTRACT - This paper gives a summary of our results on the reaction of ozone with aliphatic and aromatic compounds in aqueous solution.

Ozone oxidation of the following aliphatic compounds was investigated: oxalacetic acid, dihydroxyfumaric acid, malonic acid, tartronic acid and glyoxal. The oxidation products were quantitatively measured and the reaction mechanisms are discussed.

Reactions of ozone with the substituted aromatics naphthalene-2,7-disulfonic acid, 4-chloro-o-cresol, 2-nitro-p-cresol and p-aminobenzoic acid were studied.

In the case of 4-chloro-o-cresol, 67% of the oxidation products could be identified, namely methylglyoxal, acetic acid, pyruvic acid, oxalic acid, formic acid and mesoxalic acid.

Generally it was found that the aromatic system is destroyed by ozone attack. The heterogroups are split off and converted into their mineralized forms. The oxidation products of the compounds investigated are biodegradable.

INTRODUCTION

During the last few years many papers have been published about chemical oxidation with ozone as an advanced waste treatment process. However, only a few oxidation products have been identified. The purpose of this paper is to discuss the oxidation of aliphatic and aromatic compounds with ozone and to present the oxidation products.

First we investigated the reaction of ozone with dicarboxylic acids containing two to six carbon atoms, unsaturated, respectively substituted, and with carbonyl compounds. We have chosen these compounds because they are formed, possibly as intermediates, by the ozonation of aromatics.

By quantitative measurements of the oxidation products of these simple model compounds we can understand the reaction mechanisms, and we could then apply this knowledge to the ozonation of complicated molecules like the substituted aromatics.

EXPERIMENTAL

All reactions were conducted at room temperature. The initial concentration of the organic compounds was 1 mmole/l in water.

The samples were treated with ozonized oxygen (gas flow 24 l/hr, 10 mg ozone/min) in a 1 liter reaction flask equipped with a fritted glass disc. The ozone in the gas stream was analyzed both before and after the reaction by absorption in 2% KI solution. The end of the ozonation time was so chosen that no initial compound was detectable in the solutions.

Organic acids

The volatile acids formic and acetic acid were measured gas chromatographically by the method of Bethge and Lindström (1). Glyoxylic acid was determined by the colorimetric method of Kramer (2).

The non-volatile acids malonic, tartronic and oxalic acid were measured gas chromatographically as ethyl esters by the method of Duburque (3).

Dihydroxytartaric acid was prepared by the method of Fenton (4). In aqueous solution the free acid loses carbon dioxide by gently heating, forming tartronic acid (4). We have measured this acid after heating the ozonized solution. Dihydroxytartaric acid forms a 2,4-dinitrophenylhydrazone. We compared the R_F-values of the hydrazones of ozonized solutions with authentic samples (see thin layer chromatography section).

The identification and quantitative measurement of pyruvic acid was carried out with the help of differential pulse polarography (Princeton Polarograph, model 174). Pyruvic acid gives a signal at -1125 mV in acidic medium (pH 4.6). In basic solution (pH 10) a less intense signal was obtained at -1515 mV.

Mesoxalic acid was measured by differential pulse polarography. The acid has three signals at -365 mV, -560 mV and the most intense signal at -750 mV in acidic medium. In basic solution there is a signal at -1655 mV.

The ozonized solutions were also evaporated to dryness or extracted with organic solvents (ether, CH_2Cl_2) and the residues were separated by thin layer chromatography; R_F-values of the spots obtained were compared with those of authentic samples. We used spray reagents for the organic acids (see Stahl 5). TLC plates were cellulose (Fa. Merck) and the following solvent systems were used:

<u>Solvent systems for organic acids on cellulose TLC plates:</u>

i-propanol/acetic acid/H_2O/ethanol	60/30/10/10
ethanol/NH_3/H_2O	40/30/20
	60/20/15
dioxane/H_2O/formic acid	60/35/3
dioxane/ethanol/acetic acid	60/10/10

diisopropylether/formic acid/H_2O 90/7/3

i-propanol/NH_3/H_2O 90/10/20

Carbonyl compounds

Glyoxal was determined by the colorimetric method of Johnson (6). Mesoxalic acid semialdehyde was isolated as its 2,4-dinitrophenylosazone. The formation, the attempt at a quantitative measurement and the synthesis (7) are described in "Ozonolysis of maleic and fumaric acid" (8).

The carbonyl compounds were isolated in the form of their 2,4-dinitrophenylhydrazones. To 1 liter of ozonized solution, 50-100 ml of 4% 2,4-dinitrophenylhydrazine in 0.1N HCl was added. Fractional precipitation, depending on the oxidation products, was achieved by evaporation (30°C) step by step to smaller volumes.

TLC plates were silica gel (Fa. Merck) and the following solvent systems were used:

Solvent systems for TLC of 2,4-dinitrophenylhydrazones on silica gel:

toluene/methanol/chloroform	60/40/20
toluene/acetic acid	85/15
benzene/methanol/acetic acid	45/8/4
benzene/methanol/chloroform/acetone	60/30/10/10
dioxane/ethanol/acetic acid	60/10/10
p-xylene/chloroform/methanol	50/50/0.5
i-propanol/ethyl acetate/water	24/65/12

R_F-values of the unknown hydrazones were compared with R_F-values of the hydrazones prepared from authentic samples: pyruvic acid, mesoxalic acid, mesoxalic acid semialdehyde, glyoxal, glyoxylic acid and dihydroxytartaric acid.

Inorganic ions

Chloride, nitrate and ammonia were measured using ion-specific electrodes (Fa. Orion).

The decrease of UV-active compounds was measured using Cary 14 and Zeiss PMQ II spectrophotometers.

RESULTS

Ozonation of oxalacetic acid

The ozonation of oxalacetic acid (pH 3.4 - 3.1) leads to oxalic and mesoxalic acid as stable end products in our system. As intermediates, formic and glyoxylic acids are formed (Figure 1).

After 40 min the ozone consumption reaches a plateau at 85 mg of added ozone. In the further course of reaction the ozone consumption is zero. To oxidize 1 mmole of oxalacetic acid, 1.8 mmole of ozone is necessary.

Besides the double bond splitting [enol form (9)] (reaction products: oxalic and glyoxylic acid) which occurs to extent of 60%, we observe the splitting of a terminal carbon atom leading to mesoxalic acid. We observed similar reactions during the ozonation of fumaric acid (8).

Ozonation of dihydroxyfumaric acid

Aqueous solutions of dihydroxyfumaric acid are unstable (10), therefore the ozonation has to be started immediately after preparing the solution.

Dihydroxyfumaric acid is rapidly oxidized. The main oxidation product is oxalic acid, but we also found traces of dihydroxytartaric acid (Figure 2). Qualitatively we could identify mesoxalic acid and mesoxalic acid semialdehyde. The measured TOC of 26 mg C/l is higher than the calculated value of 21 mg C/l. That means the two not quantitatively measured compounds have concentrations below 0.05 mmole/l.

In this case the main reaction is double bond splitting. The ozone consumption was 1.4 mmole ozone per mmole of dihydroxyfumaric acid.

During this reaction the pH value changes from 3 to 3.5.

Ozonation of malonic acid

Malonic acid (a saturated compound) is well oxidized (Figure 3). The C-H bond is activated by the two carboxylic groups. Concentrations of oxalic and mesoxalic acid increase steadily. Tartronic acid, as a third oxidation product, reaches a maximum and then decreases, forming mesoxalic acid. Thus we must postulate two reaction paths. One reaction leads to tartronic acid and at least to mesoxalic acid, and the other leads to oxalic acid by splitting of a terminal carbon atom.

Therefore we found a TOC reduction of 17% (36 to 30 mg C/l). The directly measured and the calculated TOC values are in good agreement.

At the end of the reaction we determined an ozone consumption of 4 mmoles of ozone per mmole of malonic acid.

In this case we found H_2O_2 as an oxidation product. The H_2O_2 concentration reaches a maximum after 60 min and then decreases. The decrease in the H_2O_2 concentration after disappearance of malonic acid may be explained as due to the possible reactions of it with the initial oxidation products.

Ozonation of tartronic acid

In the case of tartronic acid ozonolysis we could not find any reduction in the TOC value and the formation of H_2O_2. Tartronic acid is quantitatively oxidized to mesoxalic acid (Figure 4). The pH value lies between 3.2 and 2.9 at the end of the reaction.

What about the oxidation mechanism? Figure 5 shows two possible ways, but we have not made any experiments to measure radicals. Therefore the radical reaction scheme is only a suggestion.

We cannot decide if H_2O_2 is formed, because H_2O_2 can react with the postulated hydroperoxides (Figure 5) which are the possible first intermediates.

The ozone consumption was 1 mmole of ozone per mmole of tartronic acid after 40 min.

Ozonation of glyoxal

After 50 min of ozonation the glyoxal is eliminated. With the disappearance of glyoxal, glyoxylic acid is formed. The glyoxylic acid concentration rises through a maximum and then decreases, giving rise to oxalic acid (Figure 6). Glyoxal oxidation needs 50 mg of ozone.

The TOC value decreases from 24 to 12 mg C/l. That means that a part of the intermediates is totally oxidized to carbon dioxide.

The formation of hydrogen peroxide was not observed.

Because of the different chemical properties of the compounds investigated the reaction rates are very different under our experimental conditions (Figure 7). The unsaturated compounds have the highest reaction rates and compounds in high oxidation states have the lowest rates.

Based on these results we can say the following (Table 1): The reaction of ozone with muconic acid leads to double bond cleavage, forming compounds with two carbon atoms and only traces of compounds with three carbon atoms. Also the cis-compound, maleic acid (8), is oxidized by splitting the double bond. The trans-compound, fumaric acid (8), and the substituted unsaturated dicarboxylic acids are oxidized, forming compounds with two and three carbon atoms in different yields.

The remaining TOC value is about 50% of the initial value, with the exception of tartronic and malonic acid (Table 1).

Ozonation of substituted aromatics

If we assume similar reaction mechanisms as in the case of the aliphatic compounds investigated, we then would expect the same types of oxidation products.

4-Chloro-o-cresol

With the disappearance of 4-chloro-o-cresol, chloride ion is formed. After 80 min of ozonation and complete disappearance of the initial compound, the organic chlorine is split off and appears as chloride to the extent of 100%.

As organic oxidation products, methylglyoxal, pyruvic acid, acetic acid, formic acid and oxalic acid are quantitatively measured over the course of the run (Figure 8).

Methylglyoxal is produced right from the beginning of the reaction. Its concentration reaches a maximum after 60 min (100% cresol elimination). During further progress of the reaction the methylglyoxal concentration decreases steadily. This means that its rate of formation from cresol is higher than its oxidation rate in this system. After elimination of cresol there is no further formation of methylglyoxal, and therefore methylglyoxal concentration is then decreasing.

Also, formic acid concentration rises through a maximum and then decreases, giving rise to CO_2 and water (Figure 8).

Pyruvic acid and acetic acid concentrations increase steadily during ozonation (Figure 8). Because their concentrations increase even after complete cresol elimination, they may be viewed as oxidation products of methylglyoxal.

Oxalic acid concentration increases steadily and reaches a value of 0.38 mmole/l at the end of the reaction.

When all reaction products are determined quantitatively, the sum of their carbon contents should be equal to the directly measured TOC value. But differences between the calculated and directly measured values exist (Table 2). These differences show a maximum during the course of ozonation after 40 min. This indicates that at this time (100% cresol elimination) we have non-identified intermediates. We assume that these intermediates are compounds with undestroyed ring systems. We have to investigate the properties of these intermediates, to determine whether they could be more toxic than the initial products.

The fact that the observed differences in carbon balances tend to disappear toward the end of ozonation indicates the conversion of the unstable intermediates into quantitatively determined end products. An exact agreement in the balance could not be obtained. Along with the major course of the oxidation reaction, side reactions can also take place, giving rise to small amounts of compounds other than those quantitatively measured. An example of this is the existence of mesoxalic acid and mesoxalic acid semialdehyde in trace amounts, which have been identified by comparing the R_F-values of their 2,4-dinitrophenylhydrazones with authentic samples.

Naphthalene-2,7-disulfonic acid

After 120 min of ozonation naphthalene-2,7-disulfonic acid is no longer detectable in the solution. The pH value is changed from 4.6 to 2.8 (Figure 9). Nevertheless, only 1.35 mmole/l of sulfate was measured. This means there are still organic compounds present other than naphthalene-2,7-disulfonic acid which contain the sulfonic acid group. After 300 min of ozonation almost complete desulfonation is achieved (Figure 9).

As organic oxidation products we identified and quantitatively measured formic, oxalic and mesoxalic acid. But these compounds amount to only 25% of the directly measured TOC value. Besides the identified acids, many carbonyl compounds are formed in different quantities. We have isolated the carbonyl compounds as 2,4-dinitrophenylhydrazones and separated them by thin layer chromatography (Figure 10).

Comparing the R_F-values of the hydrazones with authentic samples we could identify glyoxal, mesoxalic acid and mesoxalic acid semialdehyde (Figure 10).

At the beginning of the reaction we observed a spectrum of compounds other than those observed at the end of the reaction (300 min).

After an ozone consumption of 11 mmole/l the TOC value was reduced from 100 to 50 mg C/l. The COD value decreased from 265 to 60 and the BOD_5/COD ratio increased and reached a value of 0.8 (Figure 11). This indicates that the oxidation products are biodegradable.

4-Aminobenzoic acid

Ozonolysis of 4-aminobenzoic acid leads to the formation of ammonia and nitrate (Figure 12). After 80 min of ozonation we measured only 70% of the initial nitrogen. Obviously there are organic compounds present containing nitro groups.

As organic oxidation products, formic and oxalic acid were identified. The formic acid concentration reaches a maximum after 40 min and then decreased to zero after 80 min. The oxalic acid concentration steadily increases and reaches a value of 0.75 mmole/l after 80 min (Figure 12). The yield of oxalic acid amounts to only 45% of the directly measured TOC value. The other part of the TOC value belongs to unidentified oxidation products, some of them containing a nitro-group. After 80 min the ozone consumption was 4 mmole of ozone/mmole of 4-aminobenzoic acid. The COD value decreases from 227 to 59 mg of oxygen/l and the TOC value from 84 to 39 mg C/l. Even the BOD_5/COD value increases from zero to 0.4 (Figure 13). The oxidation products are more readily biodegradable.

2-Nitro-p-cresol

The oxidation of nitrocresol leads, after elimination of the initial product, to the formation of nitrate. At this point 90% of the nitro groups are split off (Figure 14).

LITERATURE CITED

(1) P.O. Bethge and K. Lindström, Analyst 99:137 (1974).

(2) D. N. Kramer and N. Klein, Anal. Chem. 31:250 (1959).

(3) M. Th. Duburque and I.M. Melon, "Dosage et identification de l'acide oxalique dans les milieux biologiques". Ann. Biol. Clin. 28:95 (1970).

(4) H.J.H. Fenton, "Properties and relationships of dihydroxytartaric acid". J. Chem. Soc. Transactions 73:70 (1898).

(5) E. Stahl, "Dünnschichtchromatographie". Springer Verlag, New York, Heidelberg, Berlin (1967).

(6) D.P. Johnson, F.E. Critchfield and J.E. Ruch, "Spectrophotometric determination of trace quantities of alpha-dicarbonyl compounds". Anal. Chem. 34:1389 (1962).

(7) F. Dickens and D.H. Williams, Biochem. Journal 68:74 (1958).

(8) E. Gilbert, "Über die Wirkung von Ozon auf Maleinsäure und Fumarsäure sowie über deren Oxidations-produkte in wässriger Lösung". Z. Naturforsch., in press (1977).

(9) B. E. C. Banks. J. Chem. Soc. 5043 (1961).

(10) A. Locke, "The decomposition of dihydroxymaleic acid". J. Am. Chem. Soc. 46:1246 (1924).

(11) E. Gilbert, "Chemische Vorgänge bei der Ozonanwendung. Proceedings, "Intl. Symposium on Ozone and Water" Berlin, 1977, in press.

(12) E. Gilbert, "Über den Abbau organischer Schadstoffe im Wasser durch Ozone". Vom Wasser 43:275 (1974).

(13) P. Joy, E. Gilbert and S.H. Eberle, "Reaction of ozone with p-toluenesulfonic acid in aqueous solution". Water Research (1977), in press.

Figure 1. ozonation of oxalacetic acid

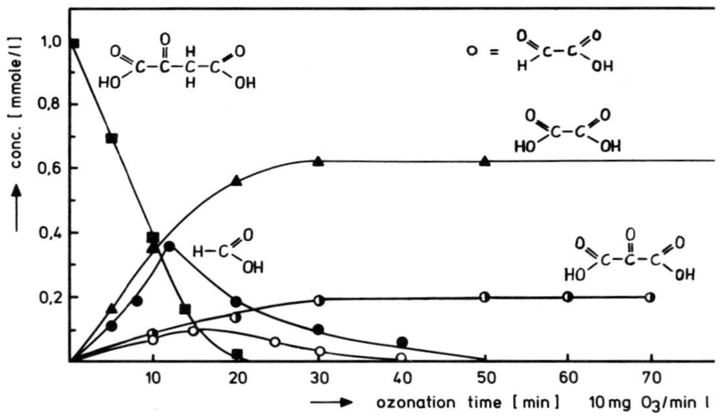

Figure 2. ozonation of dihydroxyfumaric acid

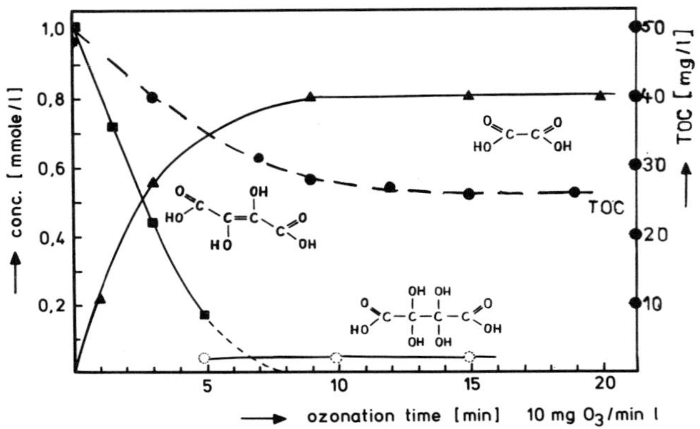

Figure 3. ozonation of malonic acid

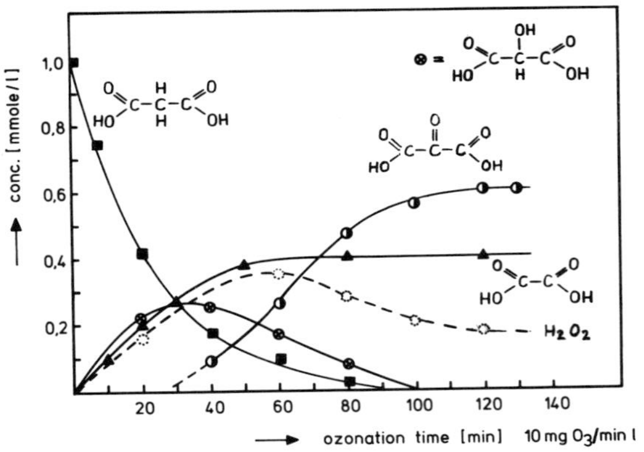

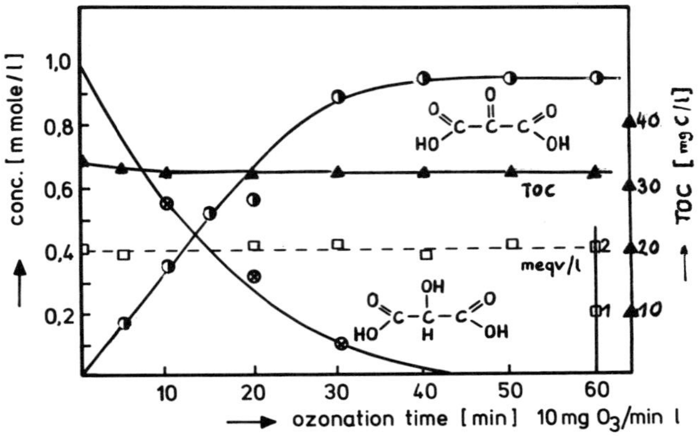

Figure 4. ozonation of tartronic acid

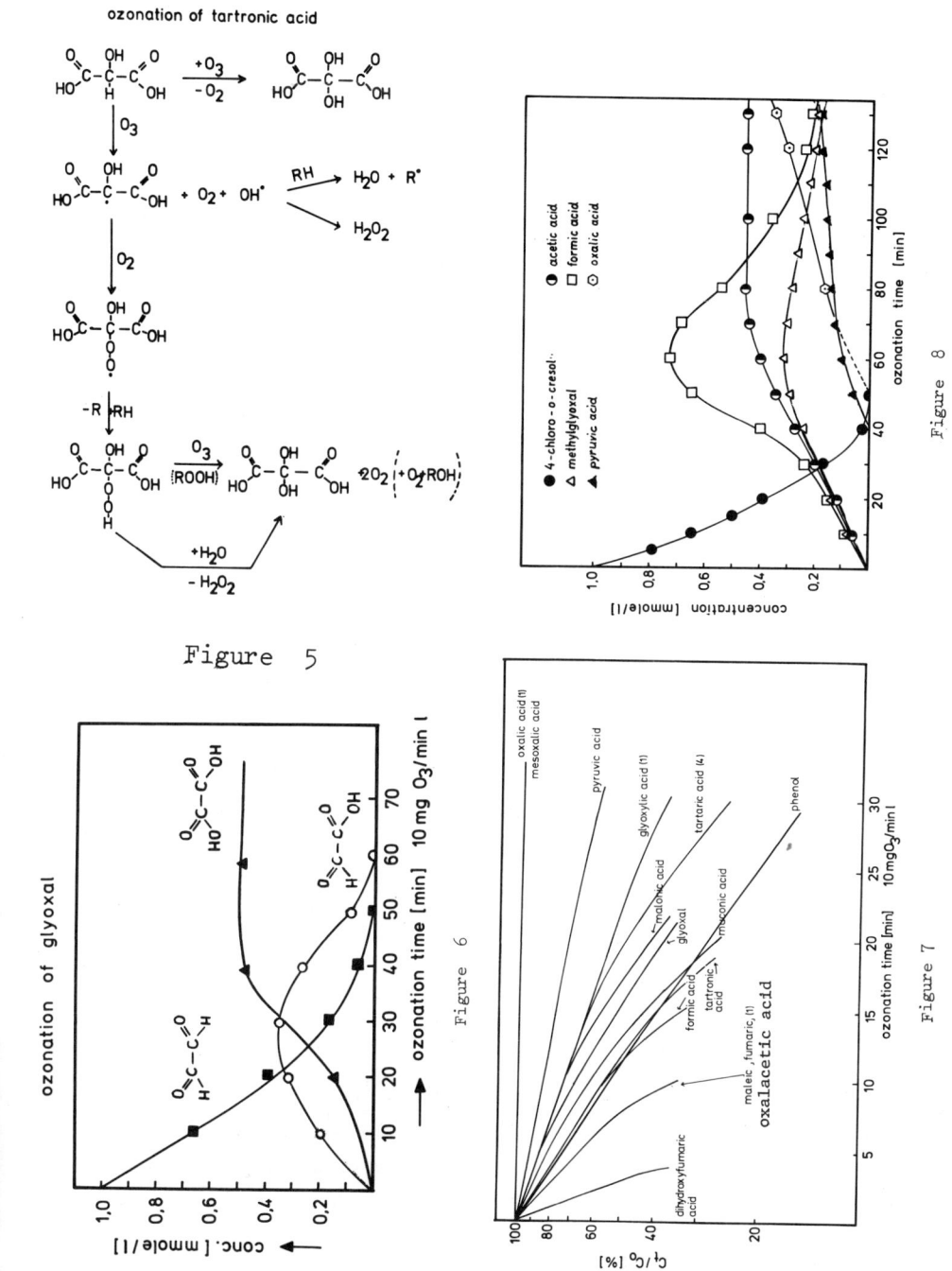

Figure 5

Figure 6

Figure 7

Figure 8

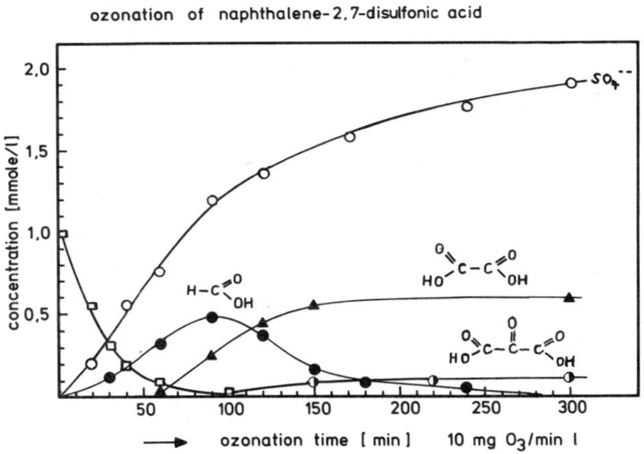

Figure 9

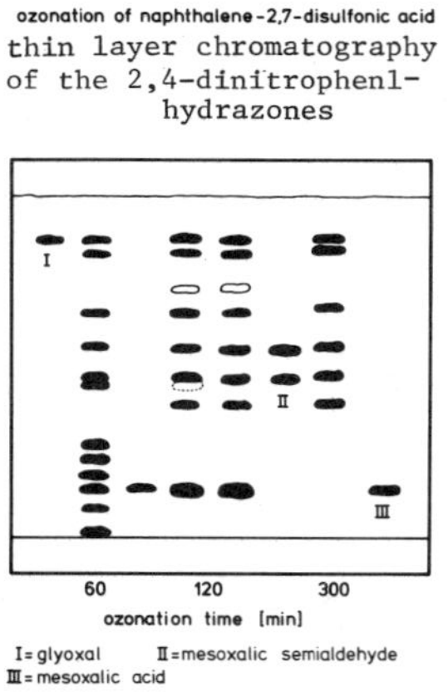

Figure 10

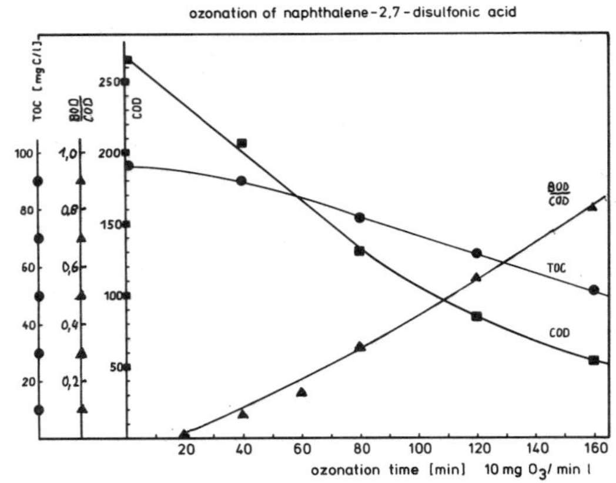

Figure 11

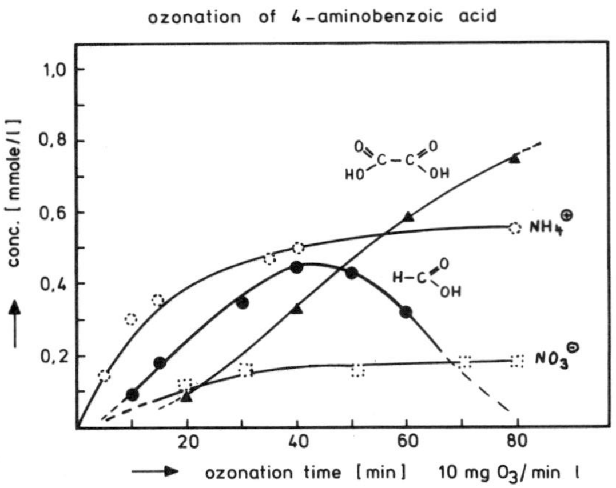

Figure 12

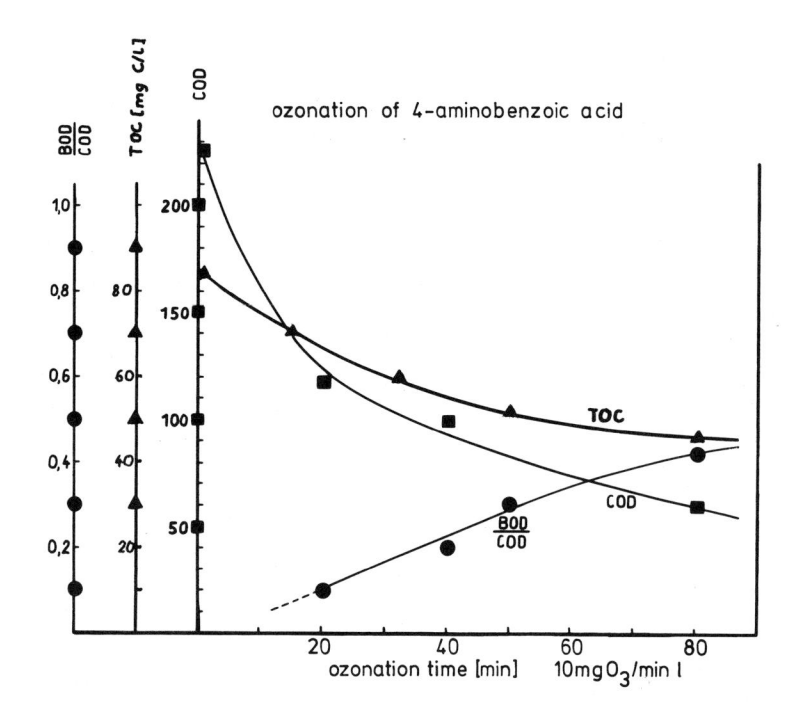

Figure 13

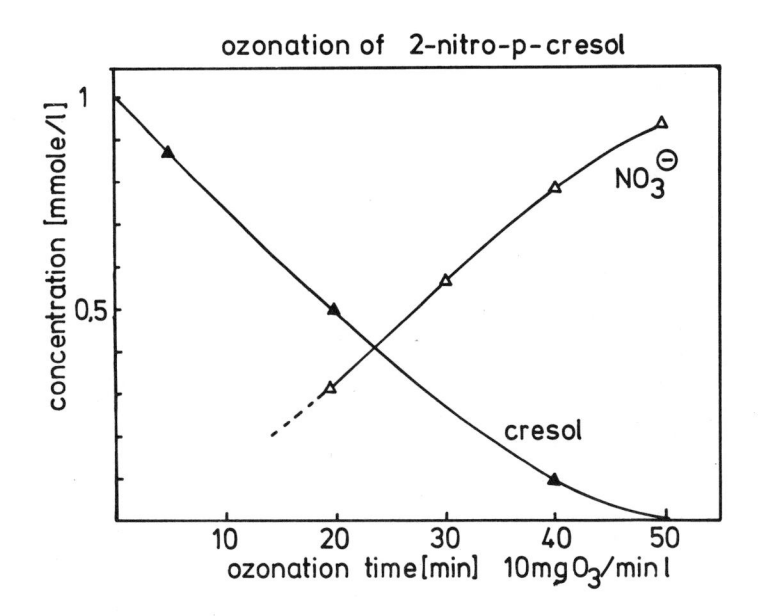

Figure 14

-241-

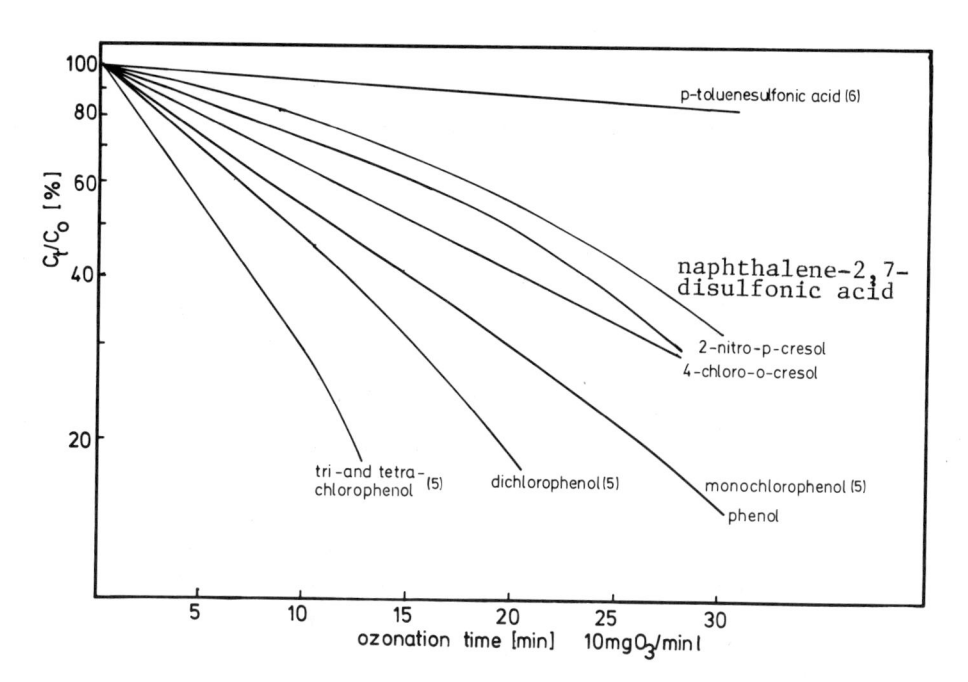

Figure 15

	double bond cleavage	C_3-compds. [%] of remaining TOC	C_2-compds. [%] of remaining TOC	remaining TOC [%] of the initial TOC
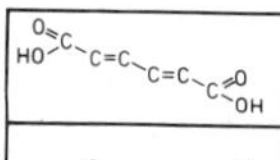 HOOC-C=C-C=C-COOH	90%	>10%	88%	47%
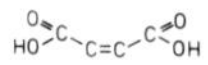 HOOC-C=C-COOH	100%	—	100%	40%
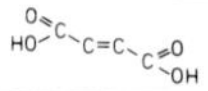 HOOC-C=C-COOH	70%	51%	49%	40%
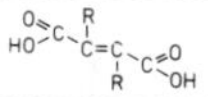 HOOC-CR=CR-COOH	60%–90%	10%–23%	50%–80%	54%–65%
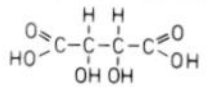 HOOC-CH(OH)-CH(OH)-COOH	—	10%	90%	50%
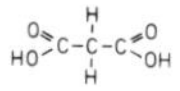 HOOC-CH(OH)-CH2-COOH	—	70%	30%	82%
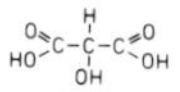 HOOC-CH(OH)-COOH	—	99%	—	99%
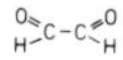 OHC-CHO	—	—	100%	46%
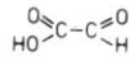 HOOC-CHO	—	—	100%	58%

Table 1

-241A-

Ozonation of 4-chloro-o-cresol, pH 5.5, TOC values directly measured and calculated from the results of the individual determinations

ozonation time [min] 24 mg O₃/min l	measured [mg C/l]	calculated [mg C/l]	difference [mg C/l]
0	86	84	—
5	83	70.5	12.5
40	73	22	51
80	62	37	25
130	55	38	17

Table 2

concentration 1 mmole/l	ozone consumption after 100% elimination	BOD_5/COD
4-chloro-o-cresol	4.5 mmole O₃/mmole	0.85
4-aminobenzoic acid	4.0 "	0.40
naphthalene-2-7-di-sulfonic acid	11.0 "	0.80
2,4,6-trichlorophenol [5]	3.6 "	0.43
p-toluenesulfonic acid [6]	16.0 "	0.65
3-chlorophenol [5]	4.5 "	0.50

Table 3

OXIDATION OF STYRENE WITH OZONE IN AQUEOUS SOLUTION

by

Floyd H. Yocum

Gulf South Research Institute
P. O. Box 26500
New Orleans, Louisiana 70186

Abstract: The oxidation of styrene with ozone in aqueous solution was studied. Ozone was found to cleave the olefinic double bond in styrene to form benzaldehyde and formaldehyde, the latter being completely oxidized to carbon dioxide and water. The benzaldehyde underwent rapid oxidation to form benzoic acid. A stirred tank reactor was operated in the semi-batch mode to identify the proper reaction regime. A model for ozone mass transfer and reaction kinetics is discussed.

Introduction

The year 1985 has been set by the Environmental Protection Agency (EPA) as the goal for zero pollution discharge into rivers, lakes, and other water sources. Small chemical plants will have difficulty justifying expensive waste treatment facilities to reach such standards. Therefore, many companies are considering regional wastewater treatment facilities. Different waste treatment methods would be incorporated to cope with the varied chemical waste products. Some of these waste products, such as styrene, are non-biodegradable. This has focused attention on pretreatment and tertiary treatment processes such as carbon adsorption, ion exchange, solvent extraction, reverse osmosis, chemical oxidation, and others which can be added to conventional waste treatment systems to remove the hard-to-oxidize (refractory) compounds. No one process has proved to be successful for removing all refractories and, therefore, a combination of chemical and physical treatments is required.

Chemical oxidation with ozone is an attractive process for a regional wastewater system. Ozone has been used for years to purify and deodorize drinking water in Europe and is recognized as one of the strongest and purest oxidants available. Unlike other oxidants, ozone can be generated as needed. It is used in the chemical industry for cleavage of carbon-carbon double bonds and has recently been applied in the industrial wastewater treatment area. The ozonation of industrial wastewater is a typical gas-liquid reaction and the effect of mass transfer on the reaction must be evaluated.

One of the objectives of this project at Gulf South Research Institute was to study the ability of ozone to oxidize styrene for application in a waste treatment plant and identify the proper reaction regime. The effects of pH and temperature on the reaction were studied. A mathematical model, based on the macroscopic view of gas-liquid reactions, was developed for mass transfer of ozone with and without chemical reaction. The model was applied to experimental data to evaluate the volumetric mass transfer coefficient and reaction rate constants.

Mathematical Model Development

A mathematical model was developed to combine with experimental data to determine the reaction rate constants, volumetric mass transfer coefficient, and the proper reaction regime for the ozone-styrene gas-liquid reaction. The model accounts for mass transfer of ozone into the liquid phase from the gas phase, ozone decomposition, and reaction with styrene and its reaction products. The model was developed from a macroscopic view of the reaction system promoted by Prengle and Barona (1). This procedure is based on a combination of Fick's law of diffusion and a material balance for each component in both phases over the entire reactor system. This can be expressed as:

$$dC_j/dt = k_L a(C_{je} - C_j) - r_j \qquad (\underline{1})$$

where:

C_j = concentration of component j

C_{je} = concentration of component j at equilibrium

k_L = liquid-phase mass transfer coefficient

a = interfacial area

r_j = reaction rate of component j

The macroscopic view is illustrated in Figure 1 with a plot of the ozone concentration in the liquid phase as a function of time. In the absence of a liquid reactant the ozone is absorbed exponentially to its solubility limit. This concentration profile of ozone in the liquid phase can be found by equating r_j to zero and solving Equation ($\underline{1}$) to get:

$$C_B = C_{Be}(1 - e^{-\phi t}) \qquad (\underline{2})$$

where:

C_B = ozone concentration in the liquid phase

C_{Be} = equilibrium ozone concentration in the liquid phase

ϕ = combination of the volumetric mass transfer coefficient and the ozone decomposition rate constant (4).

The ozone concentration profile is shown for the three most frequent gas-liquid reaction regimes in Figure 1. After an injection of liquid reactant at time equals zero, the ozone concentration will immediately drop to zero if the reaction is instantaneous, resulting in a truly mass transfer limited system as depicted by line 1. A quasi-equilibrium concentration is reached if mass transfer is controlling with a slow reaction and the ozone concentration will remain at the solubility limit if the system is reaction rate controlled as shown by profile lines 2 and 3 respectively. The reaction regime was identified for each compound ozonated by conducting a material balance on each component and comparing the data to the theoretical plot.

A brief review of the ozone-styrene reaction system is presented before developing the model for mass transfer with chemical reaction. Previous studies have shown that ozone will oxidize styrene to benzaldehyde which is easily oxidized to benzoic acid (2, 3). The ozone attack on benzoic acid should be directed to the meta positions, but actually occurs at all positions to some extent, including the carboxylic acid group. The study (4) was greatly simplified by not extending the analytical work beyond benzoic acid in search of the various reaction products. The chemical reactions studied were:

$$Ph\text{-}CH=CH_2 + 2O_3 \longrightarrow Ph\text{-}CHO + HCHO + 2O_2 \qquad (\underline{3})$$

$$HCHO + 2O_3 \longrightarrow CO_2 + H_2O + 2O_2 \qquad (\underline{4})$$

$$Ph\text{-}CHO + O_3 \longrightarrow Ph\text{-}COOH + O_2 \qquad (\underline{5})$$

$$Ph\text{-}COOH + bO_3 \longrightarrow \text{Reaction Products} \qquad (\underline{6})$$

The formaldehyde reaction (4) is instantaneous and was incorporated into the initial styrene ozonation reaction.

The successive ozone attack on styrene and its reaction products is similar to some important industrial reactions, such as the successive substitutive halogenation or nitration of hydrocarbons and the addition of alkene oxides to proton donor compounds such as amines, alcohols, water, and hydrazine. A general representation of these reactions, including the ozonation of styrene, is written as:

$$A + b_1 B \xrightarrow{k1} R \tag{7}$$

$$R + b_2 B \xrightarrow{k2} S \tag{8}$$

$$S + b_3 B \xrightarrow{k3} T \tag{9}$$

where:

    A = liquid reactant A

    B = ozone

    b = stoichiometric mole ratio of moles of ozone per mole of liquid reactant

    R, S, T = liquid reactants

    k_1, k_2, k_3 = reaction rate constants

This reaction mechanism is considered to be consecutive with respect to compounds A, R, S, and T and a parallel reaction for compound B, ozone.

The rate expressions for the consecutive-parallel reactions, assuming the reactions are irreversible, bimolecular and of constant density are:

Styrene: $\quad rA = dC_A/dt = -k_1 C_A C_B \tag{10}$

Benzaldehyde: $\quad rR = dC_R/dt = k_1 C_A C_B - k_2 C_R C_B \tag{11}$

Benzoic Acid: $\quad rS = dC_S/dt = k_2 C_R C_B - k_2 C_R C_B \tag{12}$

Levenspiel (5) has thoroughly discussed the consecutive-parallel reaction type and presents a general solution to the rate equations.

The mass-transfer equations were extended to include the chemical reactions by inserting the rate equations (10), (11), and (12) for the reaction term in equation (1). The following assumptions were made during the development to simplify the equations.

- well-mixed semi-batch stirred tank reactor operation of continuously sparged gas into a constant volume of liquid

- constant temperature and pressure

- constant volume and liquid density

- constant molar gas feed rate

- constant gas holdup

- ozone decomposes via first order reaction

- liquid reactants and products are non-volatile

The model was developed for the probable reaction regimes of liquid-phase mass transfer controlling, with an instantaneous reaction and a slow reaction, either mass transfer controlled or reaction rate controlled. The deciding factor is whether ozone penetrates the bulk liquid. In the latter case, if the reaction is extremely slow and ozone reaches an equilibrium concentration in the liquid, the reaction is rate controlled. Both models were applied to the experimental data to determine the controlling regime.

The initial model was developed for mass transfer controlling with an instantaneous reaction for the single reaction given in equation (7). The liquid reactant A represents the initial reactant which can be styrene, benzaldehyde or benzoic acid. The ozone does not penetrate into the bulk liquid for a truly mass transfer limiting system. Thus, the molar balances are simplified and the rate of disappearance of A can be found from the stoichiometry of the reaction and the limiting form of the ozone balance to give:

$$-dN_A/dt = 1/b_1 (k_L a V x_{Be}) \tag{13}$$

where:

N_A = mole of liquid reactant A

x_{Be} = equilibrium mole fraction of ozone in the liquid

Equation (13) was converted into the fraction of A remaining giving a first order linear differential equation. The equation was integrated and the integration constant evaluated for the boundary condition of $t = 0$; $N_A/N_{Ao} = 1$ giving the fraction of A remaining as

$$N_A/N_{Ao} = 1 - \left(\frac{k_L a V x_{Be}}{N_{Ao} b_1}\right) t \tag{14}$$

The equation was rewritten in terms of the ozone transferred from the gas phase by equating the gas phase molar balance to the ozone absorption rate. The ozone gas flow was regulated to starve an instantaneous reaction. Thus the amount of ozone entering the reactor was transferred to the liquid phase to react. This simplified the final equation for the fraction of A remaining to a linear function of the ozone in the feed gas.

$$N_A/N_{Ao} = 1 - \left(\frac{G y_{BF}}{N_{Ao} b_1}\right) t \tag{15}$$

Deviation of experimental data from the line projected by equation (15) indicates a slow reaction controlling. The mass transfer limiting line was calculated for each test.

If the ozonation reaction is slow, either as mass transfer controlling with a slow reaction or as an extremely slow reaction giving a reaction rate controlling regime, the ozone will penetrate into the bulk liquid. In the latter case, the ozone reaches its equilibrium value as predicted by Henry's law. For the model development, an average ozone bulk concentration, $\bar{x}_B$, was used in equation (16), which is the rate equation (10) written in mole fraction terms.

$$dN_A/dt = -k_1 V x_A x_B \tag{16}$$

Allowing $\bar{x}_B$ to be a constant transforms the reaction into the form of a pseudo-first order reaction. Rewriting equation (16) in terms of the fraction of A converted and integrating gives:

$$N_A/N_{Ao} = e^{-\omega_1 t} \tag{17}$$

where:

$$\omega_1 = k_1 V \bar{x}_B / L$$

L = total moles of liquid

The model was extended for the second reaction involving the reaction product R. Substituting equation (17) into the rate equation for R and integrating, the solution for the moles of R at any instant is expressed as

$$N_R/N_{Ao} = \frac{\omega_1}{\omega_2 - \omega_1} (e^{-\omega_1 t} - e^{-\omega_2 t}) \tag{18}$$

These models were applied to the experimental data to evaluate the rate constants.

Experimental Procedure

A stirred tank reactor was selected for the ozonation study as it provides the capability of contacting large quantities of ozone with a specified volume of wastewater. The stirred tank reactor was operated in the semi-batch mode as ozone gas was continually fed to a constant volume of wastewater. The reactor was designed to comply with standard stirred tank reactor configuration used by Westerterp and others in mass transfer studies (6,7,8). Prengle used a similar process for the ozonation of five particular refractory compounds (9). The major dimensions of the reactor are ratioed to the tank diameter for easy design and scale-up. The plexiglass reactor used at Gulf South Research Institute has a diameter of 11.5 inches and dimensions corresponding to the standard configuration shown in Figure 2. A volume of approximately 20 liters was achieved with a liquid height of 11.5 inches.

A schematic of the ozone pilot plant is shown in Figure 3. Air was used as a feed gas to the Ozone Research and Equipment Company (OREC) Model 03B4-AR ozonator. The ozonator has a rated capacity of two pounds of ozone per day. The ozone flow to the reactor was controlled by a needle valve as shown in Figure 3. The gas flow was measured with a rotameter and excess gas vented through a potassium iodide trap to the atmosphere.

The standard potassium iodide method was used to analyze the ozone concentration in the feed and exit gas as well as the liquid (10). The temperature was monitored periodically. An optional immersion heater and temperature controller were installed to operate at 15° and 25°C.

The standard operating conditions were to use a gas feed rate of 11.5 liters per minute (0.408 CFM) containing 12-15 mg/l ozone (1.0 to 1.2 wt%). A turbine speed of 700 rpm provided a sufficient volumetric mass transfer coefficient to completely consume all of the ozone if the reaction was instantaneous (4). The average test was continued for three hours. The liquid samples were analyzed for total organic carbon (TOC) with a Beckman 915 TOC analyzer and for composition via gas chromatography.

After ozonation, biological batch tests were performed to determine the enhancement of ozonation on the water biodegradability. A batch study is performed in a series of small, batch-type biological reactors. An acclimated seed is added to each of the test reactors except those in which air stripping studies are being conducted. Various amounts of ozonated sample are added to the test reactors to obtain a series of organic loadings. In addition, one reactor containing only acclimated seed and no waste is used as a baseline. Data are collected on the reactors during the three-day test period.

Results

Mass Transfer. Mass transfer tests were performed to evaluate the reactor characteristics and correlate the volumetric mass transfer coefficient with power input and gas velocity. Several turbine speeds and gas feed rates were studied. Ozone concentration profiles were plotted similar to ones shown in Figure 4 for the various turbine speeds and gas feed rates. Figure 4 is for a gas feed rate of 11.5 liters per minute (0.41 CFM). The ozone absorption rate is shown to be slow without agitation and increases as agitation is applied. Above a turbine speed of 400 rpm the system is operating in the well-mixed regime. Therefore, increasing the agitation will not appreciably enhance the mass transfer rate.

The mathematical model developed was used to evaluate the volumetric mass transfer coefficient for ozone transferring into distilled water (4).

$$C_B = C_{Be}(1-e^{-\phi t}) \qquad (\underline{2})$$

Equation 2 was utilized with the mass transfer data generated to evaluate the volumetric mass transfer coefficient. The coefficient was found to be proportional to power input raised to the one-half power when operating in the well-mixed regime. This can be expressed as:

$$K_L a \; \alpha \; (P/V)^{0.5} \qquad (\underline{19})$$

where (P/V) is the power input in hp/1000 gallons. Cooper et al. reported a similar correlation (6). Future publications will present the mass transfer study in detail.

Ozonation Reaction Study. Aqueous solutions of styrene, benzaldehyde, and benzoic acid were ozonated in the semi-batch mode. The effects of temperature and solution pH on the individual reaction rates were studied. The reaction rate constant was calculated from the initial 15 minutes of each test to reduce the effects of competing side reactions. The mass transfer limiting line was calculated for each test using equation (15). The stoichiometric mole ratio, b_1, was determined from a plot of the fraction of liquid reactant remaining versus the mole ratio of ozone to the liquid reactant. The proper reaction regime was determined by comparing the liquid reactant reduction to that predicted by equation (15) and monitoring the ozone concentration in the liquid. The reaction rate constants were determined by applying equation (17) to the experimental data.

Benzoic Acid Ozonation. Benzoic acid solutions with an average initial concentration of 300 ppm were first ozonated. The stoichiometric ratio for the benzoic acid reaction was found to be approximately four moles of ozone per mole of benzoic acid. This was used to calculate the mass transfer limiting lines.

The effect of temperature on the reaction was first studied and is displayed in Figure 5. The benzoic acid reaction at either 15° or 25°C deviates from the mass transfer limiting line. Applying equation (17) to the data, rate constants of 12.18 x 10^4 lb mole/cu ft min and 21.44 x 10^4 lb mole/cu ft min were calculated at 15° and 25°C, respectively. The increase in the reaction rate was affected by a decrease in ozone solubility at the higher temperature. The overall reduction rate was the same at both temperatures.

The benzoic acid oxidation was studied at pH values of 2, 4, and 11 at a temperature of 25°C. The effect of solution pH on the reaction was dramatic as the reaction became mass transfer limited at a pH of 11 and was due to the increased rate of ozone decomposition (11). The data are shown graphically in Figure 6 for the fraction of benzoic acid remaining as a function of time. The smallest rate constant was 6.14 x 10^4 lb mole/cu ft min for an acidic solution of pH 2. A minimum reaction rate was estimated for the mass transfer limited reaction at a pH of 11 by combining the mass transfer limiting equation (15) with the slow reaction equation (17) and solving for k. The estimated reaction rate constant to maintain a truly mass transfer limited system was 500 x 10^4 lb mole/cu ft min.

The benzoic acid reduction line for a pH of 11 begins to deviate from the mass transfer limiting line after ten minutes of ozonation, but the reaction was starved for ozone a total of 25 to 30 minutes. This indicates the high pH also increased the reaction rate of the benzoic acid oxidation products, which were competing with benzoic acid for the ozone. After 30 minutes of ozonation, ozone was detected in the exit gas, indicating that some reaction products being produced were refractory and the slower reactions were beginning to control the system. The benzoic acid reduction is significantly slowed after 85 percent of the compound was oxidized. This decrease in the rate of reaction of benzoic acid at the lower concentrations suggests the reaction is dependent on the benzoic acid concentration.

Benzaldehyde Ozonation. The benzaldehyde oxidation followed a similar pattern to that of benzoic acid. The average initial benzaldehyde concentration was 285 ppm. The stoichiometric mole ratio for benzaldehyde was evaluated to be 1.47 moles of ozone per mole of benzaldehyde by extrapolating the initial benzaldehyde reduction data. This deviates from the ratio predicted by the proposed reaction of benzaldehyde to benzoic acid and is possibly a result of benzaldehyde forming an intermediate ozonide.

A similar temperature effect was found for benzaldehyde oxidation as for the benzoic acid oxidation as shown in Figure 7. The calculated rate constant was doubled for the ten degree increment while the benzaldehyde reduction remained the same. The calculated reaction rates were 13.33×10^4 lb mole/cu ft min at 15°C and 26.23×10^4 lb mole/cu ft min at 25°C. The increase in reaction rate was again offset by the decrease in the ozone solubility in the liquid with the mole fraction decreasing from 0.57×10^{-6} to 0.24×10^{-6} at the respective temperatures. The system operated in the mass transfer limiting with a slow reaction regime, as the ozone did not reach its equilibrium concentration.

The ozonation test results for the various pH solutions are shown in Figure 8. The benzaldehyde ozonation was truly mass transfer limited at a pH of 11, but deviated after five minutes of ozonation. The calculated rate constants were 9.63×10^4 lb mole/cu ft min for a solution pH of 2 and 243.85×10^4 lb mole/cu ft min at pH of 11.

Styrene Ozonation. The styrene ozonation was difficult to evaluate due to styrene being air stripped and its relative insolubility in water. The average initial concentration was 130 ppm. In the initial data analysis the styrene reduction fell below the mass transfer limiting line which is not feasible. Therefore, air stripping tests were conducted and the styrene reaction data corrected as shown in Figure 9. The styrene oxidation was found to be truly mass transfer limited under all conditions studied. The reaction rate constant was estimated by Levenspiel's method to be 554×10^4 lb mole/cu ft min. This was verified by use of the model developed for benzaldehyde being produced from the styrene ozonation. The various rate constants determined for benzaldehyde at the different operating conditions were substituted into equation (18). A comparison of the model prediction with experimental data for a solution pH of 11 and a temperature of 25°C is shown in Figure 10. The excellent fit supports the rate constants calculated.

Biotreatability Study. The initial project objective was to improve the biotreatability of wastewater containing styrene. In order to economically monitor the water biotreatability, five-day biochemical oxygen demand (BOD$_5$) was determined on each test stream before and after ozonation. The ratio of BOD$_5$ to total organic carbon (TOC) gives the oxygen utilized for oxidizing the organic carbon per carbon atom. Comparing the experimental BOD$_5$ to TOC ratio with a theoretical value for complete oxidation gives an indication of the biotreatability.

The ratio values are given in Table 1. All the solutions had a higher BOD$_5$ to TOC ratio after ozonation than before. None of the tests reached a theoretical value. Benzoic acid and benzaldehyde were shown to be fairly biodegradable with high initial ratios. Styrene biotreatability was greatly improved as the ratio was increased from 0.47 toward the benzoic acid ratio after 90 minutes of ozonation to a final value of 2.69 after 150 minutes. The TOC was reduced from 130 to 39 ppm for a 70 percent reduction. The ozonation of styrene could be a feasible process for the treatment of styrene wastewaters.

Summary

Ozonation was found to improve the biotreatability of water containing styrene by transforming the styrene to a biodegradable compound such as benzoic acid. The BOD$_5$ to TOC ratio was improved for all the compounds studied as shown in Table 1. The styrene solution ratio was significantly improved after 150 minutes of ozonation from 0.47 to 2.69.

The complete oxidation of styrene may be costly as some reaction products react slower, but the initial oxidation of styrene to benzaldehyde was rapid. The ozonation of styrene was truly mass transfer limited with an estimated reaction rate constant of 554×10^4 lb mole/cu ft min. The rate constant was verified, as the model accurately predicted the benzaldehyde concentration from the ozonation of styrene using the various rate constants for benzaldehyde and the rate constant of styrene.

The ozonation of benzaldehyde and benzoic acid was affected by both solution pH and temperature. Benzaldehyde and benzoic acid ozonations were mass transfer limited with a slow reaction under acidic conditions. But at a pH of 11, the reactions became truly mass transfer limited. The higher oxidation rates are related to the increased hydroxyl radical formation from ozone under basic conditions. A ten degree increase in temperature produced a higher reaction rate constant which was offset by lower liquid ozone concentrations. The overall net effect was equivalent reduction rates for both temperatures. None of the reactions were truly reaction rate controlled as the ozone never reached its solubility limit. A summary of the reaction regime and rate constant for each compound is given in Table 2.

LITERATURE CITED

(1) Barona, N., and H. W. Prengle, "Design Reactors This way for Liquid-Phase Processes - Part I". Hydrocarbon Processing 52:63 (1973).

(2) Morrison, R. T., and R. N. Boyd, *Organic Chemistry*, Allyn and Bacon, Inc., Boston (1961).

(3) Subluskey, L. A., G.C. Harris, A. Maggiolo, and A. L. Tumolo, "Improved Synthesis of Aromatic Aldehydes from Ozonolysis of Olefins". *Advances in Chemistry Series No. 21, Ozone Chemistry and Technology*, American Chemical Society, 149 (1959).

(4) Yocum, F. H., "Oxidation of styrene in Aqueous Solution with Ozone," D. E. thesis, Tulane University, New Orleans, Louisiana (1977).

(5) Levenspiel, O., *Chemical Reaction Engineering*, John Wiley and Sons, Inc., New York (1967).

(6) Cooper, C. M., G.A. Fernstrom, and S. A. Miller, "Performance of Agitated Gas-Liquid Contractors", Ind. Eng. Chem. 36:504 (1944).

(7) Kawecki, W., T. Reith, J. W. Van Heuven, and W. J. Beck, "Bubble Size Distribution in the Impeller Region of a Stirred Vessel". Chem. Egr. Sci. 22:1519 (1967).

(8) Westerterp, K. R., L. L. van Dierendonck, and J. A. de Kraa, "Interfacial Areas in Agitated Gas-Liquid Contractors". Chem. Engr. Sci. 18:157 (1963).

(9) Prengle, H. W., C. G. Hewes, III, and C. E. Mauk, "Oxidation of Refractory Materials by Ozone with Ultraviolet Radiation", in Proc. Sec. Intl. Symposium on Ozone Technology R. G. Rice, P. Pichet and M. A. Vincent, Editors. Instl. Ozone Inst., Syracuse, N.Y. (1976) 224-251.

(10) Rand, M. C., A. E. Greenberg, and M. J. Taras, eds. Standard Methods for the Examination of Water and Wastewater, 14th ed., American Public Health Assoc., Washington, D.C. (1976).

(11) Adams, C. E., W. W. Eckenfelder, Jr., and R. M. Stein, "Ozonation Procedures for Industrial Waste-waters", in Proc. First Intl. Symposium on Ozone for Water and Wastewater Treatment, R. G. Rice and M. E. Browning, Editors. Intl. Ozone Inst., Cleveland, Ohio (1975), 591-613.

TABLE 1

Summary of Five-Day Biochemical Oxygen Demand to Total Organic Carbon Ratio Data for Reaction Solutions

Solution	BOD_5/TOC Theoretical	Pre Ozonation	Post Ozonation (a)
Benzoic Acid	3.05	2.46	2.89
Benzaldehyde	2.86	2.30	2.66
Styrene	3.33	0.47	2.69

(a) Ozonation performed at a solution pH of 5 and a temperature of 25°C for 150 minutes.

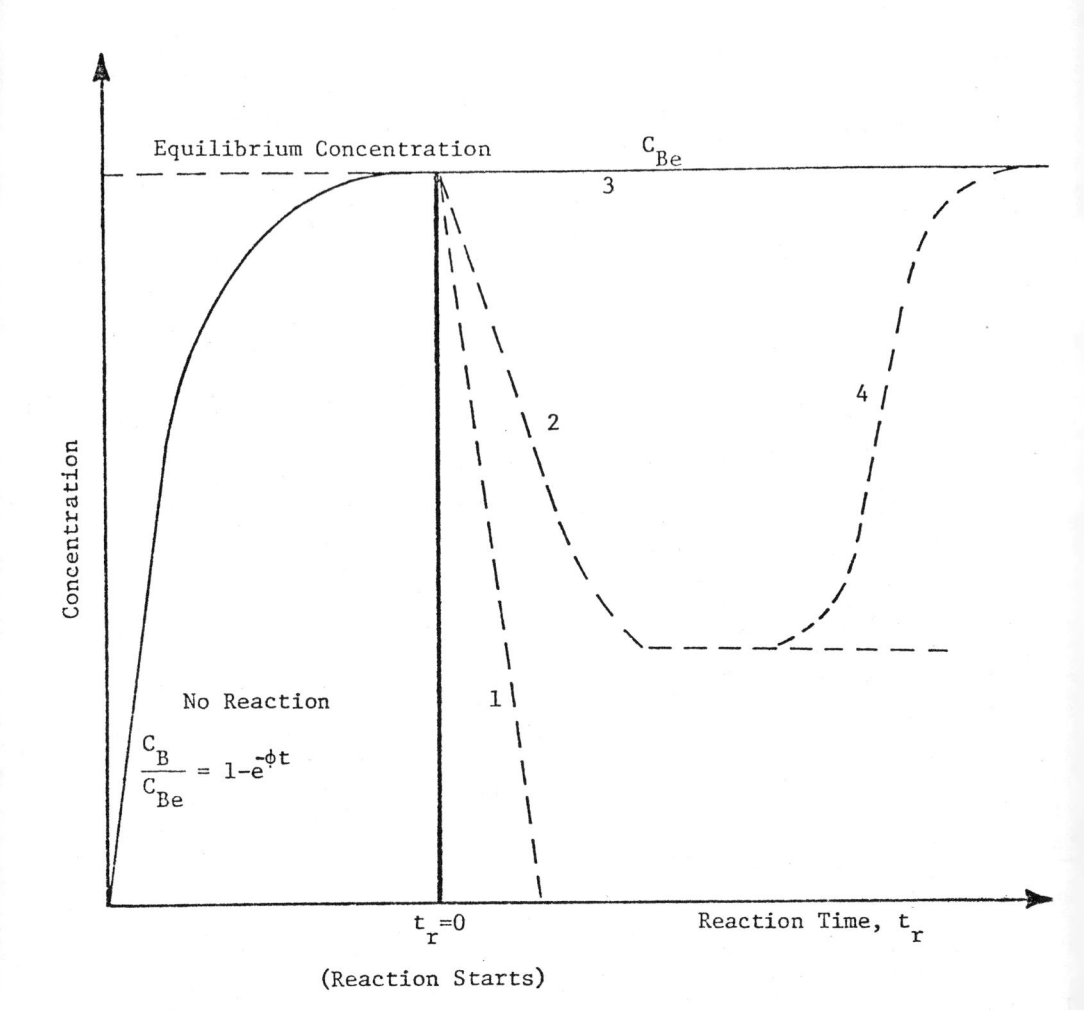

[Figure: Plot of Concentration vs. Reaction Time, t_r, showing equilibrium concentration C_{Be} approached before reaction starts at $t_r=0$ per $\frac{C_B}{C_{Be}} = 1-e^{-\phi t}$ (No Reaction region), then curves 1, 2, 3, 4 after reaction starts.]

1 – Mass Transfer Controlling – Instantaneous Reaction
2 – Mass Transfer Controlling – Slow Reaction
3 – Chemical Reaction Controlling – Very Slow Reaction
4 – After Liquid Reactant is Depleted

Figure 1   Liquid Phase Concentration of the Gas-Phase Reactant as Presented by the Macroscopic View of Interphase Mass Transfer with Chemical Reaction for Semi-Batch Operation. Probable Reaction Regimes are Illustrated.

Prengle, H.W. and N. Barona, "Make Petrochemicals by Liquid Phase Oxidation, Part 2: Kinetics, Mass Transfer and Reactor Design", Hydrocarbon Processing, 49, p. 165 (1970).

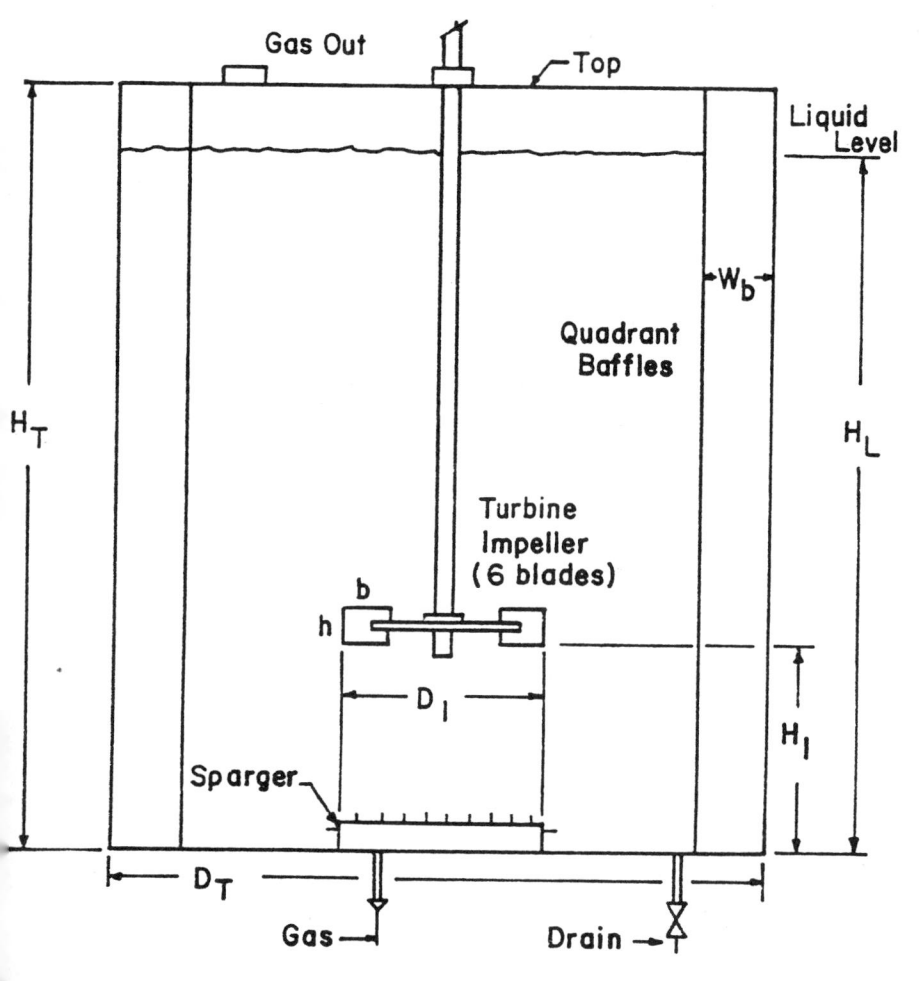

Dimensions: $H_L + D_T$, $W_b = D_T/10$, $D_I = D_T/3$, $H_I = D_I$

Sparger Holes: 2-1/6" holes/in² of sparger cross-section

Impellers: $b = D_I/4$, $h = D_I/5$

Figure 2

Standard Reactor Configuration for Stirred-Tank Reactor.

Prengle, H.W., C.G. Hewes, III, and C.E. Mauk, "Oxidation of Refractory Materials by Ozone with Ultraviolet Radiation," in R.G. Rice, P. Pichet & M.A. Vincent, Editors, Proc. Sec. Intl. Symposium on Ozone Technology. Intl. Ozone Inst., Cleveland, Ohio (1975), 591-613.

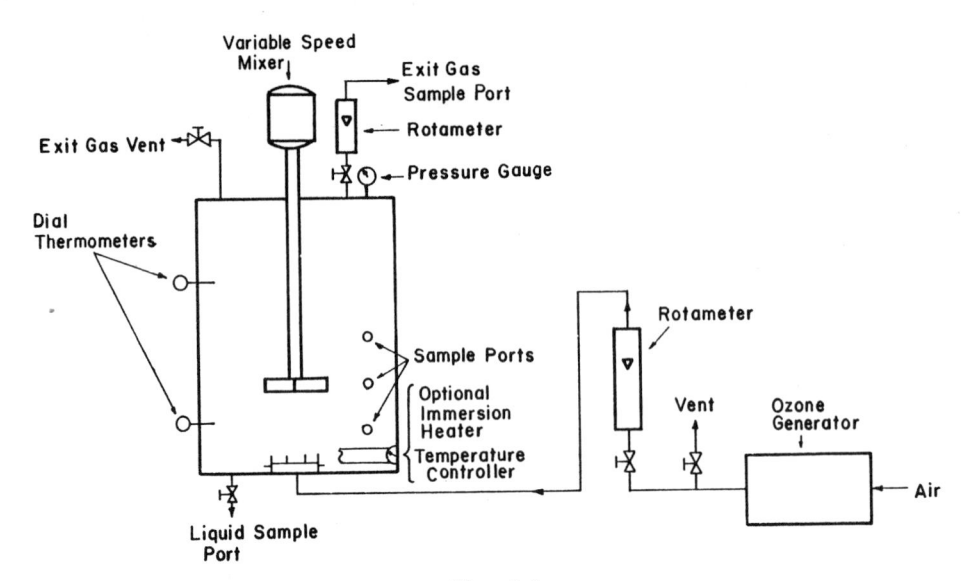

Figure 3

Ozone Pilot Plant Schematic

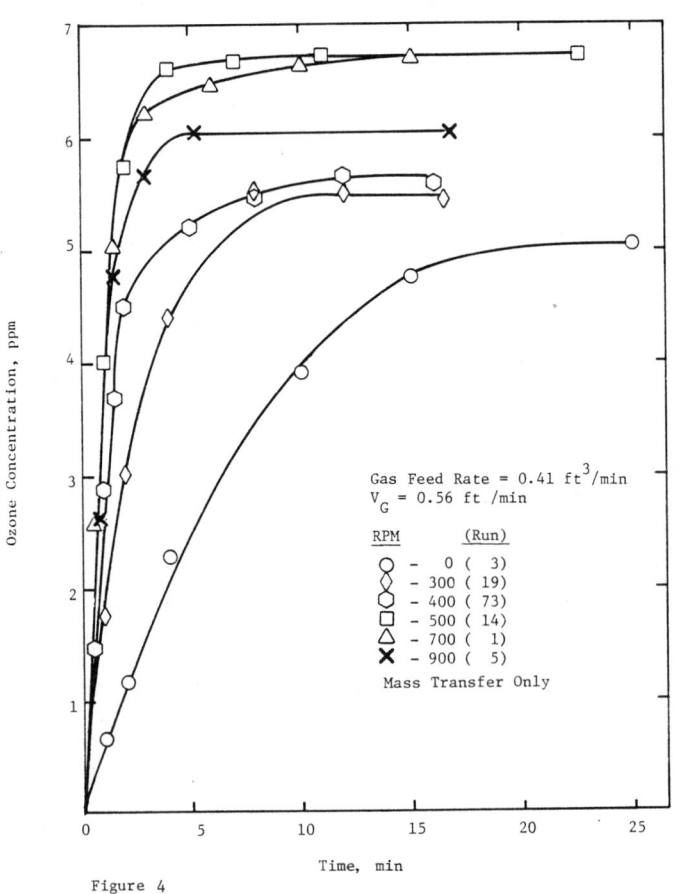

Gas Feed Rate = 0.41 ft^3/min
V_G = 0.56 ft/min

RPM	(Run)
○ – 0	(3)
◇ – 300	(19)
⬡ – 400	(73)
□ – 500	(14)
△ – 700	(1)
✕ – 900	(5)

Mass Transfer Only

Figure 4

Ozone Concentration Profile for Various Turbine Speeds at a Gas Feed Rate of 0.41 CFM and Temperature Range of 12°-15°C.

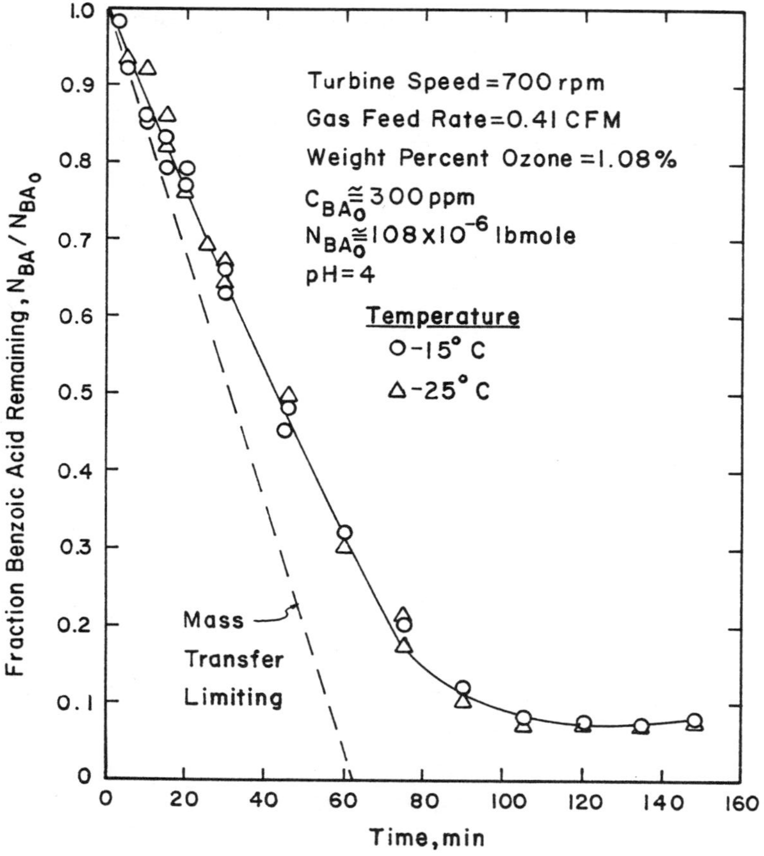

Figure 5. Effect of Temperature on the Ozonation of Benzoic Acid in Aqueous Solution.

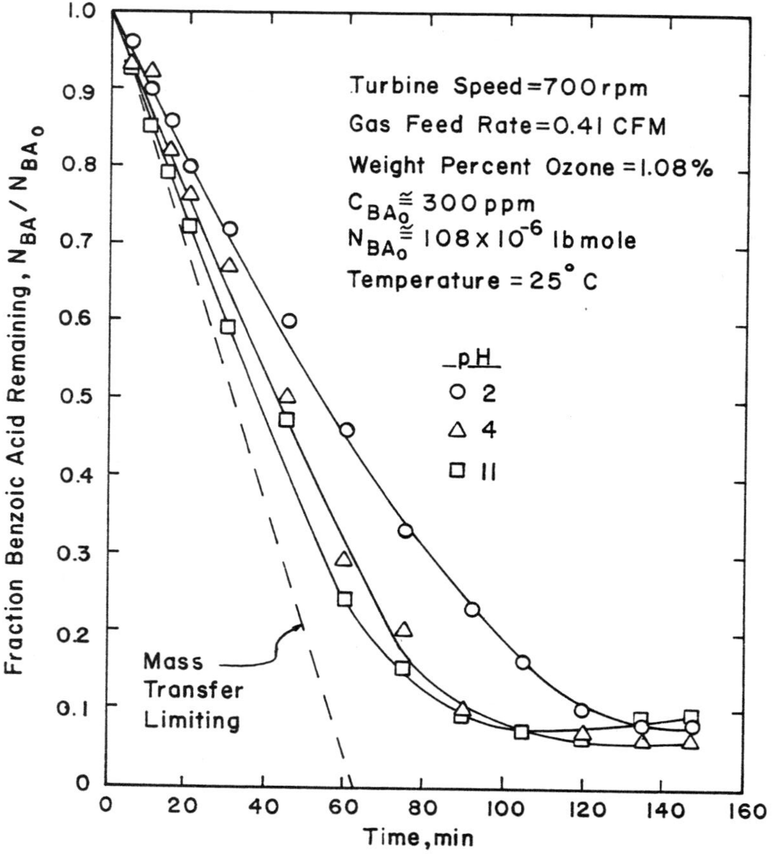

Figure 6. Effect of pH on the Ozonation of Benzoic Acid in Aqueous Solution.

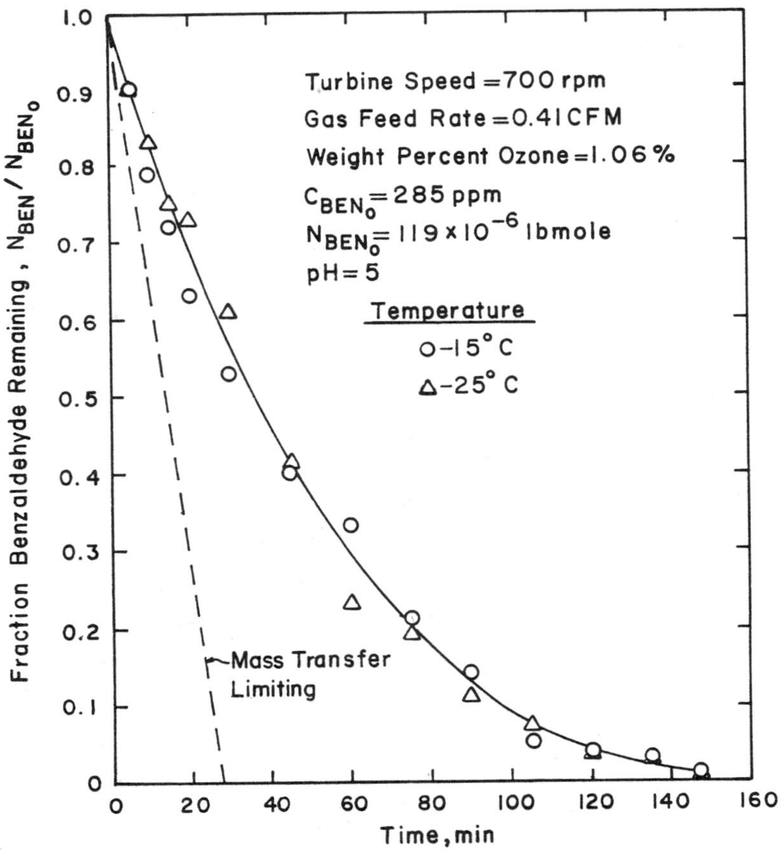

Figure 7. Effect of Temperature on the Ozonation of Benzaldehyde in Aqueous Solution.

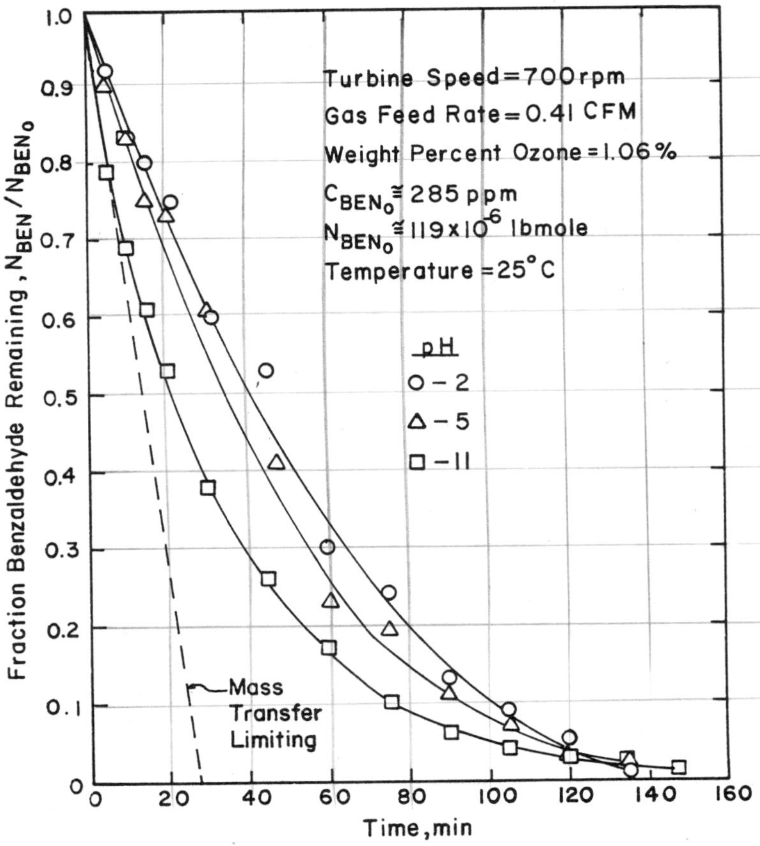

Figure 8. Effect of pH on the Ozonation of Benzaldehyde in Aqueous Solution.

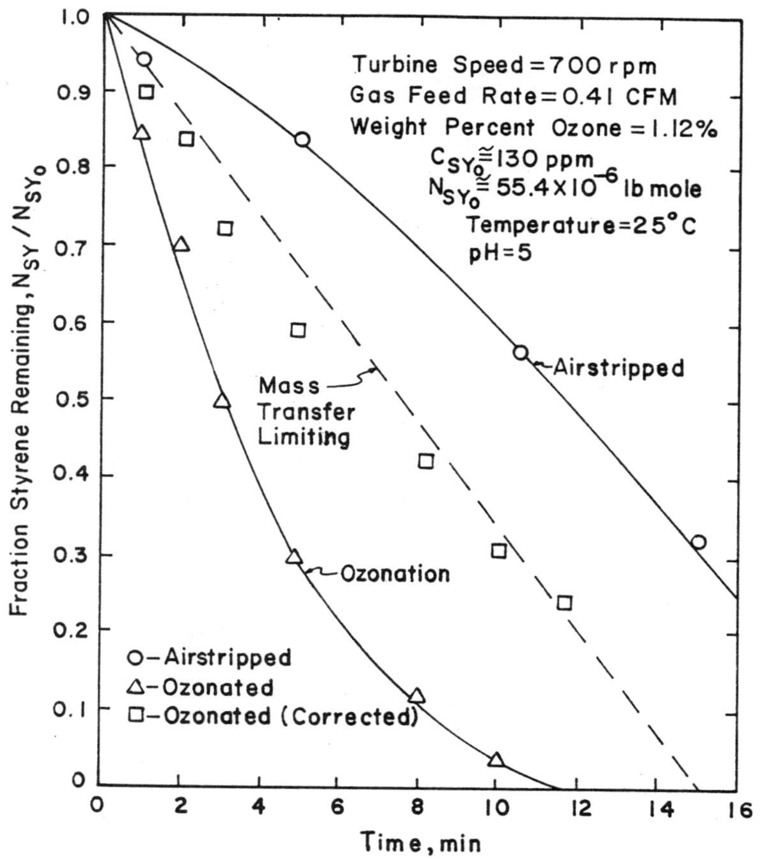

Figure 9. Ozonation of Styrene in Aqueous Solution and the Effect of Airstripping on the System.

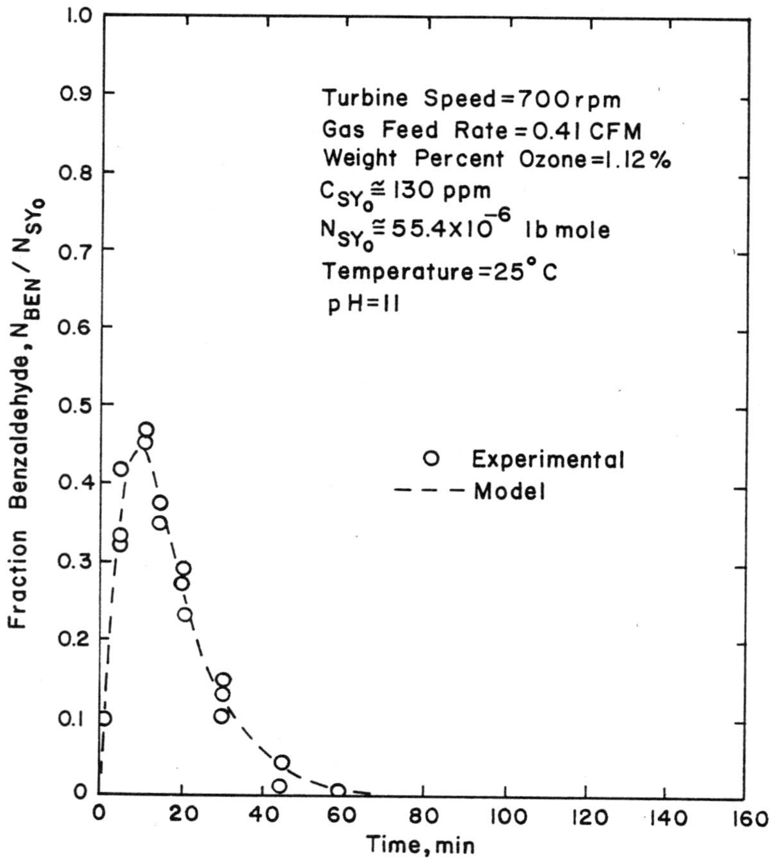

Figure 10. Application of the Model for Benzaldehyde from the Ozonation of Styrene in Aqueous Solution at 25°C and pH=11.

AN ENGINEERING APPROACH TO WATER TREATMENT PROCESS
SELECTION WITH SPECIAL EMPHASIS ON HALOGENATED ORGANICS

M. Schwartz
Assistant Chief Engineer

E. A. Lancaster
Senior Design Engineer

FENCO CONSULTANTS LTD.
1 Yonge Street
Toronto, Ontario M5E 1E7
CANADA

Bay Bulls Big Pond, an impounding reservoir, is being developed as a new source of supply for the St. John's (Newfoundland) Regional Water System. Analytical tests have shown the quality of the raw water to be very soft, acidic, low in alkalinity, low in dissolved minerals, low in turbidity, and moderate in colour (95 percent of the time the colour is less than or equal to 30 TCU). Pilot plant studies were carried out to examine the treatment process most effective and economical for colour removal. These studies included chemical precipitation of colour with alum and a polyelectrolyte, followed by filtration; and oxidation of the colour with ozone, followed by filtration.

Ozone was found to be a viable and competitive treatment process for colour removal, and bacteria kill. The results of these "First Series" pilot plant studies formed the subject of a paper presented by one of the authors at the Second International Symposium of the International Ozone Institute (1).

Since ozone was adopted as the treatment process for Bay Bulls Big Pond water, and since the "First Series" of pilot plant studies were carried out over a wide range of circumstances (to define a process), a "Second Series" pilot plant studies were undertaken to test design conditions of the ozonation process, with special consideration given to the latest developments in water treatment concepts and philosophy.

Oxidation products of organic matter have become of increasing importance and concern to the regulatory authorities and the water supply industry. Recent findings point to the potentially carcinogenic hazard of halogenated organics that may be found in chlorinated drinking water. An assessment of the risks and hazards of human ingestion of chloroform was given by Dr. R. G. Tardiff (2). Based on his extrapolation of the experimental findings with animals to humans, Dr. Tardiff estimates that chloroform in drinking water may be responsible for as much as 1.6% of the current yearly incidence of liver cancer in humans in the United States, and for as much as 1.44% of the yearly kidney cancer incidence. In order to reduce or eliminate this risk, Dr. Tardiff suggests a maximum concentration of 70 ug/l of total haloform compounds in drinking water as being a reasonable limit.

J. J. Rook(3) conducted a study which showed strong evidence that halogenated organics are produced by chlorination of humic substances, such as those which cause the colouration of natural waters (as is the case with Bay Bulls Big Pond water). Concerning water having the latter characteristics, experience has shown(4) that there is merit in using ozonation in combination with chlorination for colour removal.

Accounting for all of the above considerations, two alternate ozonation processes (and equipment) were included in the "Second Series" pilot plant studies, namely:

1. Ozonation, filtration, post-chlorination.

2. Pre-chlorination, ozonation, filtration, post-chlorination.

Each system was tested for performance, treated water quality, and the formation of halogenated organics. The results of this "Second Series" studies are reported in this paper.

PILOT PLANT DESCRIPTION

Ozonation-Filtration-Chlorination Plant

The pilot ozonation-filtration plant consisted of two independent systems operating in parallel. One system employed a positive-pressure contact column in conjunction with a high-rate gravity filter downstream of the column. The second system utilized two diffused contact columns in series with a high-rate gravity filter downstream of the columns. Flow schematics of these two systems are shown on Figures 1 and 2, respectively.

Positive Pressure Contacting System. An intake pipe extended from Bay Bulls Big Pond to a feed pump on the shoreline. The pump discharge rate was controlled at 10 gpm (0.63 l/sec) using a throttling valve and a calibrated rotameter for flow reference. The pump discharge pressure was controlled by a pressure relief valve.

Flows from the pump were discharged to the 10 ft. (3.05 m) high contact column via a positive pressure injection box located at the top end of the column. Ozone was applied to this same injection box. The gas/water mixture flowed co-currently from the injector down a central pipe. The mixture then reversed direction at the base of the central pipe and rose up a concentric circular tube, where the gas and water separated. Excess gas was vented to an ozone killer solution (made of sodium thiosulphate and potassium iodide). The ozonated water flowed over a free-fall weir to the outer shell of the column. Discharge from this column to the gravity filter was by hydrostatic pressure from an outlet at the bottom of the outer shell. The flow applied to the filter was controlled at 2 gpm (0.126 l/sec) using a pair of throttling valves, one each on the filter feed line and the line to waste, and a calibrated rotameter for flow reference.

The filter was 10 inch (254 mm) in diameter, providing a surface area of 0.5 sq ft (0.046 sq m). It was equipped with a "FLEXKLEEN" nozzle at the bottom and a 10 inch (254 mm) thick sand medium underlying a 20 inch (508 mm) deep anthracite medium. The medium size of the sand and anthracite was 0.5 mm and 1.1 mm, respectively, whereas the uniformity coefficient of these media was 1.5. The filter was operating at a positive head of 48 inches (1.22 m) minimum. This was accomplished by an adjustable outlet at the exit from the filter.

Diffused Contacting System. A second intake pipe extended from Bay Bulls Big Pond to a second feed pump on the shoreline. The pump discharge rate was controlled at 6 gpm (0.378 l/sec) using a throttling valve and a calibrated rotameter for flow reference. The pump discharge pressure was controlled by a pressure relief valve.

Flows from the pump were discharged to the top of a 4 inch (102 mm) diam 18 ft (5.48 m) high contacting column (providing a retention time of about 2 minutes). Ozone was applied to the bottom of the column through a porous diffuser. The gas and water flowed in a counter-current pattern. The ozonated water then flowed from a bottom outlet in the first column to a bottom inlet of a second contact column having the dimension of 6 inch (152 mm) diam and 18 ft (5.48 m) high (and providing a retention time of about 4 minutes). Ozone was applied to the second column through a porous diffuser at the bottom. The gas and water flowed in a co-current pattern. Ozonated water was discharged from an outlet at the top of the second contact column to a vertical pipe, which provided the hydrostatic pressure required for feeding of the gravity filter. A free fall was maintained at the contact column outlet into the vertical pipe. Excess ozone from each of the contact columns was vented to an ozone killer solution (made of sodium thiosulphate and potassium iodide).

The flow applied to the filter was controlled at 2 gpm (0.126 l/sec) using a pair of throttling valves, one each on the filter feed line and the line to waste, and a calibrated rotameter for flow reference.

The filter structure and mode of operation were identical to those described above for the positive pressure contacting column system.

Post-Chlorination. The treated water from the filters was discharged via a 50 gallon (189 l) holding tank where chlorine, in the form of sodium hypochlorite solution was added to provide a post-chlorination residual of from 0.3 to 0.5 mg/l chlorine.

Ozone Generating System. A single ozone generator, Union Carbide Model LG-2-L2, was used to feed ozone to both contacting systems. This unit is capable of producing from air 1 lb/day ozone @ 1% concentration by weight. The unit included an air preparation system comprising a compressor and air driers.

The ozone feed line (from the generator was tapped at three places and supply lines were extended to the following stations:

a) A rotameter and throttling valve to the positive pressure contacting system.
b) A rotameter and throttling valve to column 1 of the diffused contacting system.
c) A rotameter and throttling valve to column 2 of the diffused contacting system.

The total flow of ozone to the diffused contacting system was split between the two columns in such a way that contact column 2 received between 33 and 38% of the total flow, with the remaining ozone being fed to contact column 1.

Curves were developed to correlate the amount of ozone to the rate of air flow as recorded on the rotameters.

Pre-Chlorination, Ozonation, Filtration, Chlorination Plant

With certain essential modifications, the pilot plant system employed during this phase of the study was the positive pressure contacting system described previously, and the Union Carbide generating system described above. A flow schematic of the modified system is shown in Figure 3.

To provide facilities for pre-chlorination, a 200 gallon (757 l) batch tank was installed between the intake pump and the ozone contact column. This tank was provided with a recirculation pump to ensure complete mixing of the applied chlorine. All tests were based on the retention, for not less than 30 minutes, of 150 gallons (568 l) of raw water to which was added sufficient chlorine, in the form of sodium hypochlorite solution, to yield chlorine concentrations after mixing of 1.0, 2.5, or 4.0 mg/l.

The ozone generator was operated at a constant air pressure of 9 psi (0.63 kgf/cm^2) and at a constant air flow rate of 30 cfm (0.014 m^3/sec) although the latter was increased to 40 cfm (0.019 m^3/sec) for two trials at the end of the study period. The applied ozone dose was controlled by the power consumed during ozone generation and, for the purposes of the study, was decreased roughly in proportion to the increase in chlorine dose applied during pre-chlorination.

A 100 gallon (378 l) holding tank was installed between the ozone contact column and the filter column to provide a controlled and constant hydraulic feed rate to the filter. The treated water from the filter then passed into a 50 gallon (189 l) holding tank where chlorine, in the form of sodium hypochlorite solution was added to provide a post-chlorination residual of from 0.3 to 0.5 mg/l chlorine.

ANALYTICAL PROCEDURE

Analytical determinations for pH, colour, turbidity, iron, manganese, total dissolved solids, organics (as determined by the "permanganate" test in terms of oxygen uptake), and coliform organisms were carried out in all cases on daily composite samples of raw water, ozonated water from each contact column and filtered water; and of pre- and post-chlorinated water when chlorination was applied.

Analytical determinations for ozone were carried out on gas samples collected from the feed line to each contact column and the vent line from each column.

All sample points are shown on Figures 1, 2 and 3.

Water analysis was conducted in accordance with the "Standard Methods for the Examination of Water and Wastewater", 13th edition (1971), on daily composite samples. The analytical method for the determination of ozone was based on the method presented by Messrs. Birdsall, Jenkins, and Spadinger in Volume 24, page 662, of "Analytical Chemistry" (1952).

Samples for the identification and determination of halogenated organics were sent to the Organic Trace Contaminants Section of the Ontario Ministry of the Environment, in Toronto. This Section of the M.O.E. advised that, based on their own direct aqueous injection technique, the limits of detection for various haloforms are as follows:

Chloroform ($CHCl_3$) : 1 ug/l (ppb)
Dichlorobromomethane ($CHCl_2Br$): 0.5 ug/l (ppb)
Dibromochloromethane ($CHBr_2Cl$): 1 ug/l (ppb)
Carbon Tetrachloride (CCl_4) : 0.1 ug/l (ppb)
Trichloroethylene ($Cl_2C:CHCl$) : 5 ug/l (ppb)

RAW WATER QUALITY

A summary of raw water quality during the study period is as follows:

	Units	Min.	Max.	Avg.
Colour	TCU	15	30	25
pH	-	5.0	6.1	5.7
Turbidity	JTU	0.4	3.15^x	1.0
4 Hr.Permanganate	mg/l-O_2	1.4	2.7	2.0
Iron	mg/l	0.08	0.28	0.12
Manganese	mg/l	0.005	0.04	0.01
T.D.S.	mg/l	4	40	20
Alkalinity	mg/l	0.2	1.6	1.0
Trihalomethanes:				
$CHCl_3$	ug/l	ND	ND	-
$CHCl_2Br$	ug/l	ND	ND	-
$CHBr_2Cl$	ug/l	ND	ND	-
CCl_4	ug/l	ND	ND	-
$Cl_2C:CHCl$	ug/l	ND	ND	-

ND: not detected

x :   due to heavy and continuous wave action in Bay Bulls Big Pond.

STUDY RESULTS

Ozonation-Filtration-Chlorination Process

The ozonation process, in both the positive pressure contacting system and the diffused contacting system, performed efficiently in terms of colour removal.

Figure 4 represents the results obtained from the positive pressure contact system. It can be seen from this figure that at an ozone dose of 2 mg/l the water contained 5 colour units or less (conforming to the "objective" of Drinking Water Standards). The contact time provided by the system was less than 2 minutes. The ozone concentration in the air stream was about 0.5% by weight.

The results obtained from the diffused contacting system were plotted in Figure 5. It can be seen from this figure that, at a dose of about 2 mg/l ozone, the ozonated water discharged from the first column had a colour intensity of about 5 TCU. Further treatment in the second column with an additional dose of 1 mg/l reduced the colour level to less than 5 TCU. The contact time in columns 1 and 2 was 2 minutes and 4 minutes, respectively. The ozone concentration in the air stream was about 0.5% by weight.

There appears to be a correlation between the permanganate test for organics and colour. When colour removal was efficient, an increase in organics content was detected in the ozonated water. This increased organic content was eventually removed on the filters. When colour was reduced to levels of 7.5 and 10 TCU, the permanganate values remained essentially unaltered.

Measurements of turbidity (JTU) in the water before and after ozonation did not show any significant increase or decrease in this constituent. However, there seemed to be a tendency for the turbidity to drop after filtration when preceded by ozonation. This suggests that ozonation may condition the physical characteristics of the colloidal particles present in the raw water, making them more amenable to filtration.

Iron present in the raw water, if in the ferrous state, would be oxidized to the ferric state but should leave iron values, after ozonation, little changed from those in the raw water. This appeared to be true in the results obtained.

Manganese content is quite low and follows the same pattern noted above for iron.

In general, total dissolved solids concentrations did not appear to alter after ozonation. Alkalinity and pH, before and after ozonation, remained essentially unaltered.

The disinfection effect of ozone was evident in all cases, i.e., in the ozonated water from the positive pressure contact column, and the ozonated water from the first and second diffused contact column, the coliform bacteria and the faecal coliforms were found to be less than 2 per 100 ml (using the Membrane Filter method).

It is significant to note that treatment with ozone followed by filtration and post-chlorination, did not produce trihalomethane compounds.

Pre-Chlorination, Ozonation, Filtration, Chlorination Process

In the pre-chlorination process, chlorine was added to the raw water at three different concentrations, namely at 1.0, 2.5 and 4.0 mg/l. The pH of the raw water was acidic, in the general range of 5.5 to 5.8.

After 30 minutes contact time, which was preceded by recirculation to ensure complete mixing, chlorine consumption was found to be 0.4 mg/l and 0.6 mg/l at the 1.0 mg/l and 4.0 mg/l initial concentrations, respectively. This low rate of chlorine utilization, in spite of the fact that chlorine was available in the water, indicates that the pre-chlorination process is limited by chemical reaction kinetics. Specifically, the interpretation is that for the chlorine to react effectively with the organic colour in the raw water, a long contact time rather than chlorine concentration is the predominant factor. Indeed, the results of the tests confirm this observation.

Regardless of chlorine concentration, only 20% of the initial colour, some 15% of the initial iron and manganese, and 30% of the organic material (as measured by the "permanganate" test) were removed from the water. However, trihalomethanes, predominantly chloroform, were formed, and a distinct correlation was observed between the level of chloroform and chlorine utilization. At 0.4 mg/l chlorine utilization, the chloroform concentration was 22 ug/l, and at 0.6 mg/l chlorine utilization chloroform was produced in excess of 80 ug/l. These results indicate that at the pH level of the raw water, the first step in the oxidation of the organic compounds by chlorine is very slow and time dependent. Only a small portion of these organic compounds (as shown by the "permanganate", colour, iron and manganese tests), undergoes the first step of oxidation. Thereafter, chlorine is primarily utilized for the successive and relatively faster steps of oxidation which form the end product of trihalomethanes, in this case chloroform, and to a much lesser degree, dichlorobromomethane.

Figure 6 shows the effects of pre-chlorination on the raw water quality and the formation of trihalomethanes.

Ozone, employed as a follow-up to pre-chlorination, was applied at reduced dosages, roughly in proportion to the increase in applied chlorine, namely at about 2.0, 1.5, and 1.0 mg/l ozone following 1.0, 2.5, and 4.0 mg/l chlorine, respectively. The purpose of this reduction in ozone concentration was to maintain, in total, roughly equivalent overall oxidation conditions.

Ozone was found to further reduce the colour, up to 50% of the initial colour at a dose of 2.0 mg/l. Organic compounds, as measured by the "permanganate" test, were also further reduced, up to about 45% of the initial concentration at an ozone dose of 2.0 mg/l. There was no marked effect on further reduction, by ozonation, of iron and manganese.

As reported above, pre-chlorination was found to form undesirable trihalomethane compounds, notably chloroform. When subjected to ozone, the level of this chloroform was reduced, as follows:

- From 22 ug/l to 20 ug/l at an ozone does of 2.0 mg/l
- From about 50 ug/l to about 35 ug/l at an ozone dose of 3.25 mg/l

Figure 7 shows the effects of ozonation on the pre-chlorinated water quality. Trihalomethane determinations were done four days after sample collection, to simulate retention conditions in the distribution network.

Residual colour following pre-chlorination and/or ozonation appears, for the most part, to be associated with totally soluble residuals. In most cases, turbidity is reduced by filtration, but very little change in the concentration of trihalomethanes took place as a result of filtration. These latter components, on the basis of ozonation and filtration results, therefore, can be characterized as being extremely stable (refractory) and essentially present in the dissolved form.

In order to simulate conditions of post-chlorination, for purposes of obtaining a chlorine residual in the treated water to prevent organism growth in the distribution network, all filter effluents were dosed with 0.3 to 0.5 mg/l chlorine. This was applied in a batch process to the treated water in a holding tank having a retention or contact time of 30 minutes. In all cases, post-chlorination at the above applied dose and contact time had no measurable effect on colour.

Of interest, however, was the effect of post-chlorination on organic residuals and, in particular, the possible production of halogenated organics. As noted earlier, and in spite of their absence in the raw water, halogenated organics were readily formed during pre-chlorination. With ozonation alone (i.e., without pre-chlorination) no halogenated organics were formed either during ozonation or, and in particular, during post-chlorination. Whatever the mechanism, ozonation appears to be capable of breaking down the halogenated organics precursors, thus preventing the formation of these potentially hazardous materials during post-chlorination.

DISCUSSION

The two ozone contacting systems tested, i.e., the positive pressure contact column and the diffused contact columns, provided strong oxidizing conditions that effectively reduced colour. In this regard, it will be of interest to note that a single diffused column providing a contact time of 2 minutes could reduce the colour intensity to the objective level of 5 or less TCU. Similar experience has been noted elsewhere (i.e., Watchgate Treatment Plant, England; Loch Turret Pilot Plant, Scotland).

An area of interest is the possible relationship of organics to colour and its removal from drinking water. Colouration of Bay Bulls Big Pond water is caused by humic substances which are the products of plant decay. These substances include macromolecules which are condensation products of quinones and polyhydroxybenzenes and other similar complex organics of humic origin such as lignins and tannins. These complex organics tend to be highly refractory, that is, difficult to break down under normal oxidation conditions. However, under the strong oxidizing effects of ozone, these colour-producing refractory organics are effectively broken down into less refractory, more readily oxidizable organics with a corresponding reduction in the colour values attributable to the original constituents. The latter is evidenced by colour value reductions, after ozonation to 7, 5 and less than 5 TCU.

The permanganate test values before ozonation appear to measure the background oxidizable organics (not the refractory organics), and inorganics present in the reduced state (e.g., ferrous iron). After ozonation, with inorganics and the readily oxidizable organics completely oxidized, the permanganate test values appear now to be measuring that portion of the refractory organics which was only partially oxidized.

The results of the pilot plant study indicate that a portion of the refractory colour organics is oxidized to a soluble residue that is carried in the ozonated water and shows up in the permanganate test. Effective and efficient ozonation conditions will further oxidize another portion of the refractory colour organics which will coagulate with colloidal and other similar particles and be removed on filters. This phenomenon is of significant importance since it indicates that in the absence of filters, downstream of ozonation, there is the probability that organic matter, suitable as a substrate for bacterial growth, will be included in the supply water.

The pre-chlorination-ozonation process was considered to offer flexibility and economy in operation. The merit and feasibility of this process was, therefore, tested in the study.

The accepted mechanism for the reaction of chlorine with organic compounds such as humic acids (which are responsible for the colour in Bay Bulls Big Pond water) favours generally alkaline aqueous solutions. Typical of many organic reactions, this mechanism consists of a chain of sequential reactions which lead to the formation of the end product. The first reaction in these steps of reactions is known to be the slowest of them all. Once this beginning reaction is initiated, the sequential reactions proceed fast and smoothly. This phenomenon, in essence, was observed in the pilot plant study.

Due to the acidity of the raw water, the aqueous chlorination reaction with the organic compounds proceeded at a slow rate. This explains the low rate of chlorine utilization, and the low efficiency in colour and organics removal. However, once the beginning reaction was initiated, albeit at a limited rate, chlorine was utilized for the sequential reactions which formed the end product trihalomethanes, notably chloroform.

The aqueous chlorination reaction could be accelerated, and made more efficient in terms of colour removal, by either prolonging the reaction time, or increasing the pH level of the raw water to the alkaline range. However, these alternative solutions defeat the basic concepts for flexibility (simplicity) and economy in operation. Furthermore, at the alkaline range, trihalomethanes will be formed at higher concentrations, a situation we consider very undesirable.

The findings of the pilot plant study also indicate that as long as the time lapse between ozonation and post-chlorination is kept short, the organic precursor will still contain ozonides which, upon reaction with small concentrations of chlorine in the post-chlorination proces, may inhibit, and certainly minimize, the formation of trihalomethanes. This, we consider, to be a favourable situation.

CONCLUSIONS

The conclusions from the "Second Series" pilot plant study as described herein can be enumerated as follows:

a) Both positive pressure contact columns and diffused contact columns (when properly designed) are effective in treating Bay Bulls Big Pond water for colour reduction and disinfection by ozonation.

b) Filters downstream of ozonation will remove organic matter which otherwise will enter the water supply system and enhance the growth of bacteria.

c) Filters in conjunction with ozonation should be designed to operate under a positive fixed head. Also, the discharge from the ozone contact unit(s) to the filter(s) should be designed to include a free fall.

d) Pre-chlorination of the raw water for colour removal is a time-dependent reaction. Increasing the applied chlorine concentration above some (undetermined) threshold value does not produce any significant change in the reduction of colour values.

e) Ozonation following pre-chlorination provides additional colour reduction in the pre-chlorinated effluent. However, the cumulative reduction in colour resulting from pre-chlorination and ozonation appeared to be no greater than that obtained using ozonation alone, for the same dose of ozone.

f) Apart from some minor colour reduction associated with the removal or reduction of turbidity, gravity filtration following ozonation alone, or pre-chlorination followed by ozonation, produced no significant reduction in colour values. It is apparent that most, if not all of the colour present in the influent to the filter was derived from dissolved constituents.

g) Post-chlorination, following pre-chlorination, ozonation and gravity filtration, produced no measurable increase or decrease in colour.

h) Trihalomethanes are not present in the raw water in any detectable amounts, and are not formed during ozonation. Moreover, the products of ozonation do not appear to combine or react with chlorine, during post-chlorination, to form detectable quantities of trihalomethanes.

i) Trihalomethanes, notably chloroform, and to a lesser extent dichlorobromomethane, are clearly formed during the pre-chlorination of the raw water. These   to some degree to be   to decomposition by ozone. Post-chlorination following such treatment does not increase the trihalomethane concentrations.

As derived from the curves in Figure 7, the rate of chloroform decomposition by ozone follows the following expressions:

1. $CHCl_3$ (ug/l) = $10^{L-0.047D}$
   For: 30 ug/l $\leq$ $CHCl_3$ $\leq$ 90 ug/l

2. $CHCl_3$ (ug/l) = $10^{L-0.025D}$
   For: $CHCl_3$ $\leq$ 25 ug/l

Where:

L = Initial $CHCl_3$ concentration (on linear scale)

D = Ozone dose (in mg/l)

j) Since the intent is not to achieve full removal of organic residues from the treated water, post-chlorination should be considered, preferably in the form of combined chlorine or chlorine dioxide, to maintain a chlorine residual in the distribution network.

LITERATURE CITED

(1) SCHWARTZ, M. and MONCRIEFF, D.J.W. "Ozone as a Treatment Process for Colour Removal from Drinking Water". Proc. Second Intl. Symp. on Ozone Technology, R.G. Rice, P. Pichet and M.A. Vincent, Editors, Intl. Ozone Inst., Cleveland, Ohio 283-308 (1976).

(2) TARDIFF, R.G. "Health Effects of Organics: Risk and Hazard Assessment of Ingested Chloroform". Paper presented at the 96th Annual Conference, Am. Water Works Assoc., New Orleans, Louisiana, (June 20-25, 1976).

(3) ROOK, J.J. "Formation of Haloforms During Chlorination of Natural Waters." Water Treatment and Examination, 23:234-243 (1974).

(4) SANKEY, K. E. "Factors in the Watchgate Design." Water Treatment and Examination, 21: (1972).

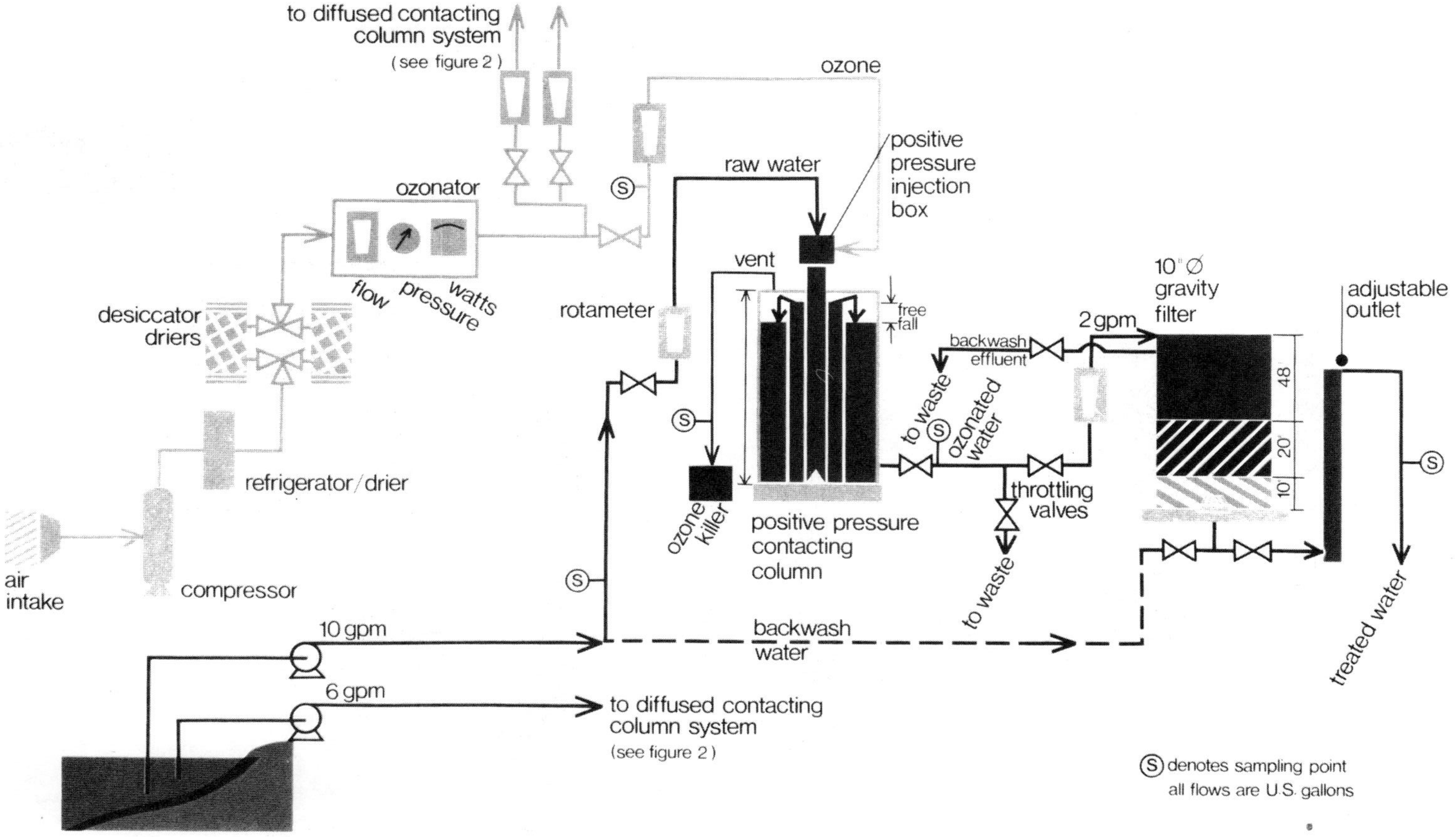

figure 1 Flow Schematic
Pilot Ozonation – Filtration Plant
(Positive Pressure Contacting System)

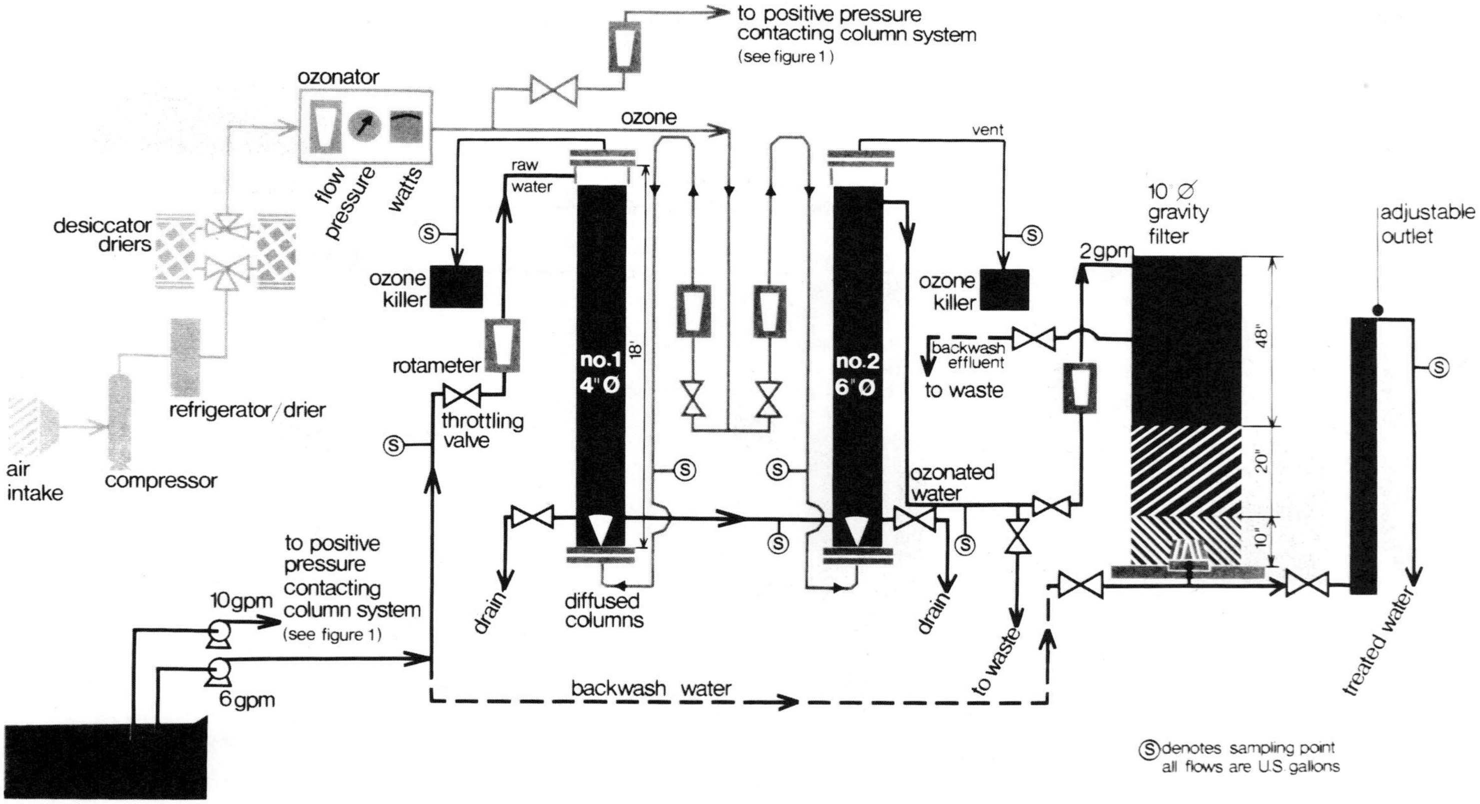

figure 2 Flow Schematic
Pilot Ozonation – Filtration Plant
(Diffused Contacting System)

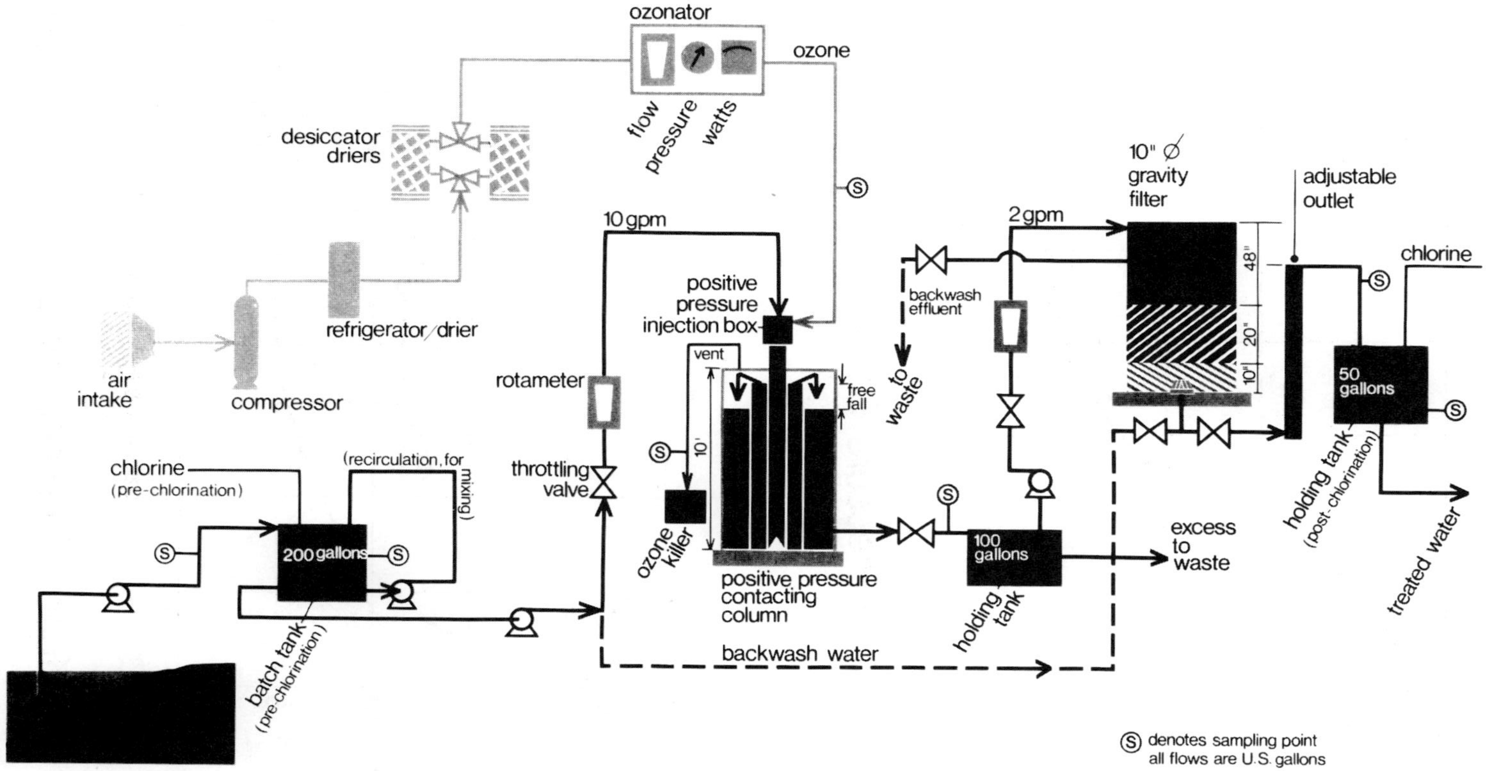

figure 3 Flow Schematic
Pilot Chlorination–Ozonation–Filtration Plant
(Positive Pressure Contacting System)

-281-

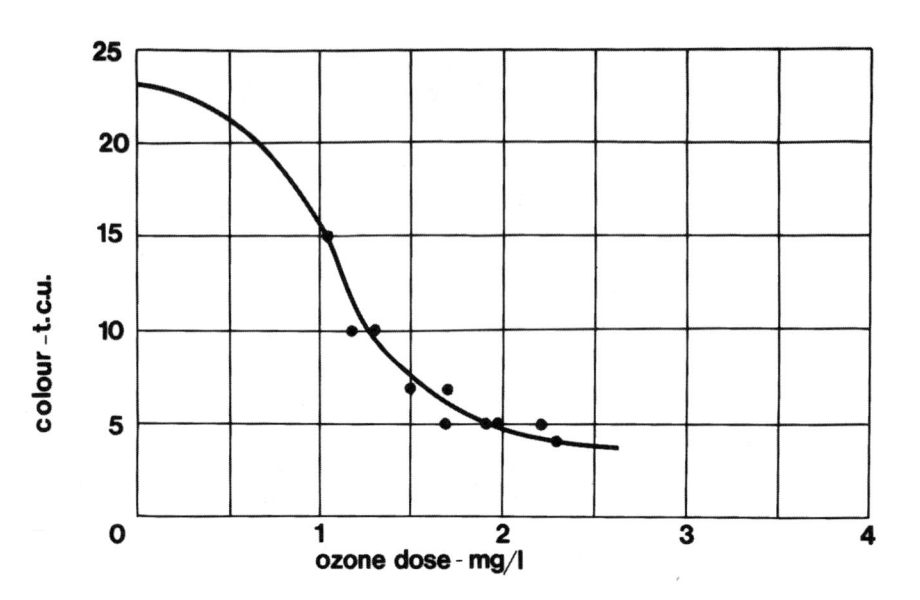

figure 4 Positive Pressure Contact System — Colour vs. ozone dose

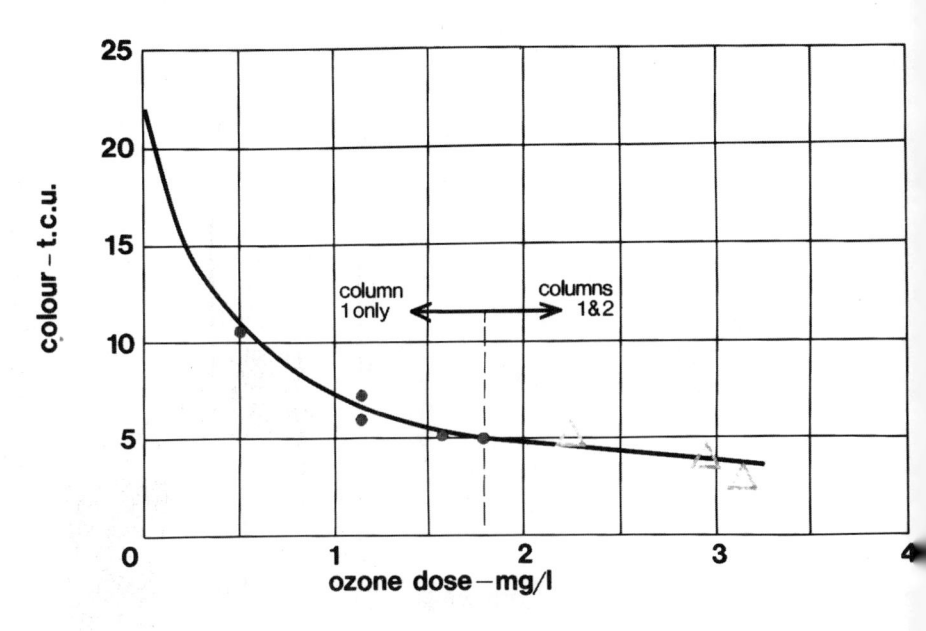

figure 5 Diffused Contact System — Colour vs. ozone dose

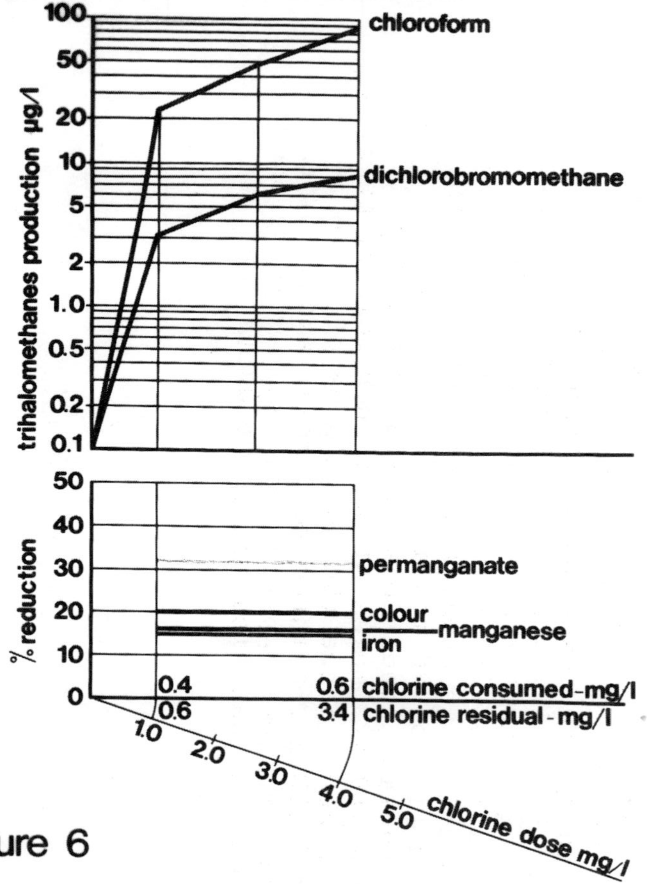

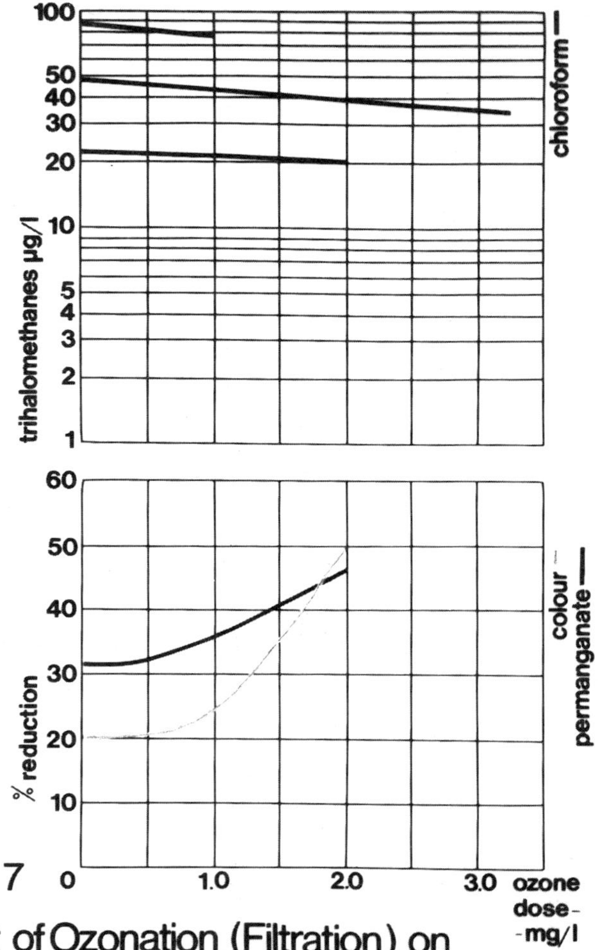

figure 7

Effect of Ozonation (Filtration) on Pre-chlorinated Water Quality

figure 6

Effect of Pre-chlorination on Raw Water Quality and Trihalomethane Formation

OZONIZATION PRODUCTS FROM CAFFEINE IN AQUEOUS SOLUTION

Robert H. Shapiro, K. J. Kolonko, P. M. Greenstein,
R. M. Barkley, and R. E. Sievers

Department of Chemistry, University of Colorado
Boulder, Colorado 80302

Introduction

A preliminary investigation into the identity of organic compounds found in the undisinfected effluent from the Upper Thompson sewage treatment plant revealed that caffeine was one of the major components. Caffeine, as well as other xanthine derivatives, had previously been found in water by others (1). We therefore initiated a study to determine the ozonized wastewater samples which contain caffeine, and to evaluate the products with respect to their biological properties.

Although the caffeine concentration in the wastewater sample was less than 1 mg/ℓ, our initial experiments were conducted at higher concentrations in order to facilitate product identification. Typical experimental conditions are a caffeine concentration of 660 mg/ℓ (0.0034 M), 0.75 ℓ volume (0.0026 mole), an ozone flow rate of 17 mg/min, and a reaction time of 90 min at 20° in pure water. Under these reaction conditions, all of the original caffeine and 4.2 moles of ozone per mole of caffeine are consumed. Since the total dose of ozone is 31.9 mmoles (1.53 g), a three fold molar excess of ozone was used in these experiments. Four major products (≥5%) and at least four minor products were generated (scheme 1).

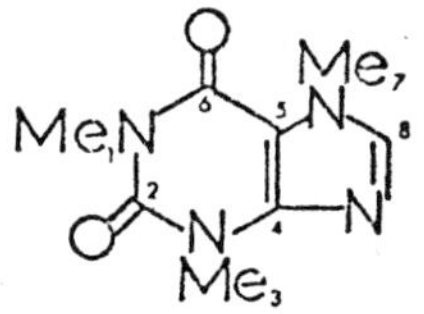

	mol. wt.	(amt)
1.	102	(1%)
2.	116	(1%)
3.	142	(34%)
4.	157(A)	(12%)
5.	157(B)	(6%)
6.	198	(44%)
7.	210	(1%)
8.	226	(1%)

Scheme 1. Molecular weights and relative amounts of the products from the ozonization of caffeine.

The number of products, their relative amounts, and their molecular weights were determined by gas chromatography-mass spectrometry (GC/MS). Both electron impact and chemical ionization mass spectrometry were performed in order to ascertain the molecular weight assignments.

Product Identifications

Molecular Weight 142.

The 142 component is readily obtained as a pure compound by allowing the ozonization to proceed for 1.5 hr followed by extraction of the aqueous reaction mixture with chloroform. The pure substance has a m.p. of 150° and shows a single peak (singlet) at 3.27 ppm in its proton magnetic resonance (PMR) spectrum (solvent $CDCl_3$). The carbon-13 magnetic resonance (CMR) spectrum of the compound dissolved in $CDCl_3$ shows three peaks at 24.9 ppm (2 carbons), 154.1 ppm (1 carbon) and 157.0 ppm (2 carbons). The upfield signal in the CMR spectrum corresponds to two equivalent methyl groups attached to nitrogen and the downfield signals correspond to two equivalent and one unique carbonyl groups. The presence of two equivalent methyls attached to nitrogen is confirmed by the PMR spectrum. The combination of the NMR and mass spectral data (i.e., five carbons, two of which are methyls, and the molecular weight of 142) indicate that the molecular formula is

$C_5H_6N_2O_3$. The only reasonable structure for this product is that of dimethylparabanic acid, which is the product of the reaction shown in scheme 2. In order to ascertain that dimethylparabanic acid is indeed the product, the independent synthesis shown in scheme 2 was performed.

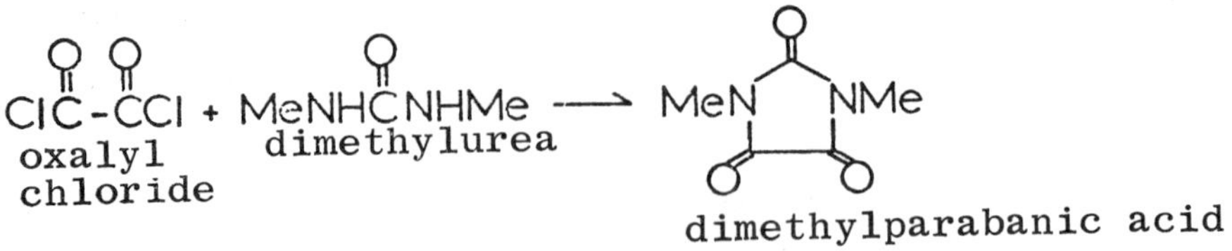

ClC-CCl + MeNHCNHMe ⟶ MeN―NMe
oxalyl   dimethylurea
chloride
                              dimethylparabanic acid

Scheme 2. Synthesis of dimethylparabanic acid.

The product from the ozonization of caffeine and the product from the independent synthesis were shown to be identical in every respect, including an undepressed mixture melting point.

Dimethylparabanic acid is a common oxidation product from caffeine; several oxidizing agents have been used to effect this conversion (2). In fact, in 1889 Leipen reported that this substance was the sole non-volatile product from the ozonization of caffeine (3). Under our reaction conditions, which are similar to those in the sewage treatment plant, dimethylparabanic acid constitutes only 35% of the product mixture. Dimethylparabanic acid was tested for mutagenicity by the Ames test (4) and was found to be non-mutagenic.

Molecular Weights 102 and 116.

These substances were not separated in the gas chromatograph. Their mass spectra were, however, quite simple even as a mixture. The major peaks of the mixture appeared at m/e 116, 102, 59, 58 and 45. From these data, it appeared that the compound with molecular weight 102 is N-methyloxalamide and the compound with the molecular weight 116 is N,N'-dimethyloxalamide. The latter compound was synthesized (scheme 3) and its mass spectrum contained major peaks only at m/e 116, 59 and 58, thus adding additional support for its presence. The m/e 59 ion arises from loss of methyl isocyanate from the molecular ion and the m/e 58 ion results from simple cleavage between the two carbonyl groups.

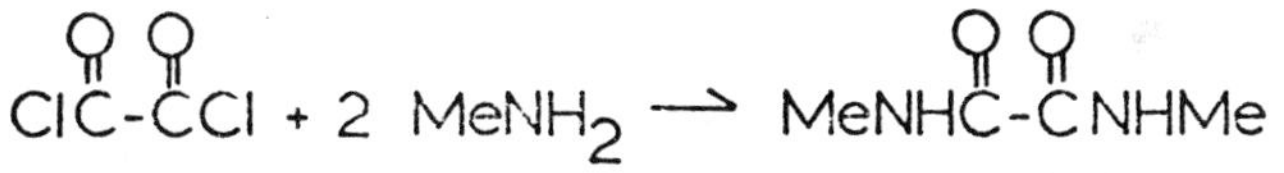

$$\text{ClC-CCl} + 2\ \text{MeNH}_2 \longrightarrow \text{MeNHC-CNHMe}$$
(with two C=O groups on each side)

oxalyl chloride   methyl amine   N,N'-dimethyloxalamide

Scheme 3. Synthesis of N,N'-dimethyloxalamide.

Molecular Weight 157.

Two compounds with molecular weight 157 are produced in the ozonization of caffeine. From the intensities of the ^{13}C isotope peaks in the mass spectra, these two products appear to be isomers with a molecular formula of $C_5H_7N_3O_3$. The major mass 157 component elutes faster from the gas chromatography column and its dominant mass spectral features are (1) strong molecular ion, (2) loss of methyliso-cyanate (M-57, m/e 100), and (3) secondary loss of formaldehyde (M-(57 + 30), m/e 70). The mass spectrum of the minor 157 component shows (1) a weak molecular ion, (2) dominant loss of an oxygen atom (M-16, m/e 141), and (3) a strong m/e 44 ion (H_2NCO^+). At the time of this writing neither of these substances have been isolated as a pure substance and therefore no NMR spectra have been obtained. Some possible structures for the mass 157 components are shown in scheme 4.

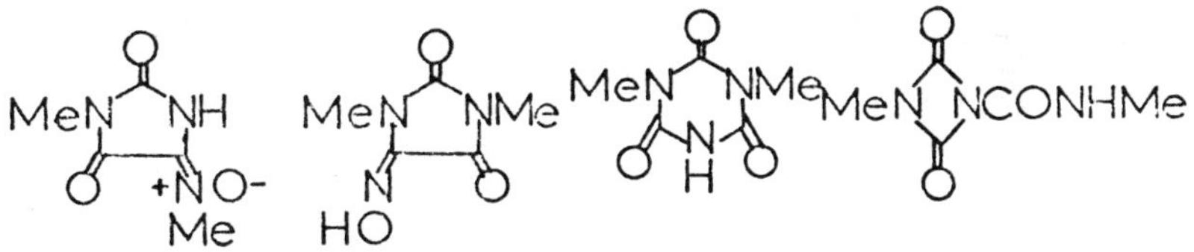

Scheme 4. Possible structures for the two mass 157 products.

Structure I is a nitrone and would be expected to expel an oxygen atom in the mass spectrometer. Structure IV is a diacyl derivative of methylurea and would be expected to lose methylisocyanate upon electron impact. Structure II is the mono-oxime of dimethylparabanic acid and Structure III is the Beckmann rearrangement product of this oxime. Work is still in progress to identify these two components.

Molecular Weight 198.

The major product from the ozonization of caffeine in aqueous solution has molecular weight 198. This substance was obtained as a pure compound by column chromatography of the reaction product mixture and showed m.p. 175-176°. The PMR spectrum of the compound dissolved in $CDCl_3$ showed four signals: (1) singlet at 3.73 ppm, (2) singlet at 3.40 ppm, (3) doublet centered at 2.97 ppm and (4) a broad peak centered at 8.05 ppm. The relative areas of the four peaks are 3:3:3:1. Addition of deuterium oxide to the PMR solution resulted in the disappearance of the broad downfield signal and the collapse of the upfield doublet to a singlet. These data were interpreted as resulting from three nonequivalent N-methyl groups, one of which being attached to an amide NH (i.e., $-CONHCH_3$).

The CMR spectrum contained seven peaks, each corresponding to one carbon atom. The combination of mass spectral and NMR data indicated a molecular formula of $C_7H_{10}N_4O_3$. The three methyl groups absorbed upfield and the four other CMR signals were observed downfield in the carbonyl region. Table 1 shows a comparison of these CMR signals with those of caffeine and dimethylurea.

Table 1

Comparison of CMR Signals from Caffeine, Dimethylurea and the compound with M.W. = 198

	2	4	5	6	8	1-Me	3-Me	7-Me
Caffeine	155.3	148.6	107.5	151.5	141.4	27.8	29.7	33.5
Dimethylurea ($Me_1HN-CO-NHMe_1$)	160.5					26.7		
Compound ($Me_1N-CO-N(Me_3)-C(=NMe)-C-NHMe_3$)	154.5	151.1	159.7	156.1		26.5	33.5	29.3

- 288 -

The NMR data, as well as the mass spectral behavior, support the structure shown in the Table. An X-Ray crystallographic study of this substance was undertaken. A sample crystallized from chloroform yielded X-Ray data which confirmed the atomic sequence of the proposed structure. A search of the literature revealed that this compound had not been described by others.

Molecular Weight 210.

This substance constituted only 1% of the product mixture. Its molecular weight is sixteen mass numbers higher than that of caffeine. Initially it was thought that C-8 of caffeine is oxidized to give 1,3,7-trimethyluric acid, but an independent synthesis of this compound showed it to be different than the ozonization product with the same molecular weight. Caffeine-N-oxide does not seem to be a likely candidate for this product, since earlier efforts by others to produce compounds of this type have been unsuccessful (5). The two double bonds present in caffeine appear to be the most probable sites of oxidation and for this reason an epoxide or oxaziridine may be suspected to be the product with molecular weight 210.

Molecular Weight 226.

This minor component corresponds to caffeine plus two oxygen atoms. The most reasonable structures for this compound seem to be a hydroperoxide or a diketone resulting from complete oxidation of the 4-5 double bond in caffeine.

Summary

Ozonization of caffeine in aqueous solution gave a complex mixture of products. It appears that ozonization of whole wastewater samples, which may contain hundreds of organic compounds, will yield a product mixture containing many more components. Therefore, it seems necessary to ozonize identified wastewater components as pure compounds and determine the structure of the products if the very complex mixtures produced from ozonized wastewater samples are to be completely analyzed.

Acknowledgements

We are grateful to Messrs. W. D. Ross and M. Wininger of Monsanto Research Corporation and Professor Ray Fall of the University of Colorado for performing Ames tests. We also wish to thank Messrs. H. Pahren and R. Andrew of the EPA for their advice and guidance. This research was supported by EPA Grant Number R-804472-01, for which we are extremely grateful.

References

(1). Identification and Analysis of Organic Pollutants In Water, Lawrence H. Keith, Ed., Ann Arbor Science Inc., Ann Arbor, Michigan, 1976.

(2). Beilsteins Handbuch Der Organischen Chemie, Vol. 26, p. 464.

(3). R. Leipen, *Monatsh.*, <u>10</u>, 184 (1889).

(4). B. N. Ames, *et al.*, *Proc. Nat. Acad. Sci. USA*, <u>72</u>, 5135 (1975); *ibid*, <u>73</u>, 950 (1976).

(5). A. A. Watson and G. B. Brown, *J. Org. Chem.*, <u>37</u>, 1867 (1972).

REACTION OF ORGANICS NONSORBABLE BY ACTIVATED CARBON WITH OZONE

W.A. Guirguis, Y.A. Hanna, R. Prober*, T. Meister and R.K. Srivastava
Cleveland Regional Sewer District
801 N. Rockwell
Cleveland, Ohio 44114

ABSTRACT

Physical-chemical treatment has been investigated at the Cleveland Regional Sewer District's Westerly Wastewater Treatment Facility since 1974. The physical chemical treatment flow scheme includes the application of ozone for disinfecting the effluent for the pilot plant. The presence of certain organic species from industrial sources has negative effects upon carbon adsorption. These organics which exhibit poor adsorption characteristics were present in significant amounts in the effluent from the carbon sorption unit process. This resulted in unacceptable effluent quality. Further, these organics reacted with ozone during the disinfection process, resulting in poor disinfection results.

Ozone pretreatment prior to the carbon adsorption process was then investigated with the objectives of improving the adsorption efficiency of the carbon.

---

*GMP Consultant

This paper presents the findings of the pilot studies and suggests a treatment process that resolves these treatment problems.

INTRODUCTION

The Cleveland Regional Sewer District is carrying out a pilot program at the Westerly Wastewater Treatment Plant to investigate physical-chemical treatment (PCT) for combined municipal and industrial wastewater, with ozone application as an integral unit process in the flow scheme. The program includes the operation of a pilot plant that simulates, on a 30 gpm scale, the sequence of unit processes in the 50 mgd capacity plant now under construction. Except for the ozonation, design criteria were developed in earlier pilot studies carried out in 1970 and 1971 (1). The objectives of the renewed pilot studies were: to investigate ozone effectiveness for disinfection and for polishing reduction of residual organics, to develop design and operational control data for the ozonation process, and to gain assurance that the proposed flow scheme would meet treatment objectives.

The pilot plant is shown schematically in Figure 1. In brief, the treatment consists of single stage lime precipitation with polymer addition as a supplemental flocculant, clarification to separate the chemical sludge, recarbonation for pH adjustment to near-neutral level, pressure filtration to further enhance solids removal, activated carbon adsorption for soluble organics removal and ozone disinfection as the final step. The pilot plant has been in operation since early 1974. In this period of nearly three years, the pilot studies have been modified in order to overcome several problems which were encountered. One of the major problems was the presence of non-sorbable organics, i.e., which exhibit poor sorption characteristics on the activated carbon. This paper presents an evaluation of the impact of non-sorbable organics on ozone application with the proposed flow scheme.

BACKGROUND

In the initial phase of the pilot plant studies, ozone disinfection was found to be very sensitive to

the performance of upstream units, and it was reported
to cause an increase in both suspended solids and
soluble BOD in the ozonated effluent (2). Both of
these results are directly related to the sorption
efficiency of the activated carbon removal of soluble
organics. At this stage, chlorine disinfection
processes could have been effectively employed, but
the question of attaining other required parameters
(COD, BOD, etc.) still would have remained unresolved.

The next phase of the pilot plant studies was
then directed to assess the limitations of the treatment processes. It was found that the physical-
chemical treatment process is sensitive to the plant
influent wastewater characteristics, particularly the
industrial components (3). The most important problem
was the inability of activated carbon to remove
certain organic species. This can be seen by the
comparison of typical gel permeation chromatography
results for the activated carbon column's influent and
effluent. Figure 2 shows that the organics removed
appear to be mainly from the lower molecular weight
fractions.* The presence of the unadsorbed organics,
particularly the higher molecular weight fractions,
deteriorates effluent quality. Further, these organics
react with ozone in competition with the microorganisms, accounting in part for the unsuccessful disinfection results observed earlier. It was then suggested that some pre-treatment be investigated to
improve the carbon adsorption efficiency.

Ozonation <u>before</u> the pressure filter and the
activated carbon contactor was found to be an effective pre-treatment step in the next phase of the study
(4). It was found definitely to improve the quality
of the carbon column effluent, particularly with
respect to soluble organics. Apparently, reactions of
ozone with organics, e.g., in addition to cleavage of
carbon-carbon double bonds or free radical oxidations,
make the influent stream to the sorption process much
more amenable to treatment. The effect of ozonation
on the gel-permeation chromatograms as shown in

---

*the relative molecular weights indicated in this
figure are those of calibrating Dextrans. The actual
molecular weights of the various peaks may differ
considerably from the apparent values shown.

Figure 3 indicates an increase in ultraviolet absorption at 254 nm, as a result of changes in the functional groups of the organics present.

Since then, pre-ozonation has been accepted as an integral unit process in the treatment scheme for the full scale plant. The next phase of the pilot plant program was directed to determine the factors which were responsible for the improved performance resulting from pre-ozonation (5). This included not only the original flow scheme modified for pre-ozonation (before the pressure filters and the activated carbon columns), but also parallel control systems to isolate effects due to: oxygen-rich conditions, biological growth and the pre-treatment itself as a function of other related variables.

The earlier results indicated that pre-ozonation enhances effluent quality to the extent that the treatment objectives can be met consistently. Further benefits not originally anticipated were achieved. These are: extended operating life and working capacity of the activated carbon with resultant high cumulative organic removal, and prevention of sulfide formation in the activated carbon column, thus eliminating the associated complications.

These improvements were found to be brought about by two factors. First, ozone pre-treatment apparently renders certain organics more biodegradable, and there is significant organics removal in the pressure filters preceding the activated carbon columns by means of biological assimilation. This reduces considerably the organic loading on the activated carbon column, which explains in part the extended operating life of the carbon. Second, the ozone pre-treatment apparently also renders other organics more readily sorbable by activated carbon. In fact, the nature of the activated carbon treatment was changed to become a steady-state process. As can be seen in Figure 4, exhaustion of the activated carbon did not occur over some 13 months of service life through September 1976. High cumulative loadings were observed in excess of one pound of soluble COD per one pound of activated carbon.

Shortly after October 1, 1976, pre-ozonation in the pilot plant was temporarily suspended. Almost immediately, breakthrough conditions occurred, as can be seen in Figure 4, (at the far right). The activated carbon column, in fact, was completely exhausted by the end of October 1976, so that the influent and effluent concentration of soluble COD were identical.

EXPERIMENTAL STUDIES

The pilot plant studies are continuing with the objectives now to develop design criteria for the modified scheme and to refine the current knowledge of all of the mechanisms cited previously. The non-sorbable organics and their reactions with ozone are singled out as the first aspect profiles measured as chemical oxygen demand (COD), as well as more detailed characterizations such as gel permeation chromatography, coupled gas chromatography-mass spectroscopy, etc. Only the COD studies are reported in this presentation.

Figure 5 shows the pilot plant as modified for the present phase of study. Pre-ozonation before the pressure filter and activated carbon column is now absent from the principal flow stream. Split off from the principal stream immediately after the pressure filters are three activated carbon columns, (C_1, C_2 and C_3), in series, allowing 10 minutes contact time each. Arbitrarily, the total of 30 minutes contact time is taken as a criterion for sorbability. That is, whatever organics remain in the effluent from column (C_3) are defined to be the non-sorbable fraction. This arrangement permits the COD concentration profile as a function of contact time. The effluent from C_3 is introduced into two further treatment systems operating in parallel. The first of these has ozonation followed by two additional activated carbon columns (OC_1 and OC_2), in series, allowing 15 minutes contact time each. There is a parallel single activated carbon column (C_4) with 30 minutes contact time, which serves as a control to isolate the effect of the additional contact time alone, i.e., without the ozonation. Copper sulfate was added to all of these systems to inhibit biological growth in the activated carbon columns. Profile measurements of dissolved oxygen and hydrogen sulfide concentrations verify that there is no biological activity within the columns.

RESULTS

For the purpose of evaluating the carbon adsorption efficiency as a function of contact time and cumulative organics removal loading, columns C_1, C_2 and C_3 are taken together as a single column, COD_f, and measurements are selected to be control parameters for process evaluation. This is due to the significant COD changes resulting from ozonation, whereas other parameters such as total organic carbon (TOC) are not affected. Results obtained by Eisenhauer (6) with pure compounds indicate the same effect.

In Figure 6, the effluent COD as a function of cumulative loading quickly reached a plateau, indicating there is no sign of a breakthrough. The cumulative loading as a function of throughput continues to increase linearly. Hence, there is still capacity available for adsorption. Thus the residual organics can be said to be those fractions defined earlier as non-sorbable organics.

Concentration profiles for COD are shown in Figure 7 plotted as cumulative frequency diagrams. (Displacement upward and toward the left indicates increasing removal of COD). There is a steady trend of decreasing COD through C_1, C_2, and C_3, but the increment removed is smaller with each succeeding column. The line for C_3 represents not only the effluent from the column, but also the data for the effluent from ozonation. That is, ozone pre-treatment affected the COD concentration at that point. However, it does have a significant effect on the sorption characteristics of certain organic species. The difference between the lines for OC_2 and C_4 is due to the ozone reactions rendering "non-sorbable" organics more readily sorbable.

The same COD data replotted in Figure 8 show that the variance (relatively flat slopes) in Figure 7 is attributable largely to day-to-day variations in the influent quality. The difference between C_3 and C_4 and OC_2 concentrations of COD are clearly seen.

CONCLUSIONS

Based on all the results, one can say that ozone pre-treatment enhances physical-chemical treatment efficiency by reactions with non-sorbable organics.

The improvement is due in part to reaction products being much more amenable to carbon adsorption. Further studies would be required to differentiate realistically between "rate and equilibrium phenomena" as the mechanics responsible for improved adsorption efficiency.

It is believed that the optimum treatment will involve a trade-off of ozone dosage vs. additional contact time.

LITERATURE CITED

(1) Zurn Environmental Engineers & Battelle Northwest, "Westerly Advanced Wastewater Treatment Facility-Process Development and Engineering Design", report prepared for the City of Cleveland, (June, 1972).

(2) Guirguis, W., et al., "Ozonation Studies at the Westerly Treatment Plant", Proc. 2nd International Symposium on Ozone Technology. R.G. Rice, P. Pichet & M.A. Vincent, Editors. International Ozone Institute, Syracuse, N.Y. (1975), p. 611-630.

(3) Guirguis, W., Melnyk, P.B., and Harris, J.P., "The Negative Impact of Industrial Waste on Physical-Chemical Treatment", presented at the 31st Purdue Industrial Waste Conference, (May, 1976).

(4) Guirguis, W.A., Jain, J.S., Hanna, Y.A., and Srivastava, P.K., "Ozone Application for Disinfection in the Westerly Advanced Wastewater Treatment Facility", Forum on Ozone Disinfection, E. Fochtman, R.G. Rice and M.E. Browning, Editors, International Ozone Institute, Syracuse, N.Y. (1977), p. 363-381.

(5) Guirguis, W.A., Cooper, T., Harris, J.P., and Ungar, A.T., "Improved Performance of Activated Carbon by Pre-Ozonation," presented at the 49th Annual Conference, Water Pollution Control Federation, (October, 1976).

(6) Eisenhauer, H.R., "Ozonization of Phenolic Wastes," J. Water Poll. Control Fed, 40:1887-1899 (1968).

DISCUSSION

<u>Howard Kwong</u>, University of Missouri at Rolla: I have three questions. One, would you indicate the capacity of the wastewater treatment plant? Secondly, with that size of plant, have you made an estimate if ozone is used as a pretreatment before the carbon adsorption, how much will the treatment cost in terms of O&M? Thirdly, yesterday and this morning there have been a few papers which indicate that at a higher pH, because of the formation of free radicals the splitting of organic chemicals would be more effective with ozone. Do you foresee that if ozone is applied before the recarbonation, that you can probably cut down on the ozone dosage to achieve the same pretreatment efficiencies?

<u>Prober</u>: Let me attempt the first two and we'll send the third to Mr. Guirguis, who is the Research Director for the Cleveland Regional Sewer District.

In terms of costs, we have not yet arrived at a figure regarding the optimum. In fact, we have a consulting engineering organization working day and night to try to refine those numbers. Would you repeat your third question?

<u>Kwong</u>: The third question is, if you apply ozone before recarbonation, would you anticipate that the dosage for carbon could be cut down?

<u>Guirguis</u>: You are speaking of the formation of hydroxyl radicals which will react with the organics?

<u>Kwong</u>: That's correct.

<u>Guirguis</u>: Okay, we must also have some equation for ozone's reaction with hydroxyl ions and the demand for ozone would increase also to satisfy the ozone-hydroxyl reaction, as stated by Stumm. So at this phase we have two ways to go: either a free radical reaction or a selective reaction in the acidic range of ozone with certain organics.

I don't believe that the technology is available with this hard data, but we are attempting to work in the acidic range and the alkaline range to see what are the reaction mechanisms at the same time.

Prober: I will add briefly, that at the pH prior to recarbonation we would indeed be into the free radical reactions, but there would be also a parasitic ozone demand due to the decomposition of ozone, and so I think, really, we would have to rely on experimentation. I would hesitate to make a guess.

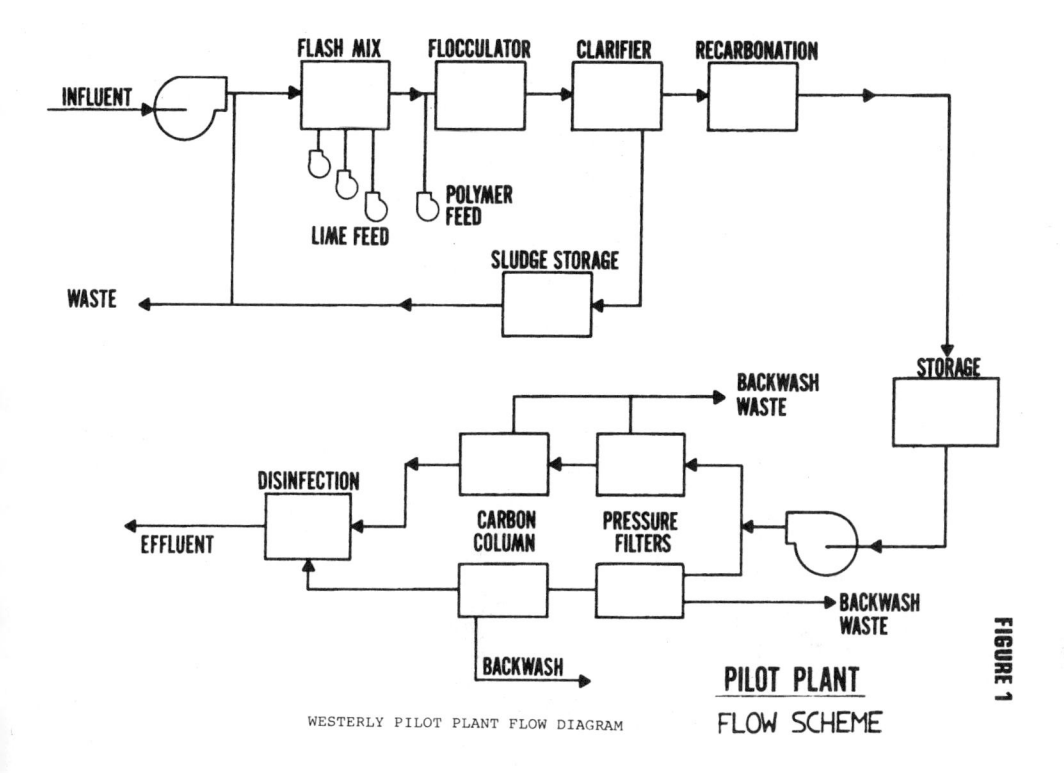

WESTERLY PILOT PLANT FLOW DIAGRAM

PILOT PLANT FLOW SCHEME

FIGURE 1

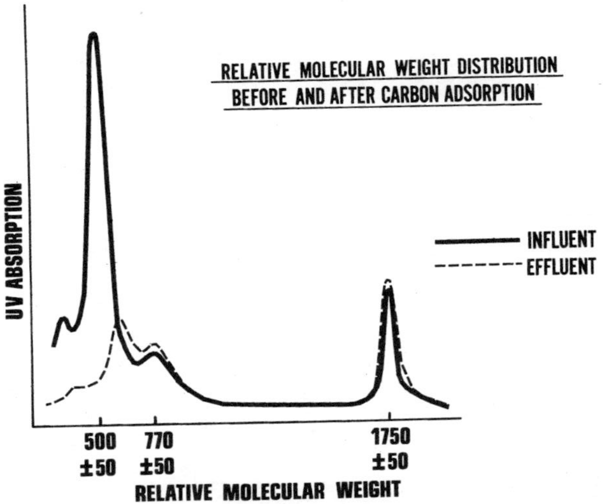

-300-

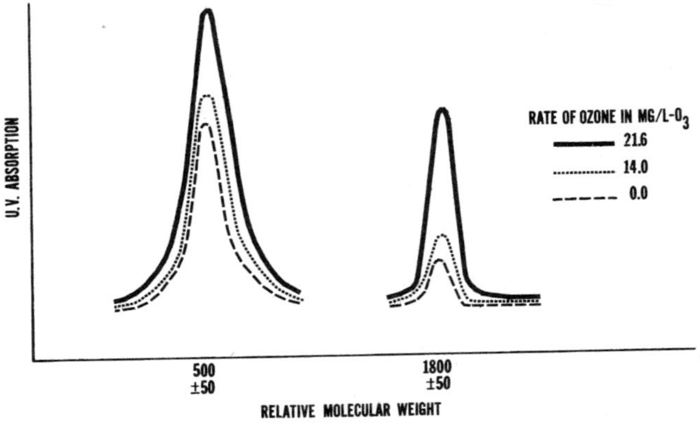

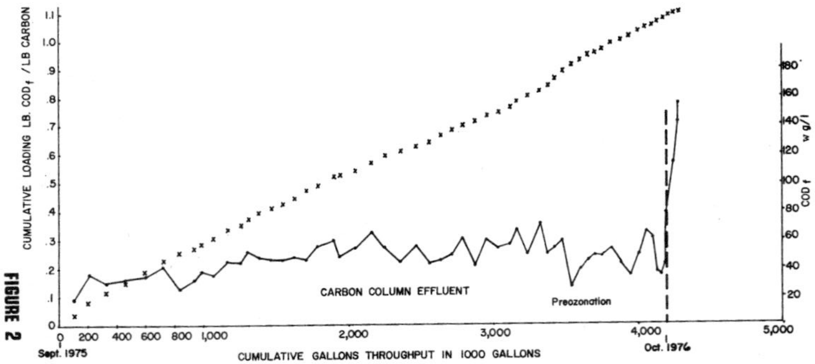

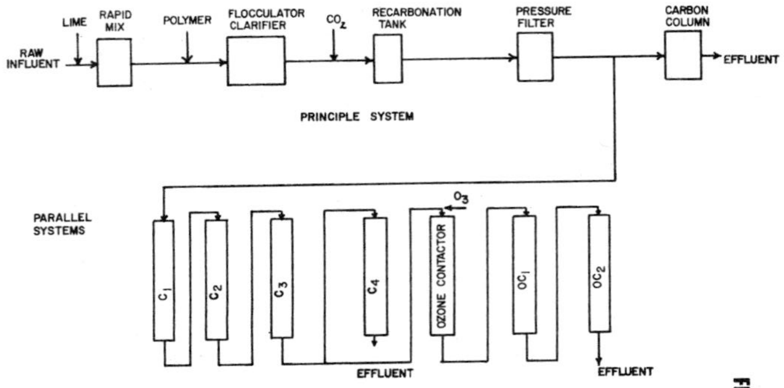

FIGURE 8

COD PROFILE OF THE CARBON COLUMN EFFLUENTS

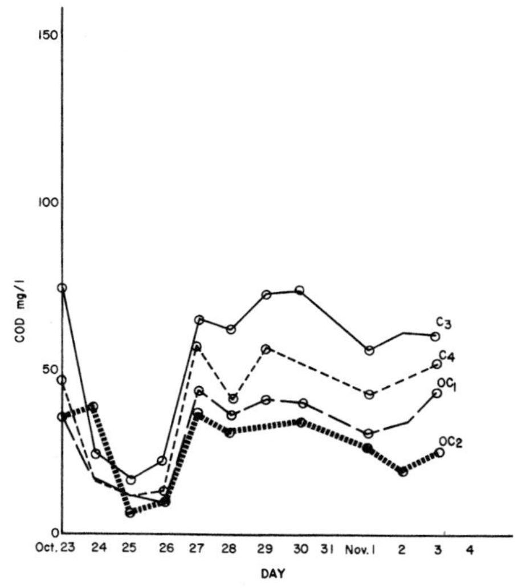

COD_f PROFILE FOR C_3, C_4 and OC_2 SYSTEMS

FIGURE 7

FREQUENCY DISTRIBUTION OF EFFLUENT COD

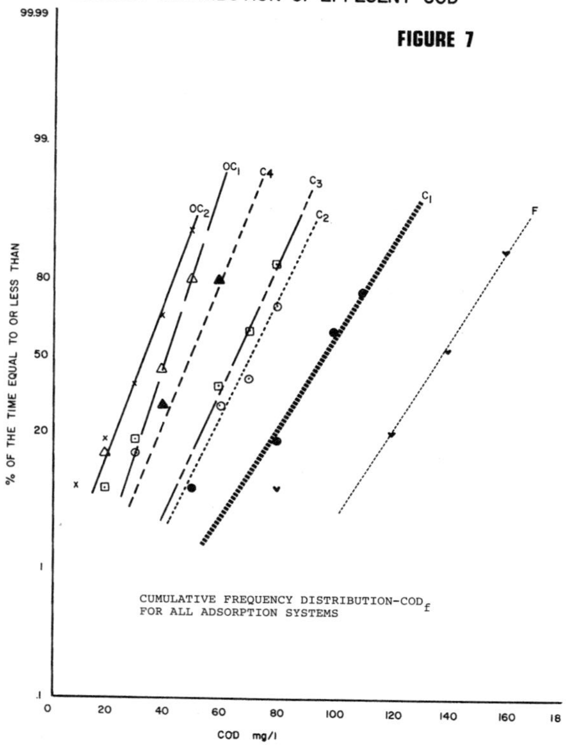

CUMULATIVE FREQUENCY DISTRIBUTION-COD_f FOR ALL ADSORPTION SYSTEMS

FIGURE 6

PERFORMANCE OF CARBON COLUMN SYSTEM C_1 C_2 C_3

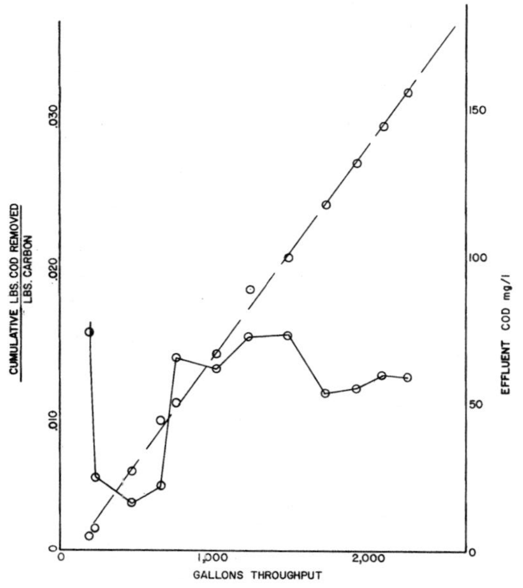

OZONE/UV OXIDATION OF PESTICIDES IN AQUEOUS SOLUTION

H. William Prengle, Jr. and Charles E. Mauk
Houston Research, Inc.
10600 Shadow Wood Drive, Suite 211
Houston, Texas 77043 USA

Abstract

This work establishes the feasibility of using ozone with ultraviolet radiation for the oxidative destruction of four pesticides: malathion, Baygon, Vapam, and DDT in aqueous solution. Each was treated by ozone with various combinations of UV intensity and temperature in a sparged stirred tank reactor. All were destroyed, as well as subsequent partial oxidation products, to below TOC detectable limit. Postulated mechanisms are presented which indicate the chemical routes from initial to final oxidation species. O_3/UV phot-oxidation of dissolved M-species occurs primarily by a combination of O_3 photolysis and M photolysis. For low level UV absorbers the oxidation rate is controlled by the former; for high level absorbers both mechanisms participate. The O_3/UV oxidation process is unequaled in destroying refractory and toxic organic species.

Introduction

In the past, the extensive use of pesticides has resulted in the pollution of water bodies with accompanying danger to both aquatic and human environments. Of particular interest are organic compounds containing: phosphorus, sulfur, and chlorine, which fall into the classifications of organo-phosphorus, carbamate, and chlorinated organics.

This paper presents oxidation rates under a wide variey of conditions, reaction mechanisms and postulated products for the following four compounds: malathion (organo-phosphorus), Baygon (carbamate),

Vapam (thiocarbamate), and DDT (chlorinated). Possible reaction mechanisms and oxidation species are proposed. In addition, a discussion is presented of the O_3/UV photo-oxidation mechanisms and a theory presented for rate controlling steps.

The work is a continuation of previously reported work (1,2,3,4) on the development of an advanced chemical oxidation water treatment system for the destruction of hazardous and refractory materials, which goes beyond oxidation with ozone alone.

Reactivity Characterization of Compounds

As a result of this and previous work, there are several ways of characterizing the reactivity of refractory type compounds in an O_3-UV reaction system, prior to actual reaction studies:
1) - by the refractory index (RFI) method (2);
2) - by the UV-spectrogram-specific absorbance as a function of wave length (180-400 nm.) (4); and
3) - by a molecular structure analysis.

For the four compounds presently considered, comments relative to the above three methods are pertinent.

The Refractory Index (RFI) method measures the difficulty of oxidation of a given compound by ozone alone. To date some 23 compounds and some seven mixed aqueous wastes have been treated. Table 1 lists most of the individual compounds: it will be noted that the four compounds in question range from refractory to very highly refractory, RFI = 37 → > 1000.

The UV-Spectrogram method provides a direct measure of the specific absorbance ($\bar{A}$) as a function of wave length in the range of interest, and also a comparison with known, good absorbers. This information is useful in selecting the best UV-radiators to be specified in a reactor design. Spectrograms for three of the four compounds studied, as well as those for pentachlorophenol, ferricyanide ion, and o-dichlorobenzene, are presented in Figure 1. Ferricyanide ion has been chosen as the reference compound on the plot and 10^3 cm^2/m.mole as a reference level, which is also the level of $\bar{A}$ for O_3 at 250 nm.

A molecular structure analysis is useful in determining reactive sites in the molecule, by comparison

with other known reactive sub-groupings. For example, the following are vulnerable to attack, by ozone enhanced with UV absorption:
1) exposed halogen atoms, e.g. chloride ion appears rapidly, electronegative atoms distort charge distribution;
2) unsaturated resonant carbon ring structures;
3) readily accessible multibond carbon atoms; on the other hand shielded multibonds, sulfur and phosphorus are much less vulnerable;
4) alcohol and ether linkages

Table 2 summarizes pertinent molecular characteristics of the compounds studied.

Oxidation of the Individual Compounds

The experimental ozonation runs were carried out in stirred batch, continuously sparged gas, liquid phase reactors. The equipment and experimental procedures have been described previously (1,2). The ultraviolet input (UV) levels referred to are given as watts of useful UV light per liter. A stoichiometric limiting line represents the theoretical maximum rate at which a compound could be oxidized for the particular O_3 rate to the reactor.

GC analyses were made to follow the disappearance of the original species; TOC analyses follow the destruction of intermediate oxidation products; and sulfate and phosphate analyses determine whether the sulfur and phosphorus atoms went to these or other forms. The pH measurements followed the formation and destruction of organic acids and the permanent formation of sulfuric and phosphoric acids.

Malathion is S-{1,2-dicarbethoxyethyl}-O,O-dimethyldithio phosphate,

$$\begin{array}{c} CH_3O \\ \diagdown \\   P-S-C-C-O-C_2H_5 \\ CH_3O \diagup | \\ HC-C-O-C_2H_5 \\ H \| \\ O \end{array}$$

with S double-bonded to P, H and O on the first carbon (O double bonded).

Figure 2 summarizes some of the malathion runs. The stoichiometric limiting line represents the theoretical rate at which TOC could be destroyed as ozone is added into the reactor, in accordance with the following chemical reaction,

$$C_{10}H_{19}S_2PO_6 + 64\ O_3 \rightarrow 19H_2O + 20CO_2 + 4\ SO_3 + P_2O_5 + 64\ O_2$$

indicating that, b=4.65 mg O_3/mg malathion. No TOC is destroyed if ozone alone or UV alone is used, but TOC is slowly destroyed if only a small amount of UV is used with ozone. The rate of TOC destruction is enhanced by elevated temperature, but greater enhancement is obtained by additional UV without elevated temperature: greatest enhancement is obtained by a combination of temperature and high UV input.

The GC data show that any ozone run has essentially destroyed malathion by the time the first sample of a run was taken, even when no UV was used. Malathion is also destroyed by UV alone, but takes about two hours for complete destruction, while ozone or O_3/UV combination requires only 15 minutes for complete destruction. Complete destruction of the malathion decomposition products, as evidenced by the TOC, takes both ozone and UV together.

Phosphorus makes up 9.4% by weight of the malathion molecule. The orthophosphate results indicate that if there is no significant destruction of organic material, as represented by TOC, there is no significant formation of orthophosphate. Where there is significant destruction of organic material, the amount of orthophosphate formed appears to correspond to the phosphorus associated with the TOC destroyed.

The malathion molecule contains two ester linkages, and when they hydrolyze, two ethanol molecules and an iso-substituted succinic acid will be formed. The succinic acid molecule is similar to two acetic molecules back-to-back. Figures 3 and 4 compare the desstruction by ozone-UV of TOC from malathion, ethanol and acetic acid (2) at room temperature and at 50°C. It is interesting to note that of the many materials studied only malathion and acetic acid show no reduction of TOC by ozone alone. As would be expected, Figure 3 shows the carbon loss from malathion at 25°C to be slowest of the three, because it must first go to ethanol and acidic forms, and the ethanol slower than acetic acid because it oxidizes to form acetic acid. However, Figure 4 indicates that at elevated temperature some other mechanism becomes more active, with malathion fastest, ethanol next, and acetic acid the slowest. One possibility is that ethanol starts to loose significant carbon through the path: ethanol to ethylene glycol to oxalic acid to carbon dioxide. In addition, malathion can lose carbon through the grouping H_3COP going to methanol, which is then further oxidized. Several simultaneous reaction mechanisms shown in Table 3 can be postulated for malathion.

Baygon is 0-isopropoxyphenyl methyl carbamate,

The data indicate that ozone or ozone-UV quickly destroys the Baygon. UV alone slowly destroys the Baygon, but accomplishes almost no destruction of the oxidation products, as represented by TOC. Figure 5 shows the destruction of TOC from Baygon and the oxidation products. The stoichiometric limiting line represents the theoretical rate at which TOC could be destroyed in accordance with the following overall chemical reaction.

$$2C_{11}H_{15}O_3N + 53O_3 \rightarrow 22CO_2 + 15H_2O + N_2 + 53O_2$$

indicating that b=6.08 mg O_3/mg Baygon. It appears that destruction of TOC proceeds very satisfactorily at room temperature with ozone and 1.32 watts/liter UV or higher, and at 50°C with as little as 0.44 watts/liter UV.

The Baygon molecule contains an ester linkage, and when hydrolyzed, a methyl amino acid is formed that is similar to glycine, NH_2CH_2COOH. Figure 6 compares the destruction by ozone-UV of TOC from Baygon and glycine (2), and shows similarity in the destruction of TOC from both molecules, but also shows that the Baygon destruction is enhanced more than glycine destruction by elevated temperature and UV.

Loss of TOC while treating with UV only is very slight. Several simultaneous reaction mechanisms, shown in Table 4, can be postulated for Baygon; none lead to toxic intermediate oxidation products.

<u>Vapam</u> is sodium N-methyldithiocarbamate dihydrate,

$$CH_3-\overset{H}{\underset{|}{N}}-\overset{S}{\underset{\|}{C}}-S-Na \cdot 2H_2O$$

Figure 7 shows the destruction of TOC from Vapam and Vapam oxidation products. The stoichiometric limiting line represents the theoretical rate at which TOC could be destroyed as ozone is added into the reactor, in accordance with the following chemical reaction,

$$2C_2H_4NS_2Na \cdot 2H_2O + 25O_3 \rightarrow 4CO_2 + 7H_2O + 4SO_3 + N_2 + 2NaOH + 25O_2$$

indicating that b=3.63 mg O_3/mg Vapam. UV alone causes no destruction of TOC and ozone alone destroys only about 70% of the TOC. Ozone with UV is required for substantial destruction of TOC.

Vapam analyses show that Vapam itself is destroyed by ozone alone or ozone-UV very quickly; essentially none is left by the time the first sample is taken at 15 minutes. UV alone will very slowly decompose Vapam, but the ozone -UV combination is required for substantial destruction of the intermediate oxidation products, as represented by TOC. The sulfate found in the samples increases as the extent of TOC destruction increases; for ozone with 1.32 watts/liter UV, the sulfate found in the final sample corresponds closely to the total sulfur in the Vapam charged to the reactor.

Figure 8 shows time profiles for Vapam, pH, TOC and sulfate for a typical O_3-UV run. The rapid drop in pH implies that oxidation of the sulfur in the molecule is an early step in the mechanism, and leads to the postulated mechanism,

$$CH_3NH(CS)SNa \xrightarrow[H_2O]{O_3,h\nu} CH_3NHCOOH + 2SO_4^= + 3H^+ + Na^+$$

followed by,

$$CH_3NHCOOH \xrightarrow{O_3} CH_3OH + CO_2 + N_2 + H_2O$$
$$O_3 \downarrow \rightarrow HCOOH \xrightarrow{O_3} CO_2 + H_2O$$

DDT is dichlorodiphenyltrichloroethane, a compound of very low solubility ($\sim$ 0.1 mg/ℓ),

[structure: phenyl–CH(CCl₃)–phenyl–Cl]

The ozone demand of DDT was calculated from,

$$2\ C_{14}H_9Cl_5 + 65 O_3 \rightarrow 28 CO_2 + 9 H_2O + 5 Cl_2 + 65 O_2$$

corresponding to b=4.4 mg O_3/mg DDT. Methanol is used as a solvent for the DDT; this combined with low solubility essentially eliminates the use of TOC to track the oxidation products.

Figure 9 shows selected data for the destruction of DDT as followed by GC analysis. The expanded scale is for ease in labeling the lines, rather than pointing up any significance in the shape of the curves. As observed previously for other compounds, O_3-UV is more effective than either ozone or ultraviolet applied separately, and 1.32 watts per liter appears to be the highest intensity from which benefit can be derived. It was observed that for the O_3-UV combination, the DDT disappeared faster than the TOC from the methanol.

As with pentachlorophenol (4), DDT by GC analysis disappears very quickly when treated with ozone-UV and less quickly when treated with UV alone. The chlorinated ring structure leads to the postulation that as with pentachlorophenol, UV light causes the chlorine atoms on the rings to go to HCl while replacing them with hydroxyls. The hydroxyls strongly activate the ring for removal of side chains (the trichloroethane bridge) and for oxidation at the ortho position. This mechanism for structure carbons would be:

$$Cl-\phenyl-\underset{CCl_3}{\overset{H}{C}}-\phenyl-Cl \xrightarrow[H_2O]{h\nu} HO-\phenyl-\underset{CCl_3}{\overset{H}{C}}-\phenyl-OH \xrightarrow{O_3}$$

$$HO-\phenyl-OH \text{ and } HO-\phenyl\overset{OH}{\underset{OH}{}} \xrightarrow{O_3}$$

Maleic and Muconic Acids $\xrightarrow{O_3}$ $HOC\overset{O}{\|}-\overset{O}{\|}COH \xrightarrow{O_3} CO_2 + H_2O$

Also, the extremely low concentration of DDT can make no detectable contribution to pH or chloride ion, and these parameters can contribute nothing new to the elucidation of reaction mechanism, which has been previously discussed for pentachlorophenol (4). The structure of the trichloroethane bridge, similar to chloroform (4) implies a similar mechanism,

$$HCCCl_3 \xrightarrow{O_3} H\overset{O}{C}Cl_3 \xrightarrow{O_3} HOO\overset{O}{C}Cl_3 \xrightarrow{H_2O}$$
(peroxide)

$$H_2O + CO_2 + H^+ + Cl^- + COCl_2 \xrightarrow{O_3} CO_2 + Cl_2$$

Because DDT will be present in such low concentration, it is believed that any phosgene formed can not be present in hazardous concentration.

Photolysis-Oxidation Mechanisms and Rates

It has been established that oxidation rates of dissolved reactant species using O_3 with UV are 10 to 10^3 fold greater than with O_3 alone; Table 5 illustrates this comparison. Consequently, several questions are pertinent: 1) what are the operable photolysis, photosensitization, oxidation mechanisms, 2) does O_2 with UV mechanisms play a significant role, and 3) which mechanisms are rate controlling?

The photo-oxidation of molecular species using O_3/UV, in aqueous media- an important contributing factor-appears to occur by a combination of several overall initiation and propagation mechanisms:
1) - by production of highly oxidizing photolysis

species formed from UV absorption by O_3 molecules;
2) - by production of free radicals and excited state photolysis species formed from UV absorption by the reactant molecules (M-species); and
3) - by reaction with water species, H^+ and OH^- as well as photolysis species formed from the water.

All produce reaction by a combination of photolysis, photosensitization, hydroxylation, oxygenation and oxidation. The extent of each mechanism in enhancing the overall reaction depends on the relative amounts of UV absorption and the quantum yield.

For comparison, Figure 1 indicates ferricyanide and O_3 have approximately equal specific absorbance values, $\bar{A} \simeq 10^3$ cm^2/m.mole, and by the theory all three mechanisms are operating. On the other hand, for compounds with low $\bar{A}$ values, the first and third mechanism are predominant.

Consider the species produced from O_3 by ultraviolet (400 nm to 180 nm) radiation (5,6):

$O_3(S_0)$ + UV < 310 nm. → $O(^1D)$ + $O_2(^1\Delta g)$

+ 260 nm. → $O(^1D)$ + $O_2(^1\Sigma_g^+)$

+ 230 nm. → $O(^3P)$ + $O_2(^3\Sigma_u^+)$ → $3O(^3P)$

+ 196 nm. → $O(^1S)$ + $O_2(^1\Delta_g)$

+ 179 nm. → $O(^1S)$ + $O_2(^1\Sigma_g^+)$

For low pressure mercury lamps as the UV source, which produce radiation distributed about 254 nm, it would appear that $O(^1D)$, $O(^2S)$, and $O(^3P)$ are the available most oxidizing species, while $O_2(^1\Sigma_g^+)$ and $O_2(^3\Sigma_u^+)$ probably also contribute. The latter raises the question of the role of dissolved O_2.

As illustrated in Table 6, our experiments indicate that some lesser amount of reaction occurs using O_2 with UV, compared to O_3 alone. Disappearance of total TOC proceeds more readily with O_3 alone than with O_2/UV. Disappearance of the original reactant species occurs to some degree using O_2 with UV as a result of a combination mechanism: producing a trace of O_3, some activated oxygen and activated oxygen and H_2O species, and activated reactant molecule species. Oxygen absorbs UV(5) giving,

$$O_2(^3\Sigma_g^-) + UV < 245 \text{ nm.} \rightarrow 2\, O(^3P)$$
$$+ \quad < 195 \text{ nm.} \rightarrow 2\, O(^3P)$$
$$+ \quad < 176 \text{ nm.} \rightarrow O(^1D) + O(^3P)$$
$$+ \quad < 134 \text{ nm.} \rightarrow O(^2P) + O(^1S)$$

But higher frequency radiation ($\lambda < 176$ nm.) would be required to produce substantial $O(^1D), O(^2P)$, and $O(^3P)$ species, comparable to O_3/UV, for oxidation.

Also since the reactions occur in aqueous media, both ionic and photolysis species (5), produced from the water, participate,

$$H_2O + UV < 242 \text{ nm.} \rightarrow H(^2S_{\frac{1}{2}}) + OH(^2\Pi)$$
$$+ \quad < 136 \text{ nm.} \rightarrow H(^2S_{\frac{1}{2}}) + OH(^2\Sigma^+)$$
$$+ \quad < 124 \text{ nm.} \rightarrow H_2 + O(^1D)$$

As an example of the photolysis process involved for a simple reactant molecule, consider the case of phenol in aqueous solution. The compound has two substantial absorption peaks at approximately 270 nm and 220 nm. The primary photochemical processes involved have been described (5) as,

$$C_6H_5OH + h\nu \rightarrow C_6H_5OH^*$$
$$\updownarrow (H_2O)$$
$$C_6H_5O^{-*} + H_3O^+$$
$$\downarrow$$
$$C_6H_5O^- + h\nu'$$
$$\updownarrow (H_2O)$$
$$C_6H_5OH + OH^-$$

illustrating the production of excited state species, free radicals and water species, which participate in the oxidation reactions.

In summary, our present theory and view hold that the overall mechanism of O_3/UV photo-oxidation of M-species in aqueous solution occurs by a combination of,

- O_3 photolysis, to produce oxidizing O-species,

- M photolysis, to produce M* and free radicals,

and

- H_2O photolysis, to produce H and OH free radicals, along with H^+ and OH^-,

which participate in a sequence of oxygenation and oxidation reactions leading to the final oxidation products, CO_2, H_2O, etc. For M-species which are low UV absorbers, the rate controlling mechanism is O-species oxidation; whereas, for high UV absorbers M-species photolysis is also very important.

Conclusions

1. The ozone-ultraviolet process has been demonstrated extremely effective in the oxidation destruction of pesticides in water, and all their intermediate oxidation products, with all carbon in the molecules going to carbon dioxide.

2. The performance of O_3-UV is unequaled by any other process with the pesticide itself being destroyed in only a few minutes of contact time with the process. In general, elevated temperature enhances the reaction rate, but not as dramatically as using additional ultraviolet light.

3. The low pressure and medium pressure mercury vapor lamps appear to have an equivalent effect based on the actual ultraviolet light intensity per liter of solution.

4. Postulated mechanisms for the O_3/UV oxidation of the four subject pesticides indicate no intermediate or partial oxidation species of significant environmental concern.

    5. The O_3/UV photo-oxidation of dissolved M-species occurs primarily by a combination of O_3 photolysis and M photolysis. For low level UV absorbers the oxidation rate is controlled by the former; whereas for high level absorbers both mechanisms participate.

    6. A properly designed multistaged reactor system, with ozone and ultraviolet light, can accomplish complete removal of pesticides from water, and based on these results, designs can now be developed for prototype units to treat potable water or effluents in the field.

LITERATURE CITED

(1) R. L. Garrison, C. E. Mauk and H. W. Prengle, Jr., "Advanced O_3-oxidation system for complexed cyanides," in Proc. First Intl. Symposium on Ozone for Water & Wastewater Treatment, R. G. Rice & M. E. Browning Editors. Intl. Ozone Inst., Cleveland, Ohio (1975), p. 551-577.

(2) H. W. Prengle, Jr., C. G. Hewes, III and C. E. Mauk, "Oxidation of refractory materials by ozone with ultraviolet radiation," Proc. Second Intl. Symposium on Ozone Technology, Intl. Ozone Inst., Cleveland, Ohio (1976), p. 224-252.

(3) H. W. Prengle, Jr., C. E. Mauk, R. W. Legan, and C. G. Hewes, III, "Ozone/UV process, effective wastewater treatment," Hydrocarbon Processing, 54(10):82-87 (1975).

(4) H. W. Prengle, Jr., C. E. Mauk, and J. E. Payne, "Ozone/UV oxidation of chlorinated compounds in water" International Ozone Institute Forum on Ozone Disinfection, Chicago, (1976).

(5) J. G. Calvert and J. N. Pitts, Jr., Photochemistry, Wiley, New York, (1966).

(6) A. U. Kahn, "Singlet molecular oxygen. A new kind of oxygen", J. Phys. Chem., 80:2219-2228 (1976).

Acknowledgement

   This paper is based on work performed under Contract DAAG-53-76-C-0089 with U. S. Army Mobility Equipment Research and Development Command (MERADCOM), Fort

Belvoir, Va.

Ultraviolet scans were made by Dr. R. A. Geanangel of the University of Houston Chemistry Department.

DISCUSSION

Maggiolo: Your paper is a fine paper and you show that now you can break things up with UV that can't be broken down alone by ozone. I like this very much. You went to free radicals, as you are showing you are doing, or something with the UV involved.

But what happens, could you comment, if you had raised the pH and used that in lieu of the UV? Could you comment as to which might be the better way? I imagine you looked at that.

Prengle: We followed pH and this previous study is reported in the IOI Montreal proceedings. In our judgment, and by evaluation of the data, pH increase is not as effective as is the UV. We steer away from pH because we work in the industrial environment and, except for minor adjustments in pH, we would like to let the pH float where it will go, rather than using it as a technique of oxidation. Now, I agree that you can expect, and there is demonstrable proof, that the pH will have an effect. I don't think, though, that pH can be compared with producing free radicals by high energy radiation.

Maggiolo: So really what you are saying is, that on a cost basis, at this time you would rather stay with the UV even though the industries could stand the cost of raising the concentration of OH and reducing it again; it is probably less costly using UV. This is your answer?

Prengle: There may be cases where you have to do pH adjustment, so that what you are suggesting would be included.

Walter Blogoslawski, National Marine Fisheries: Dr. Prengle, do you have any information on the ozone oxidation of Kepone with UV?

Prengle: No, I don't.

Andy Pincon, Ionization, Inc.: Doctor, concerning the cyanide removal, have you a cost per pound of cyanide removed with UV/ozone and the dose rates required?

Prengle: Yes, but I can't quote the cost per pound right off. If you will see me after I'm finished here I'll talk about the Hughes unit with you. As for the dose rate, there is a theoretical amount of ozone required for every substance, so when you have a mixed feed, such as is present in the case of industrial wastewater, you just determine that amount. Then, in the design, there is a factor for the percent utilization, so that combined with the theoretical amount gives us, for a particular design, the dose rate. I could give the theoretical amounts for different compounds.

Unknown: Could I ask the question in a different way? Could you tell us for the Hughes unit the ozone capacity, the UV capacity, and the peak range of cyanide that you are trying to destroy?

Prengle: Well, I'll tell you some of that. The ozone unit is a PCI Ozone unit operating on air. It produces 88 lbs/day of ozone. The inlet cyanide level is 200-300 mg/l, the effluent level is below the detectable limit in cyanide, 0.1-0.2 mg/l. The UV is nominal, we are not talking about large energy inputs. The figures are roughly in the vicinity of 1 watt per liter, which is not very much energy.

Mathilde Kland, Lawrence Berkeley Lab: I wasn't clear on the procedure that you used in ozone, oxygen and UV, and how you compared it. Did you compare rates? And were the conditions comparable?

Prengle: Yes. I will have to refer you to our previous papers in which we have described the reactor we used and the method we used to put the ozone in. It's in the Montreal paper and in the Chicago paper, and you can get that detail. The oxygen and the UV were put in in the same way, in general, that you would conduct the ozone/UV experiment.

Kland: Are the products that you get from oxygen intermediate products in the ozonation? The intermediate products, for instance, from DDT?

Prengle: We would expect similar products, but the diagram of the primary photochemical species, which is what these are---those little tables taken from Calvert and Pitts, shows that you get more of the atomic oxygen from ozone with UV in the range of UV that we are using than you do from oxygen. So it's not as good. But where it oxidizes I would expect the intermediates to be either similar or the same.

Kland: But you didn't isolate intermediates.

Prengle: No.

TABLE 1 - RFI VALUES FOR COMPOUNDS

Compound	RFI* Value	Qualitative Scale
1) Potassium cyanide (KCN)	0.41	Slightly Refractory (RFI < 1.0)
2) Phenol	0.44	
3) Chloroform	0.53	
4) Color (α-units)	0.66	
5) Complexed Cd-cyanide	0.96	
6) Pentachlorophenol	1.6	Refractory (RFI = 1 → 100)
7) Ammonium ion	8.	
8) Simulated Medical Waste	13.	
9) Glycine	19.7	
10) Palmitic Acid (NH_4 Salt)	27.3	
11) Nitrosodimethylamine	31.	
12) Baygon	37.	
13) Dichlorobutane	56.	
14) Methanol	88.	
15) Vapam	102.	Highly Refractory (RFI = 100 → 1000)
16) Glycerol	112.	
17) o-Dichlorobenzene	113.	
18) Polychlorinated Biphenyls	200 (est.)	
19) Ethanol	245	
20) Complexed Ferricyanide	270	
21) DDT	297	
22) Acetic Acid	1000	Very Highly Refractory (RFI > 1000)
23) Malathion	> 1000	

*RFI = $\dfrac{B_c^O \, t_{1/2}}{A_O}$; ($B_c^O \equiv O_3$ supplied to $t_{1/2}$; $t_{1/2}$ = hours for 1/2 conversion of A_O; A_O = initial amount of compound.)

TABLE 2 – MOLECULAR CHARACTERISTICS OF COMPOUNDS

	Compound	M. W.	$b\left(\dfrac{mgO_3}{mgA}\right)$	$\bar{A}(\lambda)$ $\left(\dfrac{cm^2}{m\cdot mole}\right)$	$\Delta\lambda$ (nm)
1)	Malathion (liquid at 25°C)	$(CH_3O)_2(PS)SCH\ COOC_2H_5$ $\quad\quad\quad\quad\quad\mid$ $\quad\quad\quad\quad\quad CH_2COOC_2H_5$ 330	4.65	1.86×10^3;	200 – 270
2)	Baygon (solid at 25°C)	$CH_3NH(CO)OC_6H_4OCH(CH_3)_2$ 198	6.08	4.23×10^3; 0.561×10^3;	180 – 270 240 – 290
3)	Vapam (solid at 25°C)	$CH_3NH(CH)SNa\cdot 2H_2O$ 129	3.63	1.42×10^3; 0.078×10^3;	180 – 210 210 – 300
4)	DDT (solid at 25°C)	$ClC_6H_4(CHCCl_3)C_6H_4Cl$ 354.6	4.40	153×10^3;	190 – 280

TABLE 3 - MALATHION REACTION MECHANISMS

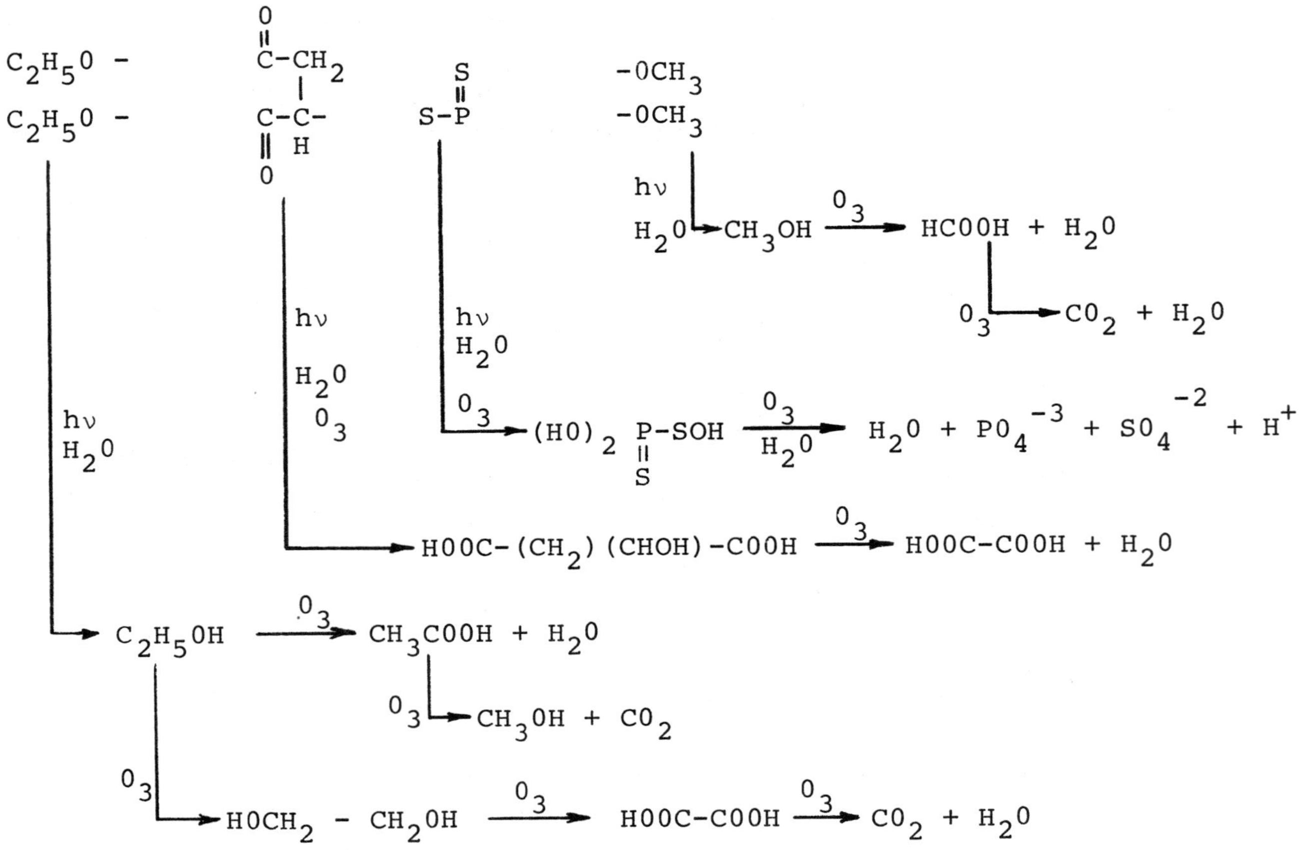

$$C_2H_5O- \underset{C_2H_5O-}{\overset{S}{\underset{\|}{\|}}} P-S- \underset{O=C-H}{\overset{O=C-CH_2}{|}} \xrightarrow[H_2O]{h\nu} \underset{C_2H_5OH}{} + \underset{O}{\overset{S=P}{\underset{\|}{\|}}} \underset{-OCH_3}{\overset{-OCH_3}{}}$$

$$\underset{O=C-H}{\overset{O=C-CH_2}{|}} \xrightarrow[O_3]{H_2O,\ h\nu} HOOC-(CH_2)(CHOH)-COOH \xrightarrow{O_3} HOOC-COOH + H_2O$$

$$\overset{S=P}{\underset{\|}{\|}} \xrightarrow[O_3]{h\nu,\ H_2O} (HO)_2\underset{S}{\overset{\|}{P}}-SOH \xrightarrow[H_2O]{O_3} H_2O + PO_4^{-3} + SO_4^{-2} + H^+$$

$$-OCH_3 \xrightarrow[H_2O]{h\nu} CH_3OH \xrightarrow{O_3} HCOOH + H_2O \xrightarrow{O_3} CO_2 + H_2O$$

$$C_2H_5OH \xrightarrow{O_3} CH_3COOH + H_2O$$

$$\xrightarrow{O_3} CH_3OH + CO_2$$

$$\xrightarrow{O_3} HOCH_2-CH_2OH \xrightarrow{O_3} HOOC-COOH \xrightarrow{O_3} CO_2 + H_2O$$

TABLE 4 - BAYGON REACTION MECHANISMS

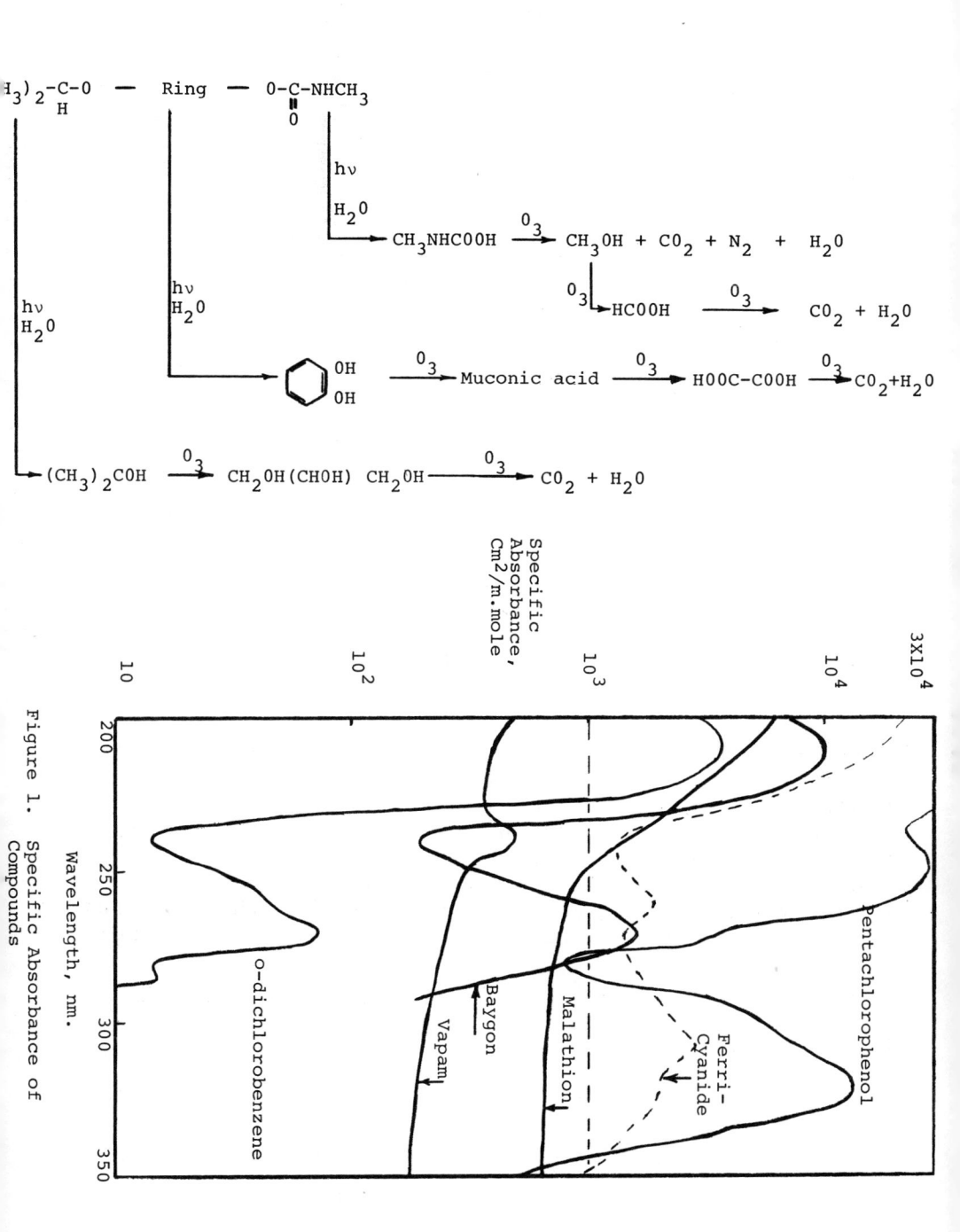

$(CH_3)_2-\underset{H}{C}-O$ — Ring — $O-\underset{\overset{\|}{O}}{C}-NHCH_3$

$\xrightarrow[H_2O]{h\nu}$ $CH_3NHCOOH \xrightarrow{O_3} CH_3OH + CO_2 + N_2 + H_2O$

$\xrightarrow{O_3} HCOOH \xrightarrow{O_3} CO_2 + H_2O$

$\xrightarrow[H_2O]{h\nu}$ benzene(OH)(OH) $\xrightarrow{O_3}$ Muconic acid $\xrightarrow{O_3}$ HOOC-COOH $\xrightarrow{O_3} CO_2 + H_2O$

$\xrightarrow[H_2O]{h\nu}$ $(CH_3)_2COH \xrightarrow{O_3} CH_2OH(CHOH)CH_2OH \xrightarrow{O_3} CO_2 + H_2O$

Figure 1. Specific Absorbance of Compounds

-318-

TABLE 5 - APPROXIMATE RELATIVE RATES OF TOC DESTRUCTION BY O_3/UV cf. O_3

COMPOUND	O_3/UV	O_3 ONLY	$K_{O_3/UV}/K_{O_3}$
1) Pentachlorophenol	20.	1.3	15.4
2) Vapam	4.0	0.02	200
3) Baygon	1.6	0.05	32
4) Malathion	1.5	0.002	750
5) Ferricyanide	1.0	0.007	143
6) DDT	0.06	0.007	8.6

TABLE 6 - PARTIAL OXIDATION BY O_3 cf. O_2/UV

Compound	Destruction of TOC by O_3	Destruction with O_2/UV* Original Species	TOC
1) Pentachlorophenol	Moderate	75%-90 min	Trace
2) Baygon	Moderate	75%-90 min	Trace
3) Methanol	Moderate		Zero
4) Vapam	Moderate	60%-180 min	Zero
5) Ferricyanide ion	Trace	Zero	
6) DDT		90%-30 min	
7) Malathion	Trace	90%-60 min	Trace
8) Acetic Acid	Trace	(Zero)	Zero

*Low pressure lamp (254 nm)

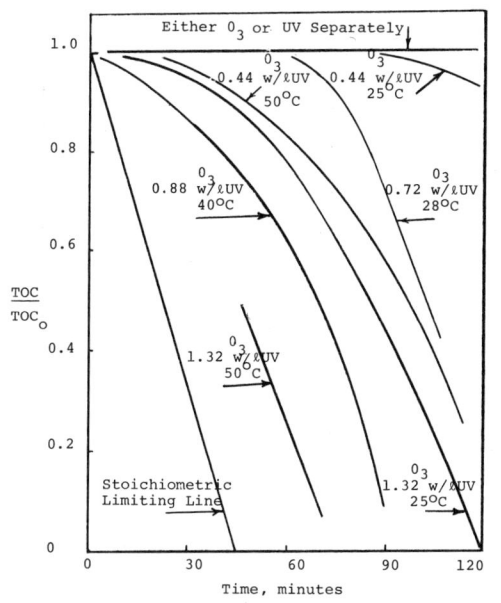

Figure 2. Destruction of Malathion TOC

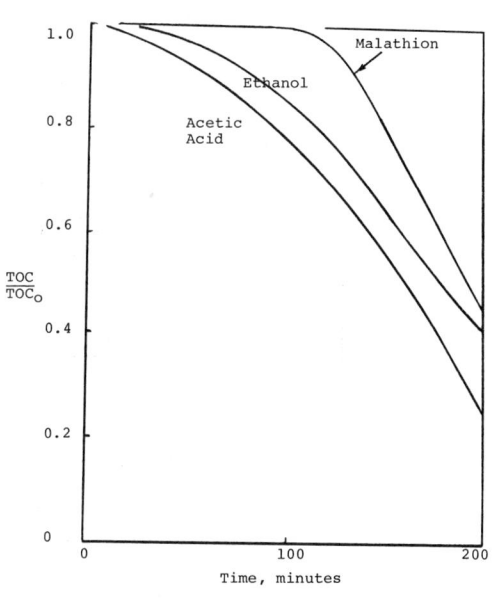

Figure 3. TOC Destruction for Malathion, Ethanol and Acetic Acid at 0.44 watts/liter UV and 25°C.

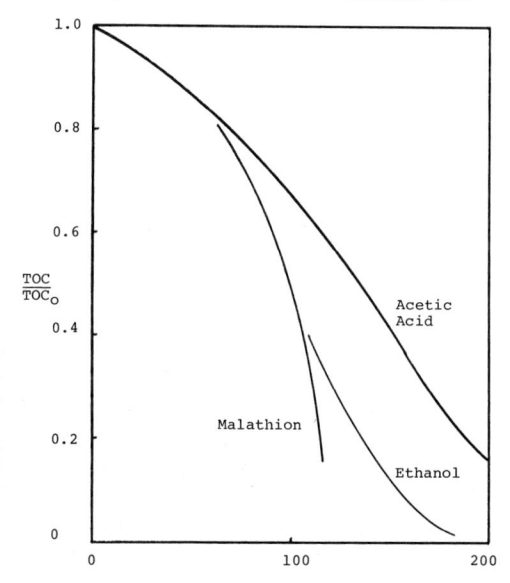

Figure 4. TOC Destruction for Malathion, Ethanol and Acetic Acid at 0.44 watts/liter UV and 50°C.

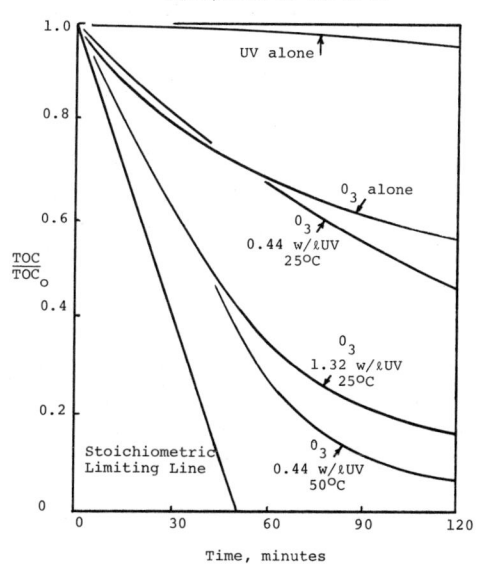

Figure 5. Destruction of Baygon TOC

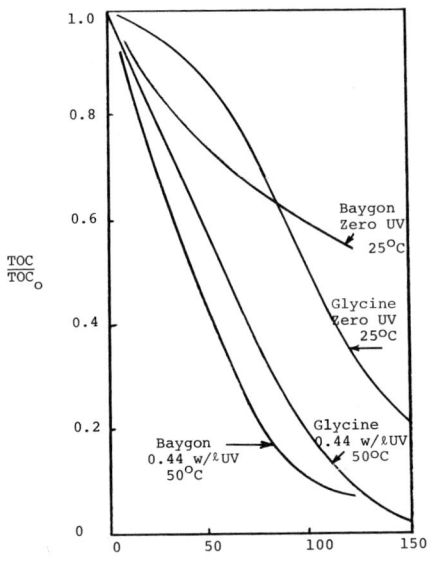

Figure 6. TOC Destruction for Baygon and Glycine

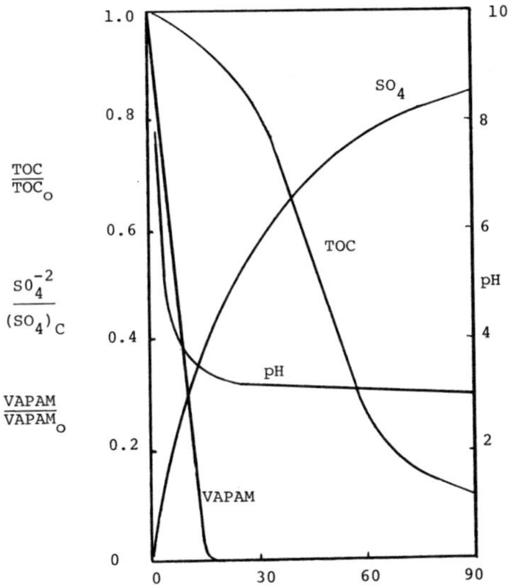

Figure 8. Related Time Profiles, Oxidation of Vapam (RE. Figure 7)

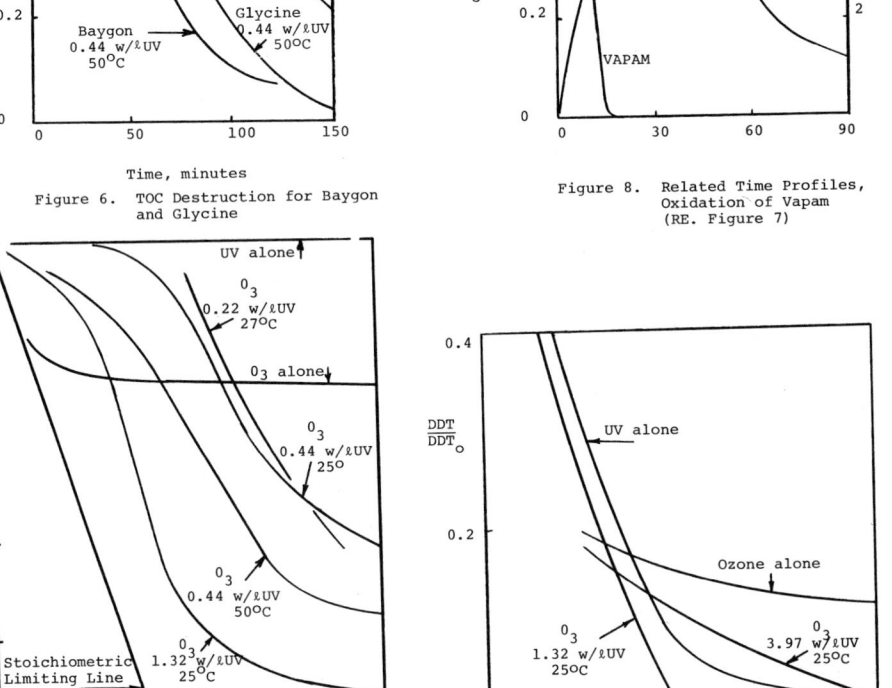

Figure 7. Destruction of Vapam TOC

Figure 9. Destruction of DDT

BY-PRODUCTS OF ORGANIC COMPOUNDS IN THE PRESENCE OF OZONE AND ULTRAVIOLET LIGHT: PRELIMINARY RESULTS

William H. Glaze, Richard Rawley and Simon Lin

Institute of Applied Sciences and Department of Chemistry, North Texas State University, Denton, Texas 76203

This laboratory has recently initiated a program sponsored under Grant R-804640-01 from the Environmental Protection Agency to investigate the efficacy of ozone combined with ultraviolet radiation for the destruction of certain refractory organics in water. The program will be carried out with the assistance of Houston Research, Inc. who have utilized O_3/UV for similar purposes for several years (1).

In the intital phases of the project, we have examined the use of ozone with and without ultraviolet radiation for the destruction of so-called "halomethane precursors" (2). This paper will focus on results obtained on Aldrich humic acid solutions, which serve as a laboratory analog of natural humus, and on a sample of Ohio River water obtained from the Cincinnati area. The results demonstrate the effectiveness of both O_3 and O_3/UV for the destruction of halomethane presursors in these two systems.

Experimental Procedure

The reactor is located at the Houston Research, Inc. laboratory and has been described in reference (1) and earlier citations therein. It consists of a stainless steel vessel which will hold 22 liters of sample. It is equipped with a high speed stirrer, a sparger for introduction of the ozone/oxygen stream, **two quartz wells for UV lamps, and ports for the** removal of aliquots or the addition of materials.

For the studies on humic acid, a 1000 mg/l stock solution of Aldrich Chemical Co., Inc. humic acid (technical grade) was prepared in 0.1 N NaOH using "organic-free" water (OFW) as the matrix (3). The rigorously cleaned reactor was filled with OFW and ozone/oxygen flow was initiated. The flow rate was 200 ± 10 ml min^{-1}(STP) containing 7.7 mg O_3/l. After the ozone concentration had reached equilibrium, 22 ml of humic acid solution was added to the reactor and the pH adjusted to approximately seven by the addition of sulfuric acid. In the case of the photochemical runs, two mercury lamps were available, one 200 watt and the other 450 watt. The lamps were turned on immediately after the introduction of the humic and sulfuric acids. In one run ozone was omitted from the gas stream, and in another both ozone and UV were omitted. Four samples were taken from the reactor at each sampling point in 120 ml septum bottles. One sample was used to measure pH and residual ozone by the iodimetric/thiosulfate titration. The results of these titrations were used to compute the quantity of sulfite ion required to quench the ozone. A fifty percent excess of sulfite was then added to the remaining bottles and they were sealed with Teflon-lined septa and aluminum crimped seals. Samples were stored on ice and shipped to NTSU for chlorination and trihalomethane analysis.

Upon receipt (no more than three hours in transit), the samples were chlorinated with an excess of chlorine in OFW so as to give a chlorine residual of 3 mg l^{-1} (amperometric titration). The samples were stored at 19 ± 2°C for eight days, after which a chlorine residual of 1.5 mg l^{-1} was determined. Halomethanes were analyzed by the liquid-liquid extraction procedure of Henderson, Peyton, and Glaze (4). Duplicate samples were analyzed in most cases and the spread is indicated for each point in Figures 2-5 and 7-10. Duplicate GC analyses for each sample were within ±2%.

Ohio River water was received from EPA/MERL, Cincinnati in a 55 gallon drum. The mixture was stirred before a portion was transferred to the reactor, without filtration. Reaction conditions were similar as above, except that ozone/oxygen flow was initiated <u>after</u> the sample had been introduced. In the case of the UV runs, the lamps were on a few minutes to warm up before gas flow was begun. Sampling was as described above, except that chlorine residual was added at 10 mg l^{-1}. It should be noted that a suspension was present in all of the Ohio River samples, and therefore it was difficult to exactly reproduce conditions from one run to another or to obtain aliquots which were exact replicates.

Results and Discussion

Chloroform and other halomethanes result from the chlorination of waters which contain various organic "precursors" in the presence of bromide and iodide ions (2,5-7); equations [1] - [4].

$$Cl_2 + H_2O \rightleftharpoons HOCl + H^+ + Cl^- \quad [1]$$
$$HOCl + Br^- \rightleftharpoons HOBr + Cl^- \quad [2]$$
$$HOCl + I^- \rightleftharpoons HOI + Cl^- \quad [3]$$
$$HOX + \text{"precursor"} \rightleftharpoons THM \quad [4]$$

(THM = trihalomethane)

In the absence of other halide ions, chloroform is probably the only halomethane produced by this reaction. Most commonly, the bromomethanes $CHCl_2Br$, $CHClBr_2$ and $CHBr_3$, along with chloroform, represent the principal species in chlorinated natural waters.

In view of the growing concern expressed by many over the presence of THM's in municipal water supplies (2), treatment technologies are under consideration either for the removal of the THM's or for the prevention of their formation (7). Treatment with ozone and ozone with ultraviolet radiation represent two approaches to this problem, and the present data relate to the efficacy of these methods.

In this work trihalomethane (THM) formation potential (2)* is measured by an 8-day chlorination regime followed by analysis of THM's by the liquid-liquid extraction method of Henderson, Peyton, and Glaze (4). The results are shown in Figures 1-5 for 1 ppm humic acid and Figures 6-10 for Ohio River Water. In the case of humic acid, chlorination

---

* The term THM formation potential is defined by Stevens and Simons (2a) as the increase in THM concentration that occurs during the storage period in the determination of the terminal THM concentration. For this work, THM formation potential refers to the concentration of THM's formed by an 8-day chlorination regime minus the blank values of THM's in the raw water. It should be noted that the method does not measure total precursor concentration (2a).

produces only small amounts of the brominated halomethanes, presumably because the water matrix was deionized. Ohio River water produces significant quantities of $CHCl_2Br$ and small quantities of $CHClBr_2$. In all cases the THM values before chlorination were subtracted from the 8-day values to obtain THM formation potentials (2).

As shown in Figures 1 and 6, ozone and ozone/UV effectively destroy THM formation potential, although the reaction conditions used were insufficient to completely remove THM precursors. It should be emphasized that the results on the humic acid solutions and the Ohio River water cannot be compared directly, since the ozone concentration at time zero was different in the two cases (see Experimental Section).* However, it is clear that ozone and ozone/UV are very nearly equal in their capacity to destroy THM formation potential in both runs. Further experimentation is currently underway to add more data on this subject.

The mechanism for the oxidation of organic compounds such as those which form the molecular structure of humic acid is still unresolved (8). One may postulate the active kinetic species to be molecular ozone (9), oxygen atoms (10), or hydroxyl radicals (11), depending on the reaction matrix. In any case, the site of attack by the oxidizing species is very likely the same type of site which undergoes haloform reactions in the presence of hypochlorous acid. The intermediate oxidation products, such as acetic (12) and oxalic (13) acid would not yield THM's upon chlorination. Thus, the rapid decrease of THM formation potential observed in this work upon O_3 or O_3/UV treatment does not imply complete destruction of organic materials. Indeed, the non-zero limit for THM formation potential shown in Figures 1 and 6 would seem to suggest the presence of some oxidation-

---

* The apparent increase in THM formation potential in the O_3/UV runs on Ohio River water at short reaction times (Figures 9 and 10) may be due to the photochemical activation of a few molecular sites toward THM formation. (The UV lamps were on for a few minutes in order to reach peak output before ozone flow was introduced.) However, the spread of the data is too severe to make a final judgment on this matter, which is under further study.

resistant product (or reactant) which yields THM's upon chlorination. Whether or not this is the case is the subject of more extensive kinetic and product studies which are currently underway.

Acknowledgment. The authors acknowledge the assistance of members of the NTSU Trace Analysis Laboratory staff and Houston Research, Inc. personnel, and the U. S. Environmental Protection Agency for a grant R-804640-01 for the support of this work.

Literature Cited

(1) H.W. Prengle, Jr., C.G. Hewes, III and C.E. Mauk, "Oxidation of Refractory Materials by Ozone with Ultraviolet Radiation," in Proc. 2nd Intl. Ozone Symp., R.G. Rice, P. Pichet and M.-A. Vincent, Editors. Intl. Ozone Inst., Cleveland, Ohio (1976), p. 224-252.

(2)(a) A.A. Stevens and J.M. Symons, "Measurement of Trihalomethane and Precursor Concentration Changes Occurring During Water Treatment and Distribution," Appendix 4 to Interim Treatment Guide for the Control of Chloroform and Other Trihalomethanes. U.S.E.P.A., Municipal Environmental Research Lab., Cincinnati, Ohio (1976); (b) A.A. Stevens, C.J. Slocum, D.R. Seeger and G.G. Robeck, ibid., Appendix 2.

(3) W.H. Glaze, G.R. Peyton and R. Rawley, "A XAD/Microcoulometric Method for the Determination of Organic Halides in Water." Environ. Sci. Tech. (in press)

(4) J.E. Henderson, IV, G.R. Peyton and W.H. Glaze, "A Convenient Liquid-Liquid Extraction Method for the Determination of Halomethanes in Water at the Parts-Per-Billion Level," in Identification and Analysis of Organic Pollutants in Water, L. Keith, Editor. Ann Arbor Science (1976), p. 105.

(5) J.J. Rook, "Formation of Haloforms During Chlorination of Natural Waters." Water Treatment and Examination 23:Part 2, 234 (1974).

(6) T.A. Bellar, J.J. Lichtenberg and R.C. Kroner, "The Occurrence of Organohalides in Chlorinated Drinking Water." Jour. AWWA 66:703 (1974).

(7) R.M. Clark, D.L. Guttman, J.L. Crawford and J.A. Machisko, "The Cost of Removing Chloroform and Other Trihalomethanes from Drinking Water Supplies," Appendix 1 to Interim Treatment Guide for the Control of Chloroform and Other Trihalomethanes. U.S.E.P.A., Municipal Environmental Research Laboratory, Cincinnati, Ohio (1976).

(8) M. Schnitzer and S.U. Khan, "Humic Substances in the Environment." Marcel Dekker, New York, N.Y. (1972), Chapter 5, Section III.

(9) P.S. Bailey, "Reactivity of Ozone with Various Organic Functional Groups Important to Water Purification," in Proc. First Intl. Symp. on Ozone for Water & Wastewater Treatment, **R.G.** Rice and M.E. Browning, Editors. Intl. Ozone Inst., **Cleveland, OH** (1975), p. 101-119.

(10) C.J. Fortin, D.R. Snelling and A. Tardif, "The Ultraviolet Flash Photolysis of Ozone and the Reaction with H_2O." Can. J. Chem. 50:2747-2760 (1972).

(11)(a) J. **Hoigné** "Comparison of the Chemical Effects of Ozone and of Irradiation on Organic Impurities in Water," in Proc. Radiation for a Clean Environment. Intl. Atomic Energy Agency, Vienna (1975), P. 297-305; (b) J. **Hoigné** and J. Bader, "Identification and Kinetic Properties of the Oxidizing Decomposition Products of Ozone in Water and Its Impact on Water Purification," in Proc. Second Intl. Symp. on Ozone for Water and Wastewater Treatment, R.G. Rice, P. Pichet and M.-A. Vincent, Editors. Intl. Ozone Inst., **Cleveland, OH** (1976), p. 271-282.

(12) F. Dobison and G.J. Lawson, "Chemical Constitution of Coal. VI. Optimum Conditions for the Preparation of Sub-humic Acid from Humic Acid by Ozonization." Fuel 38:79-87 (1959).

(13) M. Ahmed and C.R. Kinney, "Ozonization of Humic Acids Prepared from Oxidized Bituminous Coal." J. Am. Chem. Soc. 72:559-561 (1950).

DISCUSSION

<u>Arch Hill</u>, Louisiana Tech University: Do you suppose that one of the reasons that the ozone with the UV was not as good was that part of the ozone was destroyed while still in the gas bubbles themselves?

<u>Glaze</u>: I think that is <u>entirely</u> possible. In fact that's what I meant when I said that the conditions were not the same at all, when the ultraviolet radiation is on, as evidenced by the fact that there is no residual. There is a rapid, significant decrease in the ozone level simply because of the decomposition process itself. But, whether it occurs in the gas phase or in solution, I don't know how to answer that question, but maybe somebody else does.

<u>David Rosenblatt</u>, Fort Detrick. I would like to know whether any attempt has been made to see if the carbon tetrachloride which appears in the chromatogram, arises from chloroform, or whether it is formed by some other process, that is, from the same precursors from which you get the haloform derivatives.

<u>Glaze</u>: Actually, there is no carbon tetrachloride in the chromatogram, if you look at actual humic acid chlorination, except for a trace probably due to contamination. I showed that only because people like to do carbon tetrachloride analyses. That was actually a standard mixture. There is no carbon tetrachloride produced by the reaction that I am aware of; there is only a trace in all of the humic acid work. I simply show that in using those conditions one could separate carbon tetrachloride under isothermal conditions.

<u>Unknown</u>: What contact time did you use for the ozone?

<u>Glaze</u>: In this flowing reactor of the type that was described previously and with the flow rates we used, I think I showed about 200 milliliters a minute, which corresponds to about 21 milligrams per minute of ozone.

<u>Unknown</u>: What is the lamp that was used? Mercury lamps?

Glaze: Yes. Thank you. I should have mentioned that. These are indeed high pressure mercury lamps, the distribution of energies of which are published by Hanovia as a function of wavelength. They are standard high pressure mercury lamps.

Figure 1. Effect of treatment on 8-day trihalomethane formation potential of 1 mg l^{-1} humic acid solution. Chlorine dose: 3 mg l^{-1}.

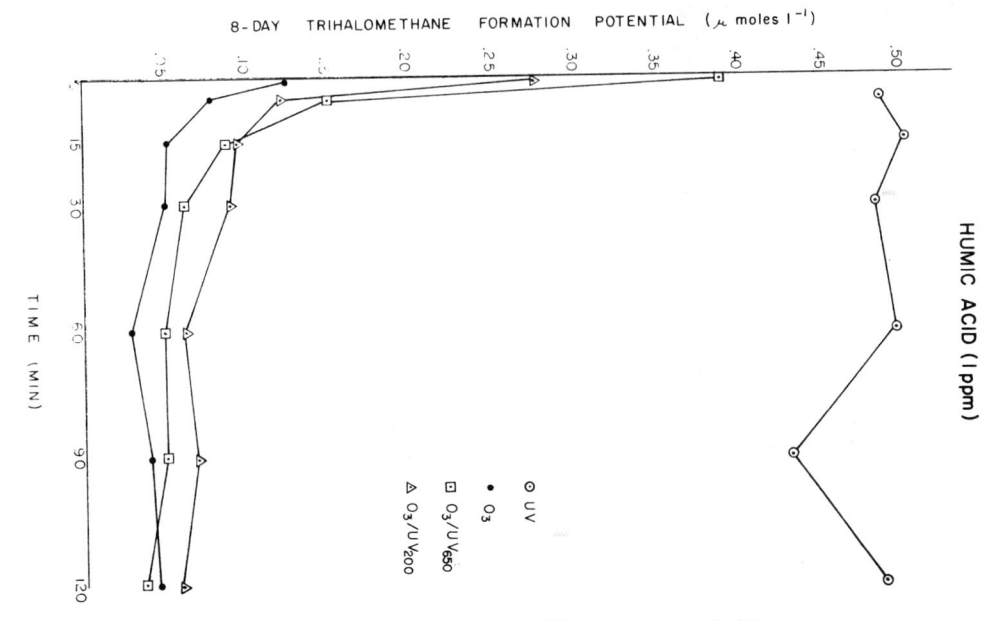

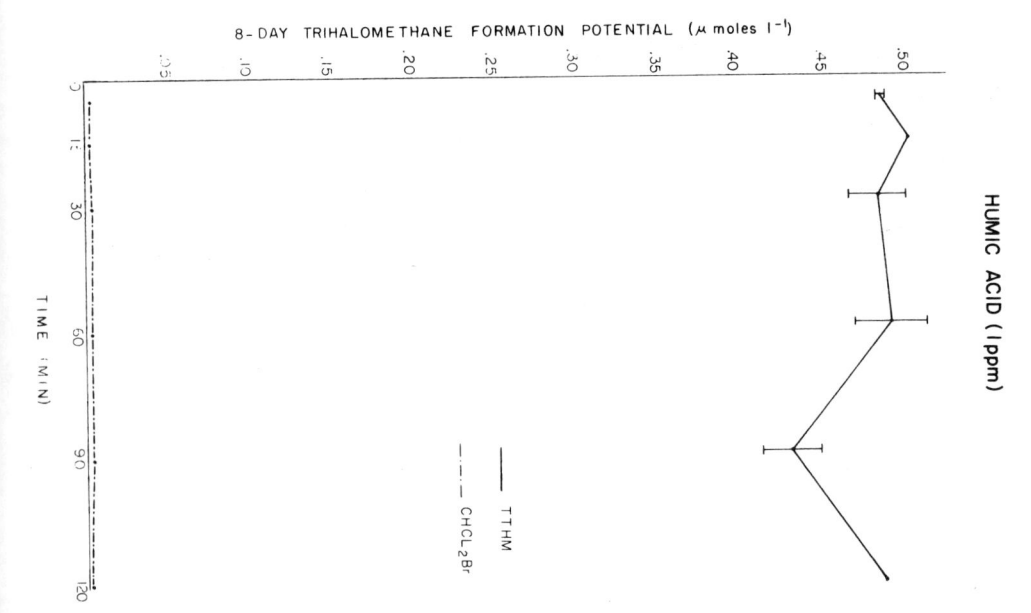

Figure 2. Effect of 650 watt UV treatment on 8-day trihalomethane formation potential of 1 mg l^{-1} humic acid solution. Chlorine dose: 3 mg l^{-1}

Figure 3. Effect of ozone treatment on 8-day trihalomethane formation potential of 1 mg l^{-1} humic acid solution. Chlorine dose: 3 mg l^{-1}.

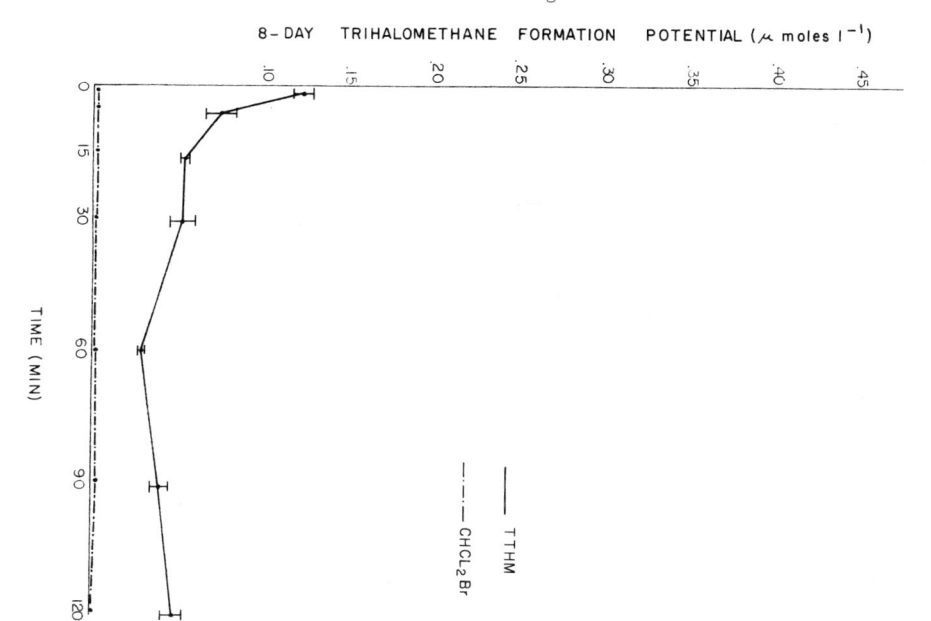

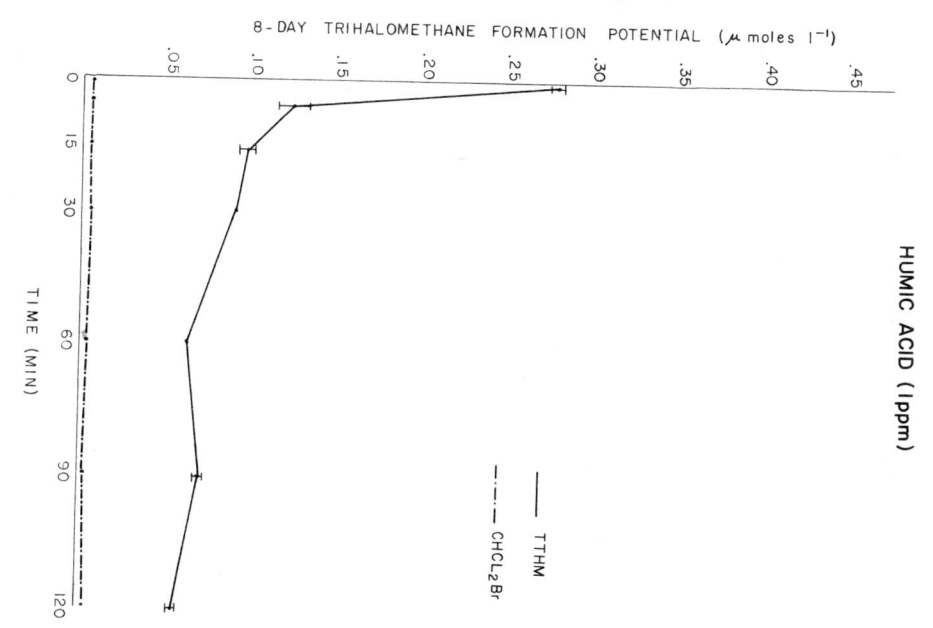

Figure 4. Effect of ozone with 200 watt UV on 8-day trihalomethane formation potential of 1 mg l^{-1} humic acid solution. Chlorine dose: 3 mg l^{-1}.

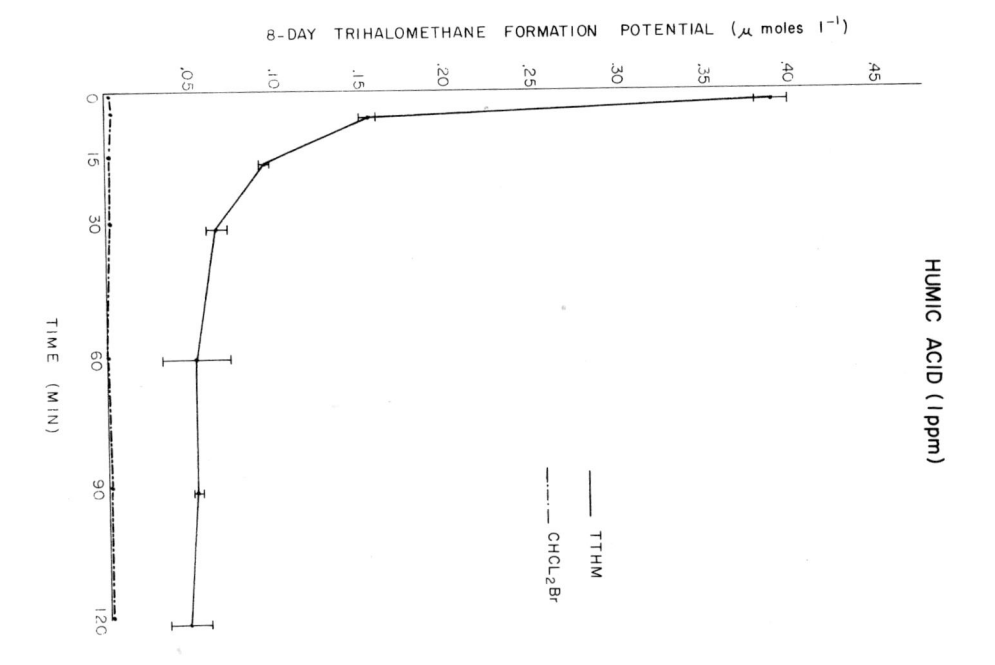

Figure 5. Effect of ozone with 650 watt UV on 8-day trihalomethane formation potential of 1 mg l^{-1} humic acid solution. Chlorine dose: 3 mg l^{-1}.

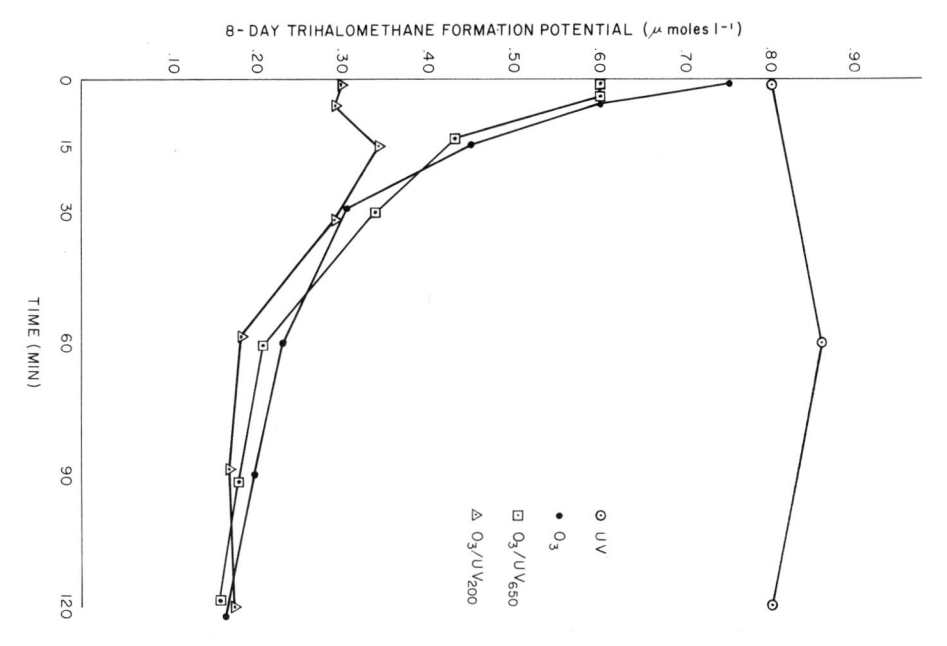

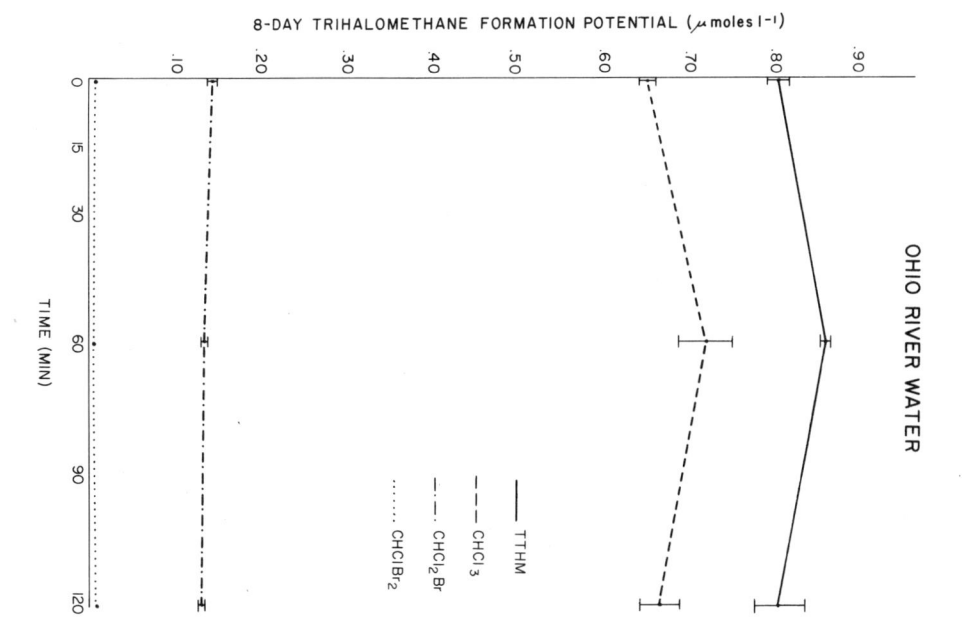

Figure 6. Effect of treatment on 8-day trihalomethane formation potential of Ohio River Water. Chlorine dose: 10 mg l^{-1}.

Figure 7. Effect of 650 watt UV on 8-day trihalomethane formation potential of Ohio River Water. Chlorine dose: 10mg l^{-1}.

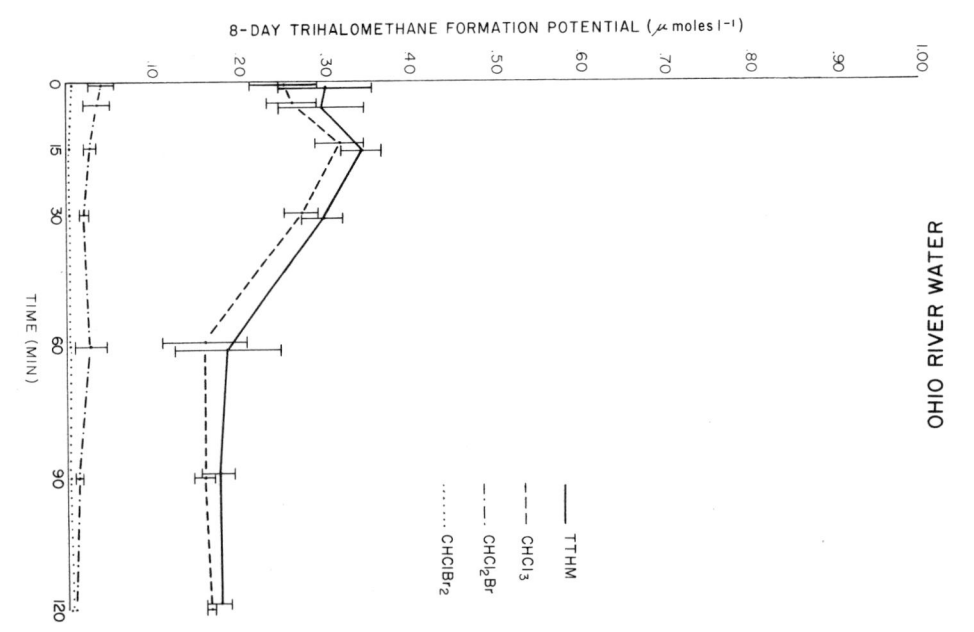

Figure 8. Effect of ozone treatment on 8-day trihalomethane formation potential of Ohio River Water Chlorine dose: 10 mg l^{-1}.

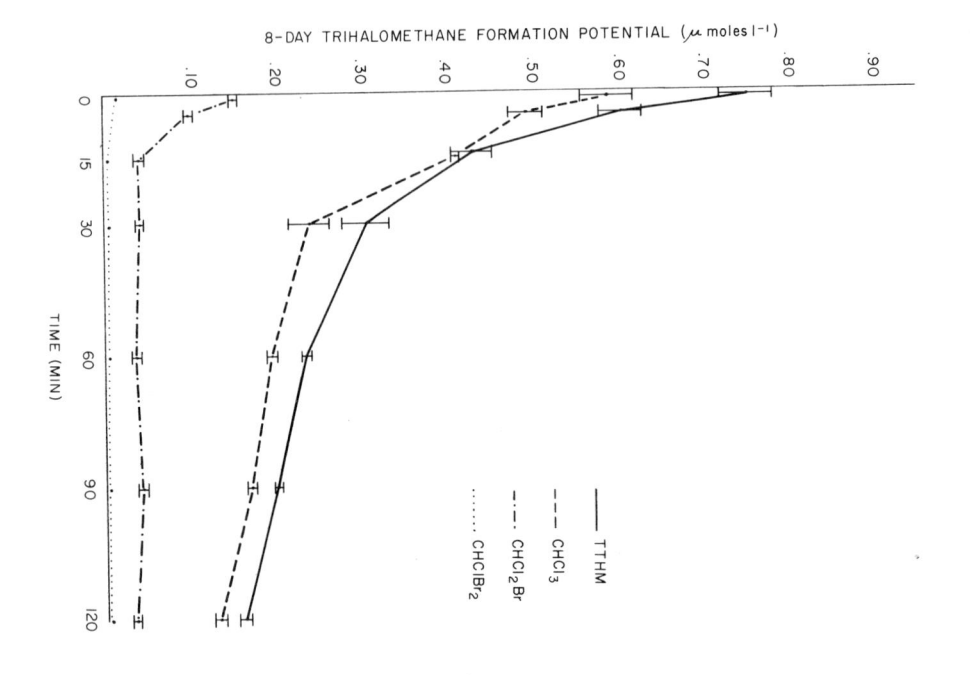

Figure 9. Effect of ozone with 200 watt UV on 8-day trihalomethane formation potential of Ohio River Water. Chlorine dose: 10 mg l^{-1}.

OHIO RIVER WATER

Figure 10. Effect of ozone with 650 watt UV on 8-day trihalomethane formation potential of Ohio River Water. Chlorine dose: 10 mg l^{-1}.

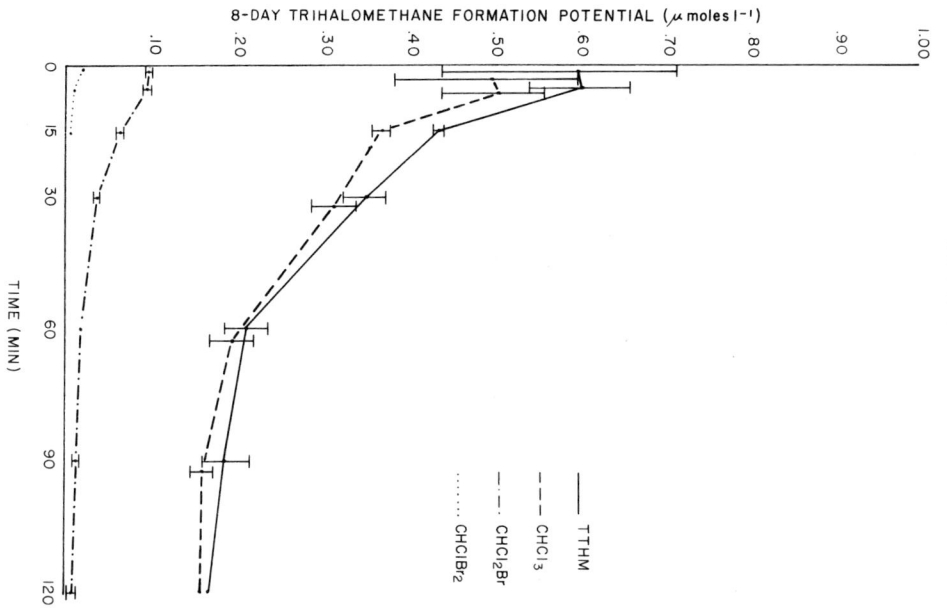

Figure 11. Effect of oxygen on 8-day trihalomethane formation potential of Ohio River Water. Chlorine dose: 10 mg l^{-1}.

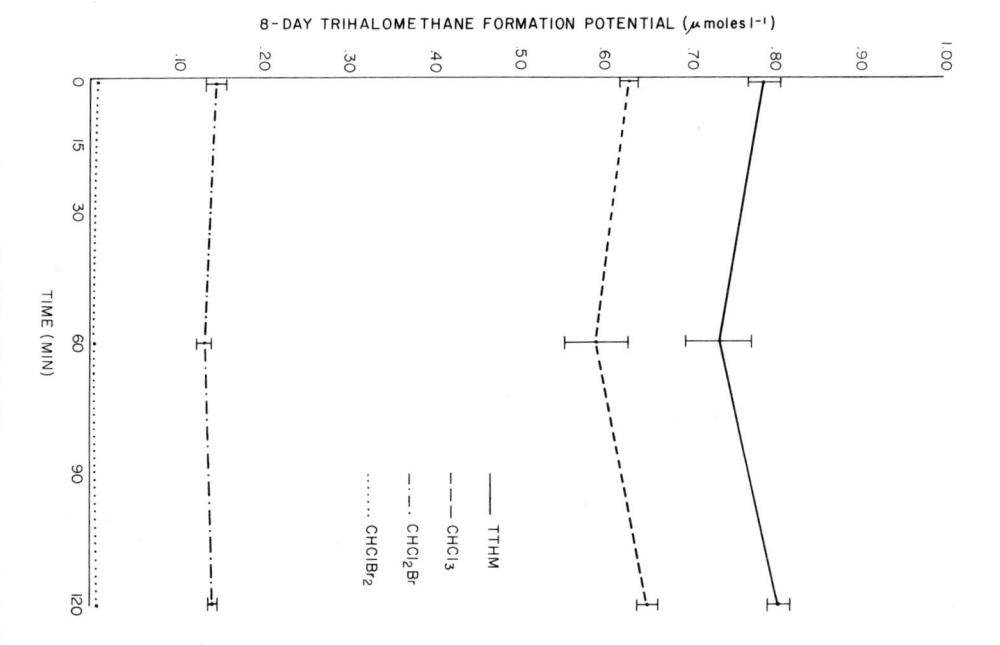

CHLORINE DIOXIDE: CHEMICAL AND PHYSICAL PROPERTIES

David H. Rosenblatt, Ph.D.

Environmental Protection Research Division
US Army Medical Bioengineering Research
and Development Laboratory
Fort Detrick, Frederick, Maryland USA 21701

Abstract

Some of the properties of ozone and chlorine dioxide (ClO_2) are compared. Physicochemical characteristics and hazards of ClO_2 vapor are presented. The synthesis of ClO_2 and the chemistry of its formation are discussed. Methods of analysis must take into account the similarities to and differences from hypochlorous acid, which may accompany ClO_2. The author's investigation of amine oxidations with ClO_2 provide an insight into the complexity and variety of interactions possible between ClO_2 and organic compounds. It is noteworthy that ClO_2 and ammonia do not react with each other.

Introduction

Our purpose in this workshop is to consider ozone and chlorine dioxide as alternatives to chlorine for the disinfection of drinking water supplies and perhaps also for the treatment of wastewaters prior to discharge.

Ozone and chlorine dioxide share certain characteristics: They are oxidants, good disinfectants, and relatively volatile. They do not appear to form toxic or esthetically undesirable residues, nor do they form haloforms by reaction with certain organic constituents of water.

They differ considerably in other respects: Ozone decomposes rapidly and spontaneously in neutral aqueous solution, whereas chlorine dioxide is quite stable -- so much so that it can usually maintain a residual longer than can an equivalent chlorine concentration. Whereas ozone is extremely reactive with a variety of organic compounds and can lower TOC, COD and BOD dramatically, chlorine dioxide, which has a considerably lower redox potential, is highly selective in its chemical reactions. A particular virtue of ozone is that the part of the molecule not incorporated into organic compounds with which it reacts is harmless or even desirable, that is, oxygen; when chlorine dioxide reacts with organic compounds, chlorite ion or other chlorine-containing inorganic species are left behind. It is to be recognized that the disinfectant qualities of ozone can suffer interference from high organic loadings, whereas the highly selective chlorine dioxide is likely to maintain its residual -- and therefore its disinfectant potentialities -- in such an environment.

I shall review here some of the highlights of chlorine dioxide chemistry, an area characterized in the past chiefly by neglect. Such information furnishes a background for safe handling of chlorine dioxide, for analytical control, for understanding the advantages and disadvantages of its use, and for making cost-effectiveness assessments.

Properties

Chlorine dioxide, a yellow gas with a chlorine-like odor, is one of the few stable non-metallic inorganic free radicals. Unlike nitrogen dioxide, it shows no tendency at all to dimerize. It is a bent molecule with a bond angle of about 118°. Chlorine dioxide melts at -59°C and boils at about 9.7°C. It is approximately ten times as soluble as chlorine in water (about 23 volumes per volume of water at 25°), and (like chlorine) forms a hydrate. The hydrate, which is a solid stable up to about 18°C, contains roughly six water molecules per molecule of ClO_2; it can be seen to precipitate when an appreciable quantity of chlorine dioxide is passed into ice water. Chlorine dioxide is considered rather toxic; its threshold limit value (TLV) for industrial inhalation exposure is 0.1 part per million (ppm), the same as that for ozone and one-tenth the TLV for chlorine. Chlorine dioxide is safe from explosions up to a concentration of roughly 4% in air, above which it is detonated by sparks; concentrations in excess of 10 or 11% in air are apt to undergo mild explosions under the influence of heat, light or shock.

The ultraviolet absorption of chlorine dioxide, with a maximum at 357 nm where the molar absorbancy is about 1250, is convenient for determining concentrations in kinetic studies, although not normally of much use for analyzing the low levels used for disinfection.

Whereas chlorine hydrolyzes rapidly and reversibly in water at any pH, chlorine dioxide is rather stable in acidic and neutral solution. Above a pH of 9-10, it disproportionates to give chlorate and chlorite ions:

$$2\ ClO_2 + 2\ OH^- \rightarrow H_2O + ClO_2^- + ClO_3^-$$

Chlorine dioxide in the vapor phase is subject to photodecomposition, especially by ultraviolet light. It is believed that photodecomposition products are involved in the explosions of chlorine dioxide-air mixtures. Mists containing decomposition products form in the air over strong aqueous chlorine dioxide solutions.

Synthesis

Chlorine dioxide may be produced by a number of methods. The reaction I prefer to use in the laboratory is convenient when an air or (preferably) nitrogen stream is used to sweep the gas from the generating solution into water, aqueous buffers, or other solvents. It may be written as follows, although this only represents the stoichiometry:

$$K_2S_2O_8 + 2\ NaClO_2 \rightarrow Na_2SO_4 + K_2SO_4 + 2\ ClO_2$$

Chlorous acid, $HClO_2$, the conjugate acid of the chlorite ion, is so unstable that one finds difficulty in measuring its dissociation constant (which is about 1.1×10^{-2} at 25°). It breaks down with a somewhat variable stoichiometry to furnish chlorine dioxide free of chlorine. At least two stoichiometries have been formulated which are concentration-dependent:

$$4\ HClO_2 \rightarrow 2\ ClO_2 + H_2O + 2\ H^+ + Cl^- + ClO_3^-\ \text{and}$$

$$5\ HClO_2 \rightarrow 4\ ClO_2 + 2\ H_2O + H^+ + Cl^-$$

A favorite method for making chlorine dioxide for water treatment involves the use of hypochlorous acid along with chlorite ion. The reaction is very slow in basic solution, and suffers from the parallel formation of chlorate ion. These two reactions are represented as:

$$HOCl + 2\ ClO_2^- \rightarrow 2\ ClO_2 + Cl^- + OH^-\ \text{and}$$

$$HOCl + ClO_2^- + OH^- \rightarrow ClO_3^- + Cl^- + H_2O$$

Unfortunately, the product distribution is not simply a result of reactant ratios. The mechanism requires that we assume a very real, even if very transient, intermediate, involved in the following series of reactions:

$$HOCl + HClO_2 \rightarrow \left[Cl - Cl \diagup^{O}_{O} \right]$$

$$2\ \left[Cl - Cl \diagup^{O}_{O} \right] \rightarrow Cl_2 + 2\ ClO_2$$

$$H_2O + \left[Cl - Cl \diagup^{O}_{O} \right] \rightarrow Cl^- + ClO_3^- + 2\ H^+$$

(and, of course, $Cl_2 + H_2O \rightarrow HOCl + H^+ + Cl^-$)

One of the consequences of this combination of reactions is that ClO_2 formation is most efficient when reactant concentrations are high, i.e., when the second-order disappearance of Cl_2O_2 (the transient species) predominates over its first-order decay to chlorate. Another consequence is that there is almost always some hypochlorite (or chlorine) left in the system unless much chlorite remains unreacted.

Akin to the preceding method (for laboratory preparation of ClO_2) is the passage of chlorine gas through a column of $NaClO_2$, from which ClO_2 emerges:

$$Cl_2 + 2\ NaClO_2 \rightarrow 2\ ClO_2 + 2\ NaCl$$

All industrial processes for generating chlorine dioxide start with chlorate salts. Since chlorite salts are made from chlorine dioxide, it is obvious that the synthesis starting with chlorate must be less expensive, and indeed is ultimately essential for the industry. The simplest process forms chlorine as a by-product:

$$4\ H^+ + 2\ Cl^- + 2\ ClO_3^- \rightarrow Cl_2 + 2\ ClO_2 + 2\ H_2O$$

By the inclusion of an appropriate reducing agent -- and there are several of these -- chlorine formation is avoided. Two such reducing agents are oxalic acid and methanol. For example:

$$2\ KClO_3 + (COOH)_2 + 2\ H_2SO_4 \rightarrow 2\ ClO_2 + 2\ CO_2 + 2\ H_2O + 2\ KHSO_4$$

Analysis

The analysis of dilute chlorine dioxide solutions is by no means a trivial problem. The simplest approach is iodometric. In neutral solution, one observes the reaction:

$$2\ ClO_2 + 2I^- \rightarrow ClO_2^- + I_2$$

Since HOCl is often present along with ClO_2, one must also consider the reaction:

$$HOCl + 2\ I^- \rightarrow Cl^- + OH^- + I_2$$

It is possible to remove the ClO_2 by sparging. The solution can be titrated before and after sparging, the difference in titers representing ClO_2. Care must be taken to avoid pH levels below about 5, since at low pH the chlorite ion (ClO_2^-) that is initially formed reacts with I^-, to give an overall reaction

$$2\ ClO_2 + 10\ I^- + 4\ H_2O \rightarrow 2\ Cl^- + 8\ OH^- + 5\ I_2$$

Instead of iodometric titration, colorimetric procedures such as the DPD (N,N-diethyl-p-phenylenediamine) method or the tyrosine method may be employed. Amperometric titration is also useful.

Alternatively, HOCl may be destroyed by a reducing agent, such as oxalic acid or malonic acid, and ClO_2 then determined colorimetrically, for instance with orthotolidine. This suffers from interference by manganese.

Finally, chlorine dioxide may be swept out of one solution and absorbed into another for iodometric determination.

Amine Reactions

For me, the most interesting part of chlorine dioxide chemistry has been the unfolding of the mechanisms of oxidation of secondary and tertiary amines. It is also highly relevant to the current interests of water chemistry, where we seek to understand and hopefully to predict the behavior of disinfectants with individual compounds. This differs from the "black box" approach, where one assumes something like:

$$O_3 + \text{Organics} \rightarrow \boxed{\text{Black Box}} \rightarrow O_2 + CO_2 + H_2O + ?$$

$$HOCl + \text{Organics} \rightarrow \boxed{\text{Black Box}} \rightarrow HCl + CO_2 + ?$$

$$ClO_2 + \text{Organics} \rightarrow \boxed{\text{Black Box}} \rightarrow CO_2 + ? + ?$$

At the time I got interested in chlorine dioxide, I was in the chemical warfare business, and others in my group had learned that the nerve agent, VX, is quickly detoxified by chlorine dioxide.

$$[(CH_3)_2CH]_2 N-CH_2CH_2S-\overset{\overset{O}{\|}}{\underset{\underset{CH_3}{|}}{P}}-OC_2H_5$$

VX

I chose triethylamine as a simple model compound for the amine part of VX and was gratified to observe a clean reaction:

$$H_2O + (C_2H_5)_3N + 2\ ClO_2 \rightarrow (C_2H_5)_2NH + 2\ ClO_2^- + CH_3CHO + 2\ H^+$$

Well, almost clean. If the pH dropped too low, ClO_2 was regenerated by the somewhat complex reaction:

$$H^+ + CH_3CHO + 3\ ClO_2^- \rightarrow CH_3\overset{O}{\underset{\|}{C}}-O^- + Cl^- + 2\ ClO_2 + H_2O$$

According to the commonly accepted kinetics of reactions involving two molecules of one substance and one molecule of a second substance, we should have been able to formulate a reaction rate law of the form (second order disappearance of ClO_2):

Rate of disappearance of ClO_2 = k [Amine] $[ClO_2]^2$

Instead, we found first order disappearance of ClO_2:

Rate of disappearance of ClO_2 = k [Amine] $[ClO_2]$

This <u>first-order</u> dependence on chlorine dioxide concentration suggested something like:

$$ClO_2 + Amine \xrightarrow{slow} First\ Intermediate$$

$$First\ Intermediate \xrightarrow{fast} Second\ Intermediate$$

$$ClO_2 + Second\ Intermediate \xrightarrow{fast} Products$$

We then had a piece of luck. We found that if we added a fair amount of sodium chlorite to such a reaction mixture, the disappearance of chlorine dioxide continued to follow first order kinetics, but at a <u>slower rate</u>. In fact, a plot of the reaction half-life against the concentration of chlorite ion was linear:

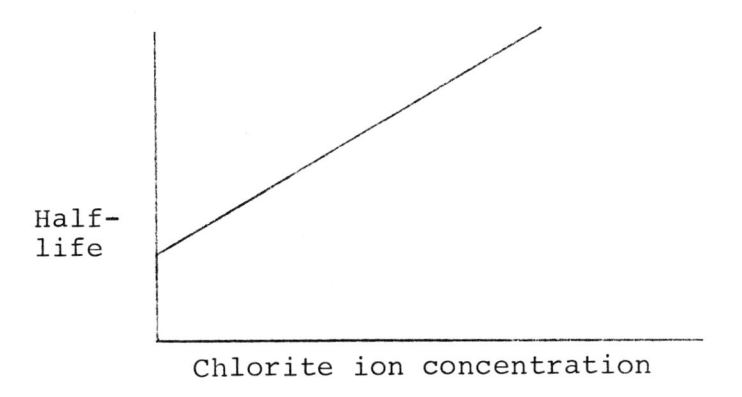

[Graph: Half-life (y-axis) vs Chlorite ion concentration (x-axis), showing a linear increase]

This permitted us to postulate a more detailed mechanism:

$$ClO_2 + (C_2H_5)_3N: \rightleftarrows ClO_2^- + (C_2H_5)_3N^{\cdot +} \text{ (cation radical)}$$

$$(C_2H_5)_3N^{\cdot +} \xrightarrow{fast} (C_2H_5)_2\ddot{N}-\underset{H}{\overset{\cdot}{C}}-CH_3 + H^+$$

$$(C_2H_5)_2\ddot{N}-\underset{H}{\overset{\cdot}{C}}-CH_3 + ClO_2 \rightarrow (C_2H_5)_2N^+=\underset{H}{C}-CH_3 + ClO_2^-$$

$$(C_2H_5)_2N^+=\underset{H}{C}-CH_3 + H_2O \rightarrow (C_2H_5)_2NH_2^+ + O=\underset{H}{C}-CH_3$$

We also showed that the rate of the first reaction depended on the density of the electric charge on the nitrogen. We did this with a series of ring-substituted benzyldimethylamines,

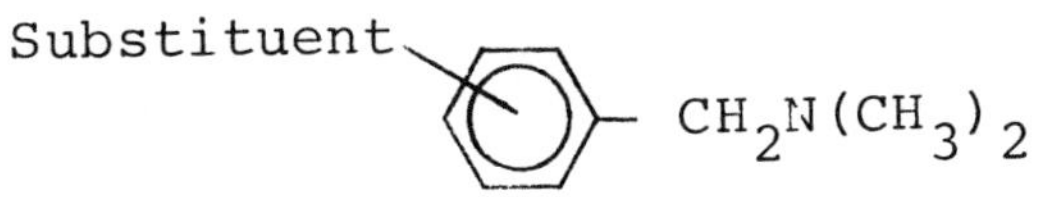

[Structure: Substituent—C₆H₄—CH₂N(CH₃)₂]

The greater the basicity of such an amine, the faster its reaction with chlorine dioxide. We were also able to solve another problem. We could distinguish between mechanisms involving removal of an electron in the first step, and those involving removal of a hydrogen atom, e.g.,

$$ClO_2 + Ar - \underset{H}{\overset{H}{C}} - \ddot{N}R_2 \longrightarrow Ar - \underset{H}{\overset{H}{C}} - \overset{+}{\dot{N}}R_2 + ClO_2^-$$

$$ClO_2 + Ar - \underset{H}{\overset{H}{C}} - \ddot{N}R_2 \longrightarrow Ar - \underset{H}{\overset{\cdot}{C}} - \ddot{N}R_2 + HClO_2$$

In fact, we observed both types of reaction, depending on the chemical structures involved, and sometimes we saw both together.

Of course, this involved making some very special molecules. Among the most interesting molecules was quinuclidine, related to the chemical warfare agent called BZ:

$$(C_6H_5)_2 \underset{OH}{\overset{}{C}} - \underset{}{\overset{O}{C}} - O - C \underset{\diagdown}{\overset{\diagup CH \diagdown}{}} \quad \text{BZ} \qquad \qquad \text{Quinuclidine}$$

Strangely enough, this was very slow in its reactions, and it gave the amine oxide, rather than the expected carbon-nitrogen cleavage. We investigated this too, but the results are a little too complicated to describe.

Finally, we looked at a relative of quinuclidine, triethylenediamine, and saw some more unusual chemistry. We called it oxidative fragmentation.

[Reaction scheme: Triethylenediamine (or "DABCO") + ClO₂· ⇌ Cation radical (red) + ClO₂⁻]

[Reaction scheme: Cation radical + ClO₂· → bis-iminium intermediate + ClO₂⁻]

[Reaction scheme: bis-iminium + 2H₂O → Piperazine + 2 H₂C=O + 2H⁺]

It is interesting to note that in this reaction second-order kinetics with respect to ClO₂ concentration is observed, i.e.,

Rate of Disappearance of ClO_2 = k [Amine] $[ClO_2]^2$

Lessons Learned

There are some general lessons that may be drawn from this discussion and from other parts of my work that I haven't had time to mention:

1. Kinetic studies constitute an important tool for determining reaction mechanisms.

2. When chlorine dioxide oxidizes electron-rich substances, such as amines, it tends to gain only one electron, forming the stable chlorite ion. This

means that possibly only one-fifth of the potential oxidizing power may be utilized. As the pH drops, chlorite becomes less stable, and more available for reaction, but the amines are more protonated, hence less reactive.

3. The structure of an amine (and presumably of the member of any class of reactive organic substances) determines its reactivity with ClO_2, and often in strange and unexpected ways. In general, tertiary amines are more reactive than secondary amines, which are more reactive than primary amines. A great advantage of ClO_2 is that <u>it doesn't react with ammonia at all</u>.

4. Chlorine dioxide is very selective about which classes of compounds it will oxidize. Often, only minimal changes occur.

5. As the equilibrium products of a redox reaction accumulate, they slow up the reaction. This is sometimes extremely important for an oxidant of low redox potential.

6. In any mixed organic system treated with ClO_2, we must look for Cl^-, ClO_2^- and ClO_3^- among the inorganic products. We must also learn a lot more about the toxicity of ClO_2^- and ClO_3^- than we now know.

7. The inorganic chemistry of ClO_2 involves some very strange species, especially the transient Cl_2O_2.

I would remind you that I have not attempted to touch at all on the reactions of ClO_2 with phenols or certain other organics, since these will be covered later in the program.

RECOMMENDED LITERATURE SOURCES

1. H.L. Robson, "Chlorine Oxygen Acids and Salts: Chlorine Dioxide," in H.F. Mark, J.J. McKetta, Jr. and D.F. Othmer, eds., Kirk-Othmer Encyclopedia of Chemical Technology, 2nd ed., Vol. 5, Interscience Publishers Div. of John Wiley & Sons, Inc., New York, 1964, pp. 35-50.

2. G. Gordon, R.G. Kieffer and D.H. Rosenblatt, "The Chemistry of Chlorine Dioxide," in S.J. Lippard, ed., Progress in Inorganic Chemistry, Vol. 15, John Wiley and Sons, Inc., New York, 1972, pp. 201-286.

3. D.H. Rosenblatt, "Chlorine and Oxychlorine Species Reactivity with Organic Substances," in J.D. Johnson, ed., Disinfection: Water and Wastewater, Ann Arbor Science Publishers, Inc., Ann Arbor, Michigan, 1975, pp. 249-276.

4. W.H. Rapson, "From Laboratory Curiosity to Heavy Chemical," Chemistry in Canada, 18:25-31 (1966).

5. W. Masschelein, "Progrès de la Chimie du Bioxyde de Chlore et ses Applications," Chimie Industrie-Genie Chimique, 97:49-61:346-354 (1967).

6. M.C. Rand, A.E. Greenberg and M.J. Taras, Standard Methods for the Examination of Water and Wastewater, 14th ed., American Public Health Association, Washington, DC, 1976, pp. 309-361.

7. H.E. Stokinger, Chairman, Documentation of the Threshold Limit Values for Substances in Workroom Air, 3rd ed., American Conference of Governmental Industrial Hygienists, 1974, pp. 46-47, 194-195.

8. J.F. Haller and S.S. Listek, "Determination of Chlorine Dioxide and Other Active Chlorine Compounds in Water," Anal. Chem., 20:639-642 (1948).

9. H.W. Hodgen and R.S. Ingols, "Direct Colorimetric Method for the Determination of Chlorine Dioxide in Water," Anal. Chem., 26:1224-1226 (1954).

USE OF CHLORINE DIOXIDE IN WATER AND WASTEWATER TREATMENT

Sidney Sussman, Olin Water Services, Stamford, CT 06904
and
James S. Rauh, Olin Water Services, Kansas City, KS 66115

Introduction

In view of the current interest in disinfection of water supplies with chlorine dioxide that has been stimulated by the EPA's recently-issued "Interim Treatment Guide for the Control of Chloroform and Other Trihalomethanes," it is appropriate at this time to summarize the present use of chlorine dioxide in the treatment of public water supplies. A similar review based upon European practice was published in 1974 by Dowling (1) and contains numerous references of interest.

Chlorine dioxide was suggested and tried as a water supply disinfectant in Paris early in this century, but only with the commercial availability of sodium chlorite about 1940 did it become practical to use chlorine dioxide for water purification on a regular basis in this country. Although its potential as a disinfectant had been long recognized and laboratory and plant scale tests were initiated in the early forties in order to develop its use as a disinfectant, the accidental discovery of its effectiveness in preventing the formation of chlorophenol tastes and various odors in waters quickly diverted interest in the use of chlorine dioxide for potable water treatment from disinfection to taste and odor control.

Since then, chlorine dioxide has been used by a considerable number of water utilities, primarily for the prevention or control of tastes and odors. To a smaller extent, it has been used for control of algae in reservoirs and for removal of manganese. In 1958, Granstrom and Lee (2) reported the responses to a questionnaire sent to 150 American water treatment plants believed to be using chlorine dioxide. Fifty-six plants reported that they were using it, in almost all cases for the control of phenolic or algal tastes or odors. Additional reasons for its use, in most cases along with taste or odor control, were algae control in 7 plants, iron or manganese removal in 3 plants, and disinfection in 15 plants.

The total number of water treatment plants currently using chlorine dioxide is not reliably known, but Table 1 gives a partial listing along with the purpose of the treatment in each case.

Disinfectant Properties

Present interest in the use of chlorine dioxide for water treatment is, of course, focussed on its disinfectant properties. In a series of papers published in 1947-1949, Ridenour and co-workers (3-6) established the fact that chlorine dioxide was as effective a biocide as chlorine in water. This work, carried out using as test organisms E. Coli, E. Typhus, S. Disenteriae, mouse-adapted poliomyelitis virus, the spore-former S. Subtilus and other organisms, established not only that chlorine dioxide was an effective bactericide but that it was also a potent viricide and sporicide. In addition, the effects of pH and temperature on the biological efficiency of chlorine dioxide were studied, developing the important point that chlorine dioxide is much less pH sensitive than chlorine with its biocidal effectiveness considerably higher than that of the latter at high pH values. (Figures 1, 2, 3)

More recent work by Bernarde and co-workers at Rutgers University (7-9) confirmed the findings of the earlier workers on disinfection by chlorine dioxide and established the mechanism as resulting from an abrupt inhibition of protein syntheses. (Figures 4, 5)

Despite this background and its adoption by some European cities for disinfection, alone or together with taste and odor control, no American water supply used chlorine dioxide solely for disinfection until its recent adoption by Hamilton, Ohio (10).

Admittedly, chlorine dioxide suffers a cost disadvantage when compared with chlorine as a disinfectant on a pound for pound basis. However, the unique properties of chlorine dioxide under special conditions were sufficiently well known for some years to attract operators of several types of industrial water systems to its use as a disinfecting agent where these special properties offered unique advantages or where chlorine dioxide became the superior disinfectant on a cost-effective basis. Interestingly enough, this included the use of chlorine dioxide in food processing plant waters which are normally considered to be simply potable waters. Chlorine dioxide has been in use for disinfecting flume waters and similar waters in food processing plants for better than fifteen years (11). During the past two to three years, there has been a rapid adoption of chlorine dioxide for disinfection of industrial cooling waters, particularly those which have a high chlorine demand (12).

Ridenour and Ingols (4) demonstrated the superiority of chlorine dioxide over chlorine for the disinfection of E. Coli in sewage effluent (Figure 5) and the lesser reactivity of chlorine dioxide with many nitrogen-containing compounds suggests that chlorine dioxide may have advantages for disinfection in municipal wastewater treatment. At this time, we are not aware of any use of chlorine dioxide for this purpose, although there have been several applications for disinfection in industrial wastewater treatment.

Physical and Chemical Properties

Chlorine dioxide is a gas whose color changes from yellow-green to orange-red as the concentration increases. It condenses to form a red, highly unstable liquid which freezes at $-59°C$ and boils at $+11°C$. The gas has a disagreeable, irritating odor which resembles both chlorine and ozone. At 17 ppm in air, the odor of the gas becomes evident and at 45 ppm is quite irritating.

Chlorine dioxide gas is sensitive to pressure and temperature so that it cannot be shipped in bulk, but must be generated and used on site. The gas can be handled safely by dilution with sufficient air or nitrogen to keep the ClO_2 concentration below 10%. In the pulp and paper industry, which requires reasonably high concentrations of the gas in solution, it is routinely handled in this way.

At the lower concentrations needed for water treatment operations, it is safer and more convenient to handle solutions of chlorine dioxide. It is fairly soluble in water, dissolving at 30mm partial pressure to form a solution containing 2.9 grams per liter at room temperature or more than 10.0 grams per liter in chilled water. Unlike chlorine, chlorine dioxide does not react with the water, but forms a true solution.

The amount of chlorine or chlorine-yielders present in a water is commonly reported as "available chlorine" which, in water works practice, usually refers to compounds which hydrolyze in water to form hypochlorous acid or a hypochlorite. By this definition, chlorine dioxide does not produce available chlorine. In general chemical usage, however, the term "available chlorine" is used to define the oxidizing capacity of any compound relative to chlorine, measuring the oxidizing capacity by the release of free iodine from potassium iodide. By this definition, chlorine dioxide has an available chlorine content of 263% compared with 100% for chlorine itself as shown by the following:

$$\frac{\text{(Electron changes)} \times 35.5}{\text{Molecular Weight}} \times 100 = \% \text{ Available Chlorine} \quad (1)$$

	Chlorine	Chlorine Dioxide
Molecule	Cl_2	ClO_2
Valence of Cl in compound	0	+4
Electron change to chloride ion	2	5
% available chlorine (Eq.1)	100.	263.

The degree to which this oxidizing capacity can be utilized in water treatment depends upon the pH of the system and the nature of the reacting substances in the water. For example, the superiority of chlorine dioxide over chlorine in avoiding the development of tastes when treating waters contaminated with certain industrial wastes depends upon the fact that chlorine forms a series of chlorophenols noted for their distinctive tastes at very low concentrations whereas chlorine dioxide breaks the ring structure of the phenolic compounds producing tasteless products. In waters contaminated with ammonia and many amines, chlorine reacts to form a family of chloroamines which are much weaker oxidizing agents than either chlorine or chlorine dioxide, consuming a considerable amount of chlorine before there is a free available chlorine residual. In contrast, chlorine dioxide does not react with ammonia and most amines. Finally, we have already mentioned the fact that chlorination of many surface waters produces chloroform, whereas treatment of these waters with chlorine dioxide does not.

Chlorine Dioxide Generation

The large volumes and higher concentrations of chlorine dioxide required for bleaching of pulp and paper have led this industry to generate its chlorine dioxide requirements by reduction of sodium chlorate by Equation 2 or by similar reactions with other reducing agents.

$$2\ NaClO_3 + SO_2 \longrightarrow 2\ ClO_2 + Na_2SO_4 \quad (2)$$

The safety hazards inherent in handling chlorine dioxide gas, together with the relatively low concentrations of chlorine dioxide required for water treatment, make it safer and more convenient to generate solutions of chlorine dioxide for this purpose from sodium chlorite solution. This preparation can be accomplished by either of two simple and easily controlled reactions.

At treatment plants where gaseous chlorine is already in use or is preferred for economic reasons, a sodium chlorite solution may be proportionately introduced into the strong chlorine solution downstream of the chlorinator. A solution of chlorine dioxide is generated in accordance with Equations 3 and 4, and this is then injected into the main water stream at an appropriate point or points in the treatment plant. For efficient generation of chlorine dioxide, the chlorine solution should contain at least 500 mg/l Cl_2.

$$Cl_2 + H_2O \longrightarrow HOCl + HCl \tag{3}$$

$$2\ NaClO_2 + HOCl + HCl \longrightarrow 2\ ClO_2 + 2\ NaCl + H_2O \tag{4}$$

If chlorine gas cannot be used for any reason, chlorine dioxide solution can be generated using a hypochlorite solution. The sodium chlorite solution and an acid are both proportioned into a flowing stream of hypochlorite solution. The acid is required in order to reduce the pH to the level necessary for complete conversion of the chlorite to chlorine dioxide. Equation 5 summarizes this reaction when using hydrochloric acid. Sulfuric acid may be used equally well.

$$NaOCl + 2\ HCl + 2\ NaClO_2 \longrightarrow 2\ ClO_2 + 3\ NaCl + H_2O \tag{5}$$

Mechanically, the preparation of chlorine dioxide for water works use is simple. Diaphragm-type chemical feed pumps are used to introduce the sodium chlorite solution and, if necessary, the acid into the chlorine solution or hypochlorite solution. Rapid and thorough mixing of the reactants is effected by passing them through a relatively short packed column. Both column and packing may be constructed of plastic, ceramic, or glass. If the column itself is not transparent, a transparent section of piping is included immediately downstream of it. This facilitates control by providing an inspection point where the yellow-green color of the chlorine dioxide solution may be readily seen by the operator. Absence of any color is an obvious indication that no chlorine dioxide is being generated. Although an experienced operator can estimate from the color of the effluent stream whether the chlorine dioxide concentration is close to that desired, a sample cock at this location is also desirable so that the concentration can be checked by an iodometric titration. The concentration range of chlorine dioxide at this point is usually 1,000-3,000 mg/l ClO_2. From this point, the chlorine dioxide solution is introduced into the main water stream at an appropriate location in the same way that a chlorine solution would be introduced.

Figure 6 shows a typical chlorine dioxide generator and auxiliaries for use with chlorine and Figure 7 shows the corresponding generator and auxiliaries for use with a hypochlorite.

It is clear from this description and the generator layouts that the preparation of chlorine dioxide presents no unusual mechanical and operational problems for a water treatment plant that is already equipped for chlorinating water either with chlorine or with a hypochlorite.

Treatment Control

Similarly, the control of treatment with chlorine dioxide has no mysteries for operators who have already been testing for available chlorine. By most chlorine test procedures, chlorine dioxide reacts the same as free available chlorine.

The conventional orthotolidine (OT) test produces a yellow color with chlorine dioxide just as it does with both free and combined available chlorine. This test cannot be used to distinguish between chlorine and chlorine dioxide. The orthotolidine arsenite (OTA) procedure also responds to chlorine dioxide, but does not permit distinguishing between ClO_2 and free available chlorine from other sources.

Another modification of the orthotolidine test, the orthotolidine-oxalic acid (OTO) method can be used to distinguish between chlorine and chlorine dioxide. Free and combined residual chlorine are eliminated by addition of oxalic acid. Any color then developed by the subsequent addition of orthotolidine results from the presence of chlorine dioxide. It is less intense than the color developed with a like concentration of chlorine. In addition to the usual interferences with the OT method, the presence of excess chlorite can cause high readings and high calcium waters may cause turbidity by precipitating some of the oxalate reagent.

Another colorimetric test is based upon the development of a bluish-pink color on adding H-acid (1-amino-8-naphthol-3,6-disulfonic acid) to samples acidified to pH 4.1-4.3 and in the presence of ferric chloride. This test also measures chlorine.

The DPD (N,N-diethyl-p-phenylenediamine) method can also be used for measurement of free available chlorine resulting from the presence of either chlorine dioxide or chlorine.

When using conventional color comparators or titration methods calibrated for mg/l chlorine, multiplying the results by 1.9 will convert them to mg/l chlorine dioxide.

Safe Handling

A number of water works operators have expressed concern about the hazards of handling sodium chlorite and chlorine dioxide. Actually, these are no more hazardous than any other oxidizing agent. In more than 30 years of American experience with the use of chlorine dioxide in water works, we have had no reports of accidents. Most reports of the hazards involved with the use of chlorine dioxide appear to stem from its application in pulp and paper plants where the less stable chlorine dioxide gas is separated at some stage in the process.

However, the sodium chlorite used to generate chlorine dioxide is a powerful oxidant and, if mishandled, can be as hazardous as other oxidants, such as the hypochlorites. The prime hazards derive from mixing of the sodium chlorite as the solid or strong solutions with acids so that large amounts of chlorine dioxide can evolve unexpectedly, or from contamination of the solid with combustible organic materials or reducing agents so that a fire is caused.

When using sodium chlorite for generating chlorine dioxide, the principal hazard is fire caused by permitting the drying out of a spill on a wood floor, clothing or other organic matter. Prompt flushing of any spill with ample amounts of water completely avoids this hazard.

Since traces of chlorine dioxide in the air are irritating, all areas subject to possible contamination should be well ventilated. This is, of course, exactly the same precaution required when using chlorine.

Operators handling the materials used for preparing chlorine dioxide in water treatment plants require only the normal personal protective equipment: splash-proof goggles or face shield, rubber or plastic gloves, and rubber or plastic apron. As with chlorine, gas masks should be available for emergency use. Operators should not smoke when handling chemicals, and should always wash well before eating.

Conclusions

Chlorine dioxide has been used for more than 30 years for the treatment of potable and non-potable waters for disinfection, taste and odor control, and iron and manganese removal. Its effectiveness and techniques for its generation and measurement are well known. The simplicity and low cost of the generating equipment, together with the familiarity of most water works operating personnel with the required equipment, testing methods, and chemicals handling procedures, make possible a rapid changeover from disinfection with chlorine to disinfection with chlorine dioxide for most water treatment plants.

REFERENCES

1. L. T. Dowling. Chlorine dioxide in potable water treatment, Water Treat. & Exam. 1974, 23, 190-204.

2. M. L. Granstrom and G. F. Lee. Generation and use of chlorine dioxide in water treatment, J. Am. Water Works Assn. 1958, 50, 1453-1466.

3. G. M. Ridenour and R. S. Ingols. Inactivation of Poliomyelitis Virus by "Free" Chlorine, Am. J. Pub. Health 1946, 36, (6) 639 ff.

4. G. M. Ridenour and R. S. Ingols. Bactericidal properties of chlorine dioxide, J. Am. Water Works Assn. 1947, 39, 561-567.

5. G. M. Ridenour and E. H. Armbruster. Bactericidal effects of chlorine dioxide, J. Am. Water Works Assn. 1949, 41, 537-550.

6. G. M. Ridenour, R. S. Ingols, and E. H. Armbruster. Sporicidal properties of chlorine dioxide, Water & Sewage Works 1949, 96 (8) 1.

7. M. A. Bernarde, B. M. Israel, V. P. Olivieri, and M. L. Granstrom. Efficiency of chlorine dioxide as a bactericide, Appl. Microbiol. 1965, 13 (5) 776-780.

8. M. A. Bernarde, W. B. Snow, and V. P. Olivieri. Chlorine dioxide disinfection temperature effects, J. Appl. Bacter. 1967, 30 (1) 159-167.

9. M. A. Bernarde, W. B. Snow, V. P. Olivieri, and B. Davidson. Kinetics and mechanism of bacterial disinfection by chlorine dioxide, Appl. Microbiol. 1967, 15 (2) 257-265.

10. H. W. Augenstein. Use of chlorine dioxide to disinfect water supplies, J. Am. Water Works Assn. 1974, 66 (12) 716-717.

11. J. L. Welch and J. F. Folinazzo. Use of chlorine dioxide for cannery sanitation and water conservation, Food Technology 1959, 13 (3) 179-182.

12. W. J. Ward. Chlorine dioxide - A new development in effective microbio control. Paper presented at Cooling Tower Institute Annual Meeting, Houston, Jan. 1976.

Table 1. Partial List of
Public Water Supplies Treated with Chlorine Dioxide

State or Country	Location and Function of Treatment*	
GA	Atlanta (Mn)	
KY	Covington (T/O)	
MI	Escanaba (T/O) Gladstone (T/O)	Menominee (T/O)
OH	Bowling Green (T/O) Columbus (T/O) Defiance (T/O) Elyria (T/O) Hamilton (D) Lorain (T/O) Napoleon (T/O)	Newark (T/O) Oregon (T/O) Sandusky (T/O) Toledo (T/O) Youngstown (T/O) (Mahoning Valley Sanitary District)
NY	Binghamton (T/O) Buffalo (T/O) (Erie County Water Authority)	Niagara Falls Tonawanda (T/O)
PA	Philadelphia (A)	Pittsburgh Vic. (T/O) (Western PA Water Co., Hays Mine Plant)
TX	Midland (T/O)	
WI	Green Bay (T/O)	
WV	Wheeling (T/O)	
England	Liverpool (D)	
Italy	Adria (D) Apulia (D) Florence (D)	Palermo (D) Turin (D)
Switzerland	Basel (D) Berne (D)	Zurich (D)

*Letters in parentheses: (A) = Algae control, (D) = Disinfection, (Mn) = Manganese removal, (T/O) = Taste and odor reduction.

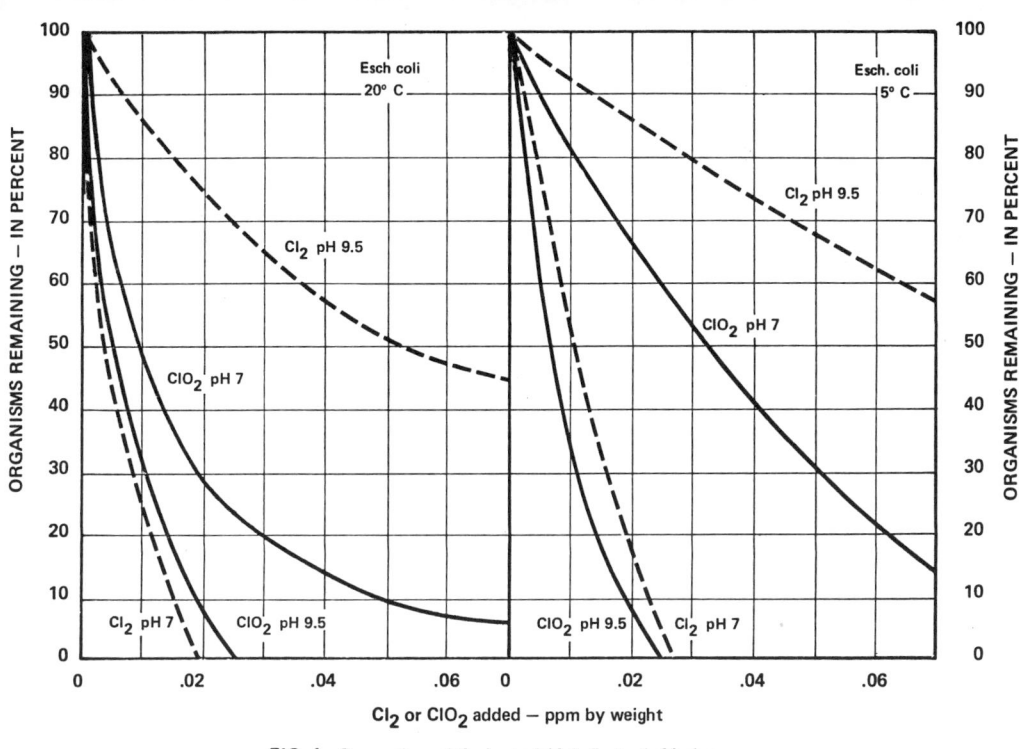

FIG. 1. *Comparison of the bactericidal effects of chlorine and chlorine dioxide on Esch. coli at different temperatures and pH with five minute contact*

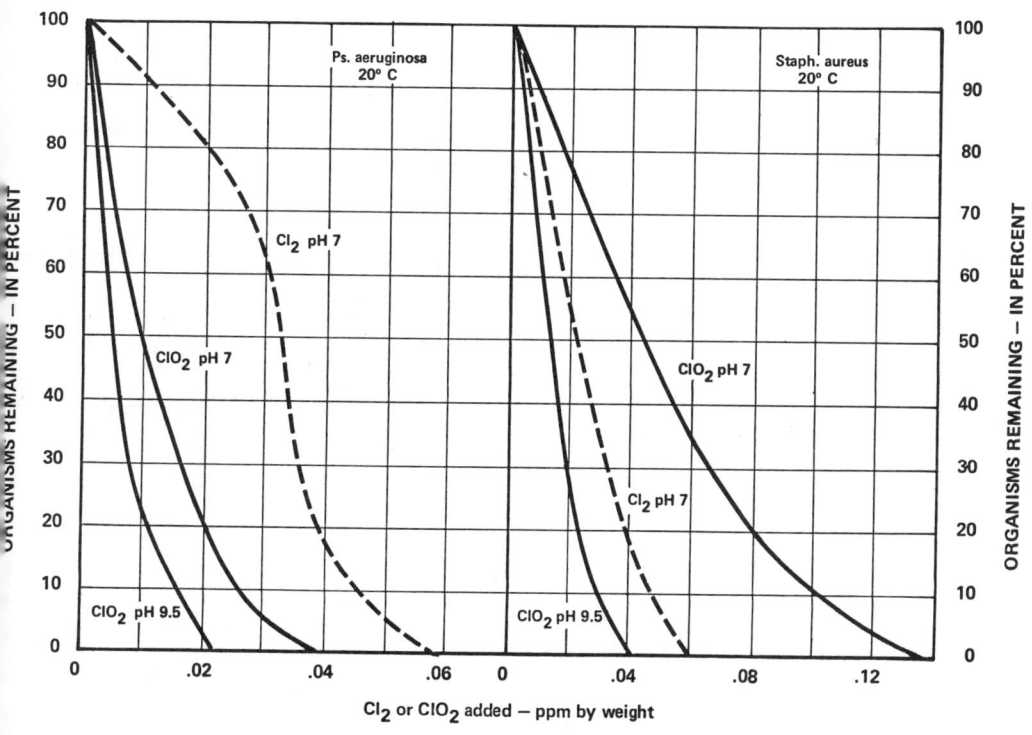

FIG. 2. *Comparison of the bactericidal effects of chlorine and chlorine dioxide on Ps. aeruginosa and Staph. aureus at different pH with five minute contact*

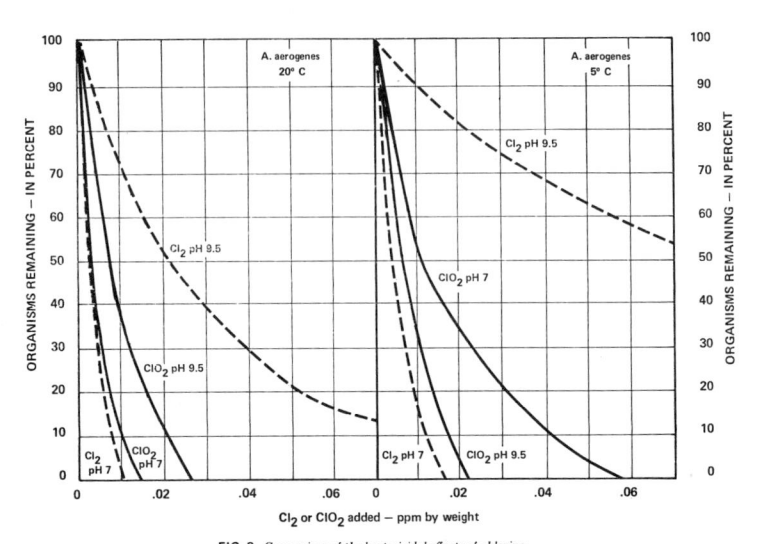

FIG. 3. *Comparison of the bactericidal effects of chlorine and chlorine dioxide on A. aerogenes at different temperatures and pH with five minute contact*

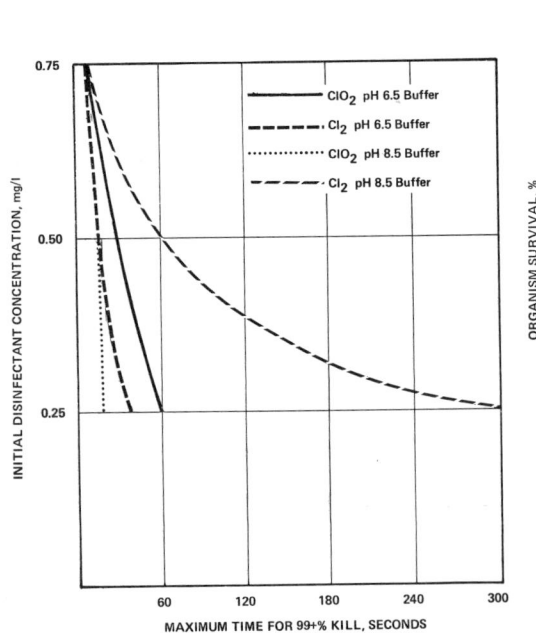

FIG. 4. *Effect of pH on kill*
Disinfection of E. coli in organic free buffer, 15,000 cells/ml at 24°C

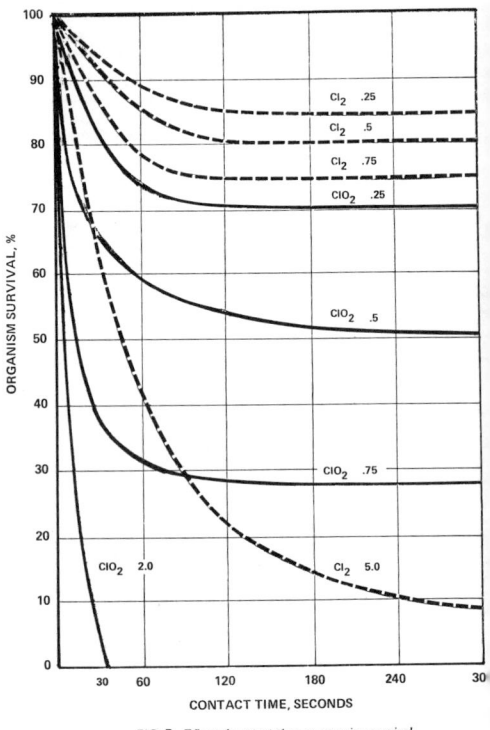

FIG. 5. *Effect of contact time on organism survival*
Disinfection of E. coli in sewage effluent   Chlorine dioxide vs. chlorine pH 8.5, 15,000 cells/ml at 24°C   Dosages in mg/l

FIG. 6. Chlorine Dioxide Generator -- Using Chlorine

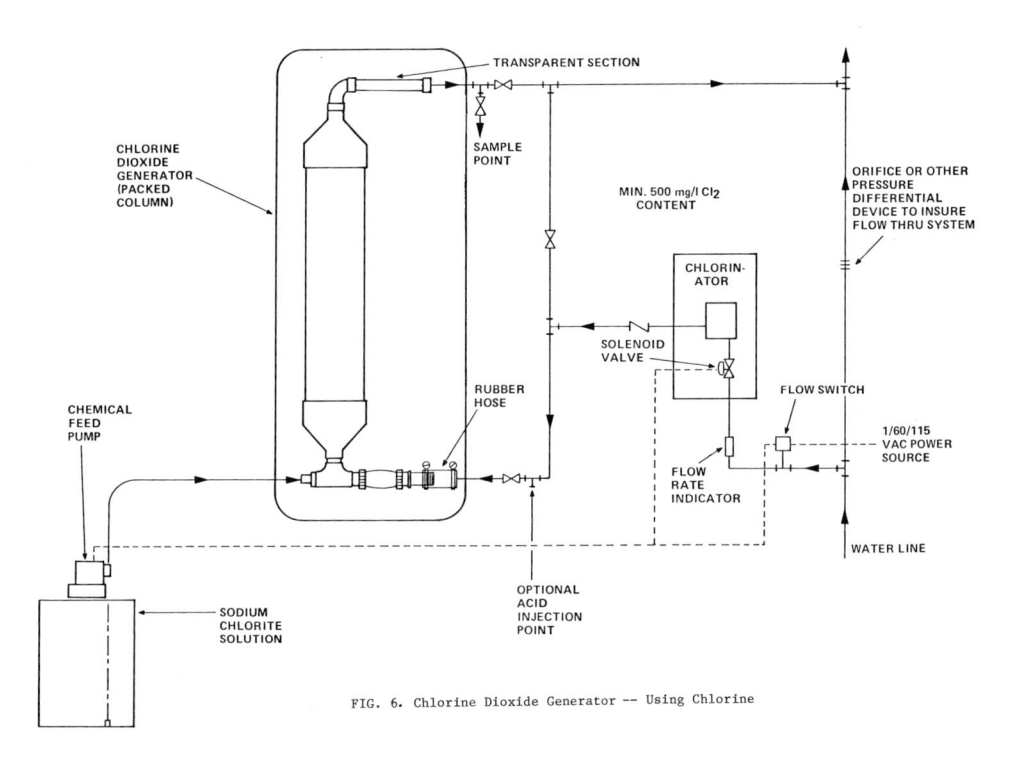

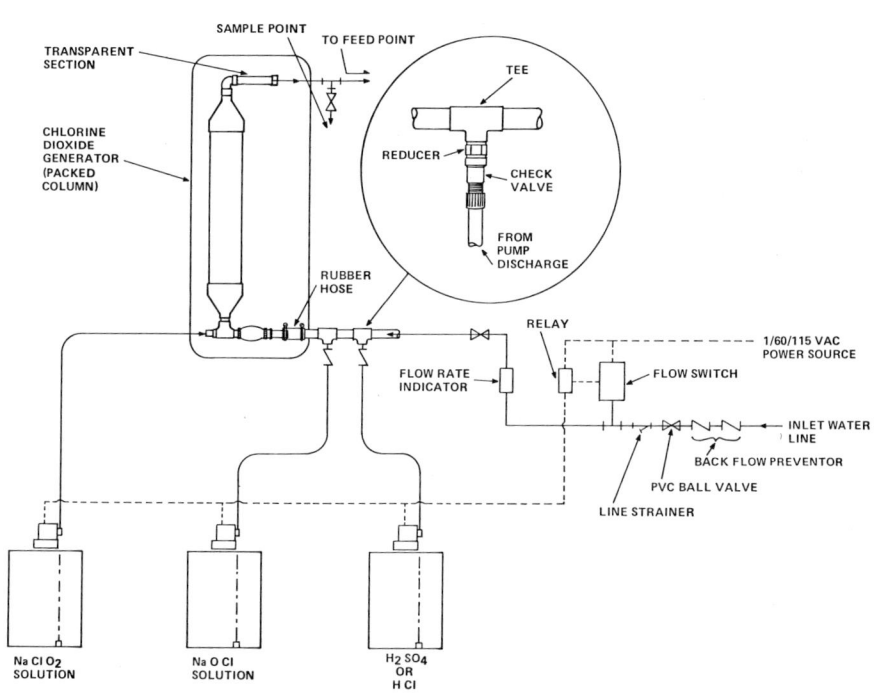

FIG. 7. Chlorine Dioxide Generator -- Using Hypochlorite

-355-

CHLORINE DIOXIDE

AN OVERVIEW OF ITS PREPARATION, PROPERTIES AND USES

R.J. Gall - Research and Development Department

Hooker Chemicals & Plastics Corp.
P.O. Box 8
Niagara Falls, New York 14302

Properties And Uses

Many thanks for the opportunity to share with you some thoughts on a most interesting and unique chemical, chlorine dioxide.

The purpose of this paper is to present an overview of the preparation, properties and uses of chlorine dioxide and to familiarize you with Hooker's experience and contributions to ClO_2 technology.

History

Chlorine dioxide, a greenish-yellow gas, was first discovered by Sir Humphrey Davy in 1811 (1) when he reacted potassium chlorate with hydrochloric acid. He called the gas produced "euchlorine", which is a mixture of chlorine dioxide and chlorine. Later in 1834 (2) Watt and Burgess, the inventors of alkaline pulping, mentioned the use of "euchlorine" as a suitable bleaching agent for soda pulp in their original patent.

In the early twenties (3), Eric Schmidt, a German botanist used a chlorine dioxide solution to dissolve lignin from very thin wood specimens so that he could microscopically examine the residual carbohydrate structure. Schmidt noticed the residue was white and since "chance favors the prepared mind", he recognized that he had discovered a new method for bleaching pulp. He obtained a patent on the treatment of pulp with chlorine dioxide in dilute acetic acid solution.

Schmidt's discovery remained a laboratory curiosity for a number of years because there was no simple, safe and economical way of producing ClO_2.

Then in the nineteen thirties (4), the Mathieson Alkali Works developed the first commercial process for generating ClO_2 from sodium chlorate. The ClO_2 was used to produce sodium chlorite, a recognized bleaching agent. By 1939, sodium chlorite was established as a commercial product.

In 1943, ClO_2 generated from sodium chlorite, was used for the first time in the Niagara Falls water treatment plant as a disinfectant and for taste and odor control. The use of ClO_2 for disinfection overcame the objectionable chlorophenols that had been developed when the phenol contaminated Niagara River water was treated by conventional chlorination processes. By 1945 the practice was officially adopted and its use by other plants with similar problems was begun.

In the summer of 1946 three pulp mills; Canadian International Pulp Co. at Temiskaming, Quebec, Mo och Domsjo AB, Husum Sweden and Stora Kopparbergs Bergslags AB, at Skutskar Sweden, independently began to bleach wood pulp with ClO_2 (5). Four years later two more mills; CIP in Canada and International Paper at Natchez Mississippi, started to generate ClO_2 by the Rapson-Wayman process. In 1951 the Harmac, B.C. mill of MacMillan and Bloedel Ltd. started to use ClO_2 and its growth has continued rapidly from that time. Today, some 60+ U.S. mills are generating a total of 500+ tons per day for pulp bleaching.

Co-currently with this development, the use of ClO_2 for taste and odor control, and more recently as the primary disinfectant for potable water, has grown and is now used in several hundred plants throughout the world.

Physical And Chemical Properties (6,7)

Chlorine dioxide is a yellowish-green gas which approaches an orange-red as its concentration increases. It is soluble in water as a true gas and is normally used as an aqueous solution. Its chemical formula is ClO_2 and structural formula $\overset{OO}{Cl}$. The molecular weight of ClO_2 is 67.5; ClO_2 freezes at -59°C forming a red, highly unstable liquid. It boils at +11°C. The density of ClO_2 is 1.62 at the boiling point and 1.765 at -56°C.

ClO_2 has a disagreeable, irritating odor somewhat similar to chlorine. The odor is evident at about 17 ppm in air and at about 45 ppm, is quite irritating. Its toxic effects are similar to chlorine gas. Concentration of ClO_2 vapor at about 11% in air may give mild explosions or "puffs" and concentrations of over about 4% in air may be set off by an electric spark to sustain a decomposition wave (in commercial practice, this potential hazard has been recognized and fail-safe systems have been engineered to overcome it).

ClO_2 may be readily decomposed on exposure to UV light. Therefore, it is always stored in the dark. The sensitivity of gaseous and liquid ClO_2 to temperature and pressure has mitigated against its bulk shipment. Thus, it has been generated and used on-site. ClO_2 can be handled safely in the gas phase by diluting with air or nitrogen to keep its concentration below about 10-11%.

The water solubility of ClO_2 depends on temperature and pressure. This may be seen in Figure 1 (8). At room temperature, 25°C and at 40 mm partial pressure, it is soluble to the extent of 2.9 grams/liter. In chilled water its solubility increases to over 7 grams/liter and at 80 mm partial pressure, the solubility increases to 6.2 and 15 grams/liter, respectively. ClO_2 is five times more soluble in water than chlorine and does not react with water as does chlorine. This difference in solubility is the basis on which the two gases are separated in some manufacturing processes. ClO_2, however, is extremely volatile and may be readily stripped from aqueous solutions by a minimum of aeration. Therefore, solution transfer systems should be designed to minimize or eliminate this potential loss. Chlorine dioxide is paramagnetic and thus may be differentiated from sodium chlorite in the in-line analysis of process streams.

Chlorine dioxide forms a hydrate with water and

has been the basis of a proposed method of transportation. The chlorine dioxide hydrate is made by contacting concentrated ClO_2 in air with water at near its freezing point. The slush of crystals formed may then be frozen, coated with ice and stored below -18°C.

ClO_2 does not contain "available chlorine" per se since it does not hydrolyze to form hypochlorous acid as does chlorine (9). In chemical terms, "available chlorine" is used to define the oxidizing potential of any compound relative to chlorine. When viewed in this light, ClO_2 contains 263% available or equivalent chlorine. The chlorine atom in ClO_2 has a valence of +4. When it is reduced to chloride, it loses five electrons and its available or equivalent chlorine may be calculated as follows:

$$\overset{+4}{ClO_2} \longrightarrow \overset{-1}{Cl} \quad -5e$$

$$\frac{5 \times 35.5}{67.5} \times 100 = 263\%$$

In a comparative way, when chlorine is reduced it loses two electrons as per the following reaction:

$$\overset{0}{Cl_2} + H_2O \longrightarrow \overset{+1}{HOCl} + \overset{-1}{HCl} \quad -2e$$

$$\frac{2 \times 35.5}{71} \times 100 = 100\%$$

From this it may be seen that ClO_2 possesses more than 2 1/2 times the oxidizing capacity of chlorine. This potential of ClO_2 is not always fully utilized. It depends in large measure on the pH of the system and the nature of the reactants involved. For example at high **pH** values and in the presence of reducing agents, only 20% of its oxidizing capacity is utilized. However, on the acid side, all of its oxidizing capacity will be used.

Preparation Or Generation

Chlorine dioxide may be generated from sodium chlorate or sodium chlorite. Since sodium chlorite is commercially produced by the reaction of ClO_2 with H_2O_2 in NaOH solution, it becomes apparent that for large scale operations the sodium chlorate route is the most economical.

Chlorine dioxide is produced commercially for the pulp and paper industry by the reduction of sodium chlorate in the presence of a strong acid. Sodium chloride, hydrochloric acid, sulfur dioxide and methanol are the reducing agents normally used. Sulfuric is usually the acid of choice.

Four different processes are currently used in North America. They are generally named after the company which developed them or the reducing agent used. They are:

1. The Mathieson or SO_2 Process
2. The Solvay or Methanol Process
3. The Hooker R-2 Process
4. The Hooker SVP(R) Process

The Mathieson Or SO_2 Process (10,11)

SO_2 is used as the reducing agent in this process and the overall reaction may be expressed as:

$$2NaClO_3 + H_2SO_4 + SO_2 \longrightarrow 2ClO_2 + 2NaHSO_4$$

Side reactions also take place and include:

$$2NaClO_3 + 5SO_2 + 4H_2O \longrightarrow Cl_2 + 3H_2SO_4 + Na_2SO_4$$

Figure 2 illustrates the Mathieson process. Two reaction vessels are operated in series and are termed primary and secondary generators.

The primary generator is filled to overflow with a solution containing H_2SO_4, $NaClO_3$ and chloride. A solution of $NaClO_3$, about 45%, is fed into the top of the generator and discharged below the surface of the reaction mixture. Co-currently, concentrated H_2SO_4 is fed is a similar manner. The chlorate feed is regulated to control the ClO_2 production and the H_2SO_4

feed is regulated to maintain the desired acid concentration of about 10N.

This process is run at atmospheric pressure. Air is sparged into the generators to mix the reaction media, strip the ClO_2 and provide a diluent for the ClO_2 to limit its concentration to about 10 to 12%. It is also used to carry the SO_2 reducing agent into the reaction mixture at a level of 6-8% in air.

The reaction mixture in the first generator continuously overflows into the secondary generator and is sparged with additional SO_2-air mixture. Here the reaction is completed. The acid concentration increases in the second generator through the formation of H_2SO_4 from the inefficient reaction of SO_2 with $NaClO_3$.

The spent acid overflows to a stripper where the dissolved gases are removed and then to storage. The acid, sulfur and sodium values are recovered for use in mill operations.

The ClO_2 is absorbed in chilled water, 2-5°C, in a packed column and yields a 6-8 g/l solution. The air containing a trace of ClO_2 passes through an induced draft fan, which maintains a slight negative pressure in the generator, and then discharged to the atmosphere. The negative pressure is used to prevent loss of ClO_2 to the atmosphere through the pressure relief valve vent and to minimize the possibility of gas compression which can cause ClO_2 decomposition.

Since the overall process is exothermic, the generators are jacket cooled. The primary generator is operated at about 40°C and the secondary at 45°C.

The Solvay Or Methanol Process (12)

Methanol is used as the reducing agent in this process.

The overall reaction may be written as:

$$2NaClO_3 + CH_3OH + H_2SO_4 \rightarrow 2ClO_2 + HCHO + Na_2SO_4 + 2H_2O$$

An intermediate reaction is:

$$2NaClO_3 + CH_3OH + 2H_2SO_4 \rightarrow ClO_2 + 1/2\ Cl_2 + CO_2 + 2NaHSO_4 + 3H_2O$$

The reaction efficiency may be determined by analyzing the exit gas or ClO_2 solution for the relative quantities of ClO_2 and Cl_2. Since Cl_2 should not be present as a reaction product, its level is an indirect measure of reaction efficiency. This procedure may not be applied to the Mathieson system; however, since any SO_2 escaping will destroy the Cl_2 and give erroneous results.

The flow sheet for the Solvay process is illustrated in Figure 3. The operation of the Solvay process is quite similar to the Mathieson except that the methanol is added to the incoming stream of the $NaClO_3$ solution, about 45%, immediately ahead of the generator. Concentrated sulfuric is diluted to about 75% and cooled prior to use. The process streams are regulated to provide the optimum reaction conditions of about 9 normal.

In this process, methanol and H_2SO_4 are fed to the secondary generator and drive the reaction to completion. As a result, the acidity is increased.

The ClO_2 is recovered by the same technique used in the Mathieson process.

The reaction rates in the Solvay process are considerably slower than in the Mathieson and therefore, high operating temperatures are required. The primary generator is run about 60°C and the secondary at about 63°C. Some cooling may be required in the primary while the secondary may require jacket heating with hot water.

The Hooker - R-2 Process (13)

The Hooker - R-2 process may also be called the "chloride reduction" process since it uses sodium chloride as the primary reductant. The reaction to produce ClO_2 in this process may be summarized as:

$$NaClO_3 + NaCl + H_2SO_4 \rightarrow ClO_2 + 1/2\ Cl_2 + Na_2SO_4 + H_2O$$

A competing reaction and common to all processes is the formation of Cl_2 when there is a low level of acid. This is shown in the following equation:

$$NaClO_3 + 5NaCl + 3H_2SO_4 \rightarrow 3Cl_2 + 3Na_2SO_4 + 3H_2O$$

This reaction is favored somewhat in the R-2 process in order to achieve the optimum conversion of $NaClO_3$. In practice, this is accomplished by feeding "R-2 solution" to the generator. The "R-2 solution" is an aqueous mixture of NaCl and $NaClO_3$ and provides a simple method for reactant addition. The feed rates of the 'R-2 solution" and concentrated H_2SO_4, are regulated to yield high acid reaction mixture. It should be noted that two thirds of the H_2SO_4 is used to adjust generator acidity.

The R-2 process is very efficient and yields of about 95% ClO_2 have been obtained.

A flow diagram for the R-2 process is illustrated in Figure 4.

Air sparging is also used in the R-2 process for mixing, stripping and ClO_2 - Cl_2 dilution.

Cl_2 is produced as a major by-product. As insignificant amount is absorbed with the ClO_2 and the remainder is reacted with dilute NaOH in a scrubbing tower to yield sodium hypochlorite. The recovered "hypo" is then used in the pulp bleaching process.

The reaction rates and efficiency of the R-2 process are such that only a single generator is required. Here too the spent acid is air stripped.

The reactions which produce ClO_2 in this process do not produce much heat.

The Hooker SVP(R) Process (14)

The Hooker SVP(R) process is a patented process and subject to license.

The reactions and feed solutions used in this process are similar to those illustrated for R-2. The equipment and operating conditions; however, vary considerably from R-2 and the other processes.

In the SVP process, ClO_2 generation, evaporation and crystallization of sodium sulfate take place in a evaporator-crystallizer.

The generator is operated at reduced pressure and the acidity and temperature conditions are maintained to produce an anhydrous sodium sulfate that may be readily recovered. The exact conditions are proprietary and only available under license.

Figure 5 illustrates the flow sheet for a typical SVP process wherein the co-product Cl_2 is recovered as weak Cl_2 water.

R-2 solution is fed to the circulated reaction mixture from the generator-evaporator-crystallizer (A) just ahead of the circulating pump (B) and heat exchanger (C). Heat is required to evaporate the water in the system.

Sulfuric acid is injected into the circulating stream. If the SVP process is installed to increase capacity, the spent acid from an existing conventional ClO_2 plant may also be used in a cascaded fashion.

The ClO_2, Cl_2 and water vapors formed in the generator, are fed to an indirect condenser where a substantial amount of heat is removed. The gases plus condensate flow to the ClO_2 absorber where ClO_2 and some Cl_2 are absorbed. The resultant solution contains a preponderence of ClO_2. The Cl_2 and non-condensables are passed to the Cl_2 absorber, wherein the Cl_2 is absorbed. The Cl_2 may also be absorbed in NaOH.

The non-condensables are discharged to the atmosphere.

The solid by-product, sodium sulfate, is continuously removed and filtered. The filter cake is washed with a small amount of water. The combined filtrates are returned to the generator.

The feed rate of H_2SO_4 and R-2 solution is controlled to yield the desired production rate and reaction-mixture-concentration.

A small amount of air is bled into the system to maintain the ClO_2 concentration below 12%.

Comparison Of the ClO_2 Processes (15)

Typical comparative consumption figures are shown in Table 1. It may be observed that the SVP is the most efficient and requires the least amount of reactants.

The ClO_2 produced along with the co-products of the various processes is detailed in Table 2.

Since these processes have been designed for use in the production of bleached pulp, most of the by-products may be recycled and used in other phases of pulp production. When this is possible, a credit is taken for amounts used, in calculating the economics of ClO_2 production.

The comparative economics of these ClO_2 generating processes are detailed in Table 3. It may be seen that the SVP produces the lowest cost ClO_2.

The Mathieson, Solvay and Hooker R-2 generators are made of lead lined steel while the SVP is fabricated from fiberglass reinforced polyester.

Several of these processes may be amenable to scale down for water and wastewater treatment depending on volume requirements, the treatment criteria and the allowable level of by-products entering the treated stream.

Additional Chlorate Routes to ClO_2 Generation

A number of other lesser known routes have been developed to produce ClO_2 from sodium chlorate. The Canadian International Paper Process (16), although it was the first process used in a North American pulp mill, has fallen into disuse because of its somewhat lower efficiency, about 70%. However, for substantially lower production rates, it may well be a process of choice when compared to producing ClO_2 from the more expensive sodium chlorite.

In this process, the sulfuric acid required is generated in situ by the over-reduction of sodium chlorate. The overall reaction may be written as follows:

$$2NaClO_3 + SO_2 \rightarrow 2ClO_2 + Na_2SO_4$$

A flow diagram is illustrated in Figure 6.

In this process, SO_2 is introduced into the base of a packed column and flows counter-current to a stream of concentrated sodium chlorate solution. The flow of chlorate solution is relatively low in contrast to the volume of gas. There is very little holdup in the column and thus, the process may be started and stopped quickly.

A significant improvement of this process has been developed by Jaszka of Hooker (17). In this process, chlorine is introduced with the SO_2 as shown in Figure 7. This reduces the chlorate requirement to oxidize the SO_2 to sulfuric acid and therefore, increased efficiences are obtained.

Both of the above processes yield ClO_2 with a small amount of Cl_2. With proper design and operation, the Cl_2 may be eliminated. A spent acid stream containing sulfuric acid and sodium bisulfate is discharged. Some sodium chloride will also be in the effluent from the Jaszka process.

The Société Universelle de Produits Chimiques et d'Appareiliages (18) of Paris has also described a ClO_2 generator based on the sodium chloride reduction of sodium chlorate in sulfuric acid solution. This unit has a capacity of up to 1,000 pounds per day.

Sodium Chlorite Route To ClO_2

For those industrial applications requiring lower volumes of ClO_2, (i.e., textile and flour bleaching and water treatment) the ClO_2 is conveniently generated from sodium chlorite. Three methods are generally employed. These include the treatment with acid and the reaction with Cl_2 or hypochlorite.

From Acidification

This method is generally preferred for textile and flour bleaching. The reaction may be expressed as:

$$10NaClO_2 + 5H_2SO_4 \rightarrow 8ClO_2 + 5Na_2SO_4 + 2HCl + 4H_2O$$

If the ClO_2 is prepared in the presence of reducing materials such as unbleached textiles or is continuously removed from the reaction media, 85 to 90% of the oxidizing power of chlorite appears as ClO_2 and 10% or less as chloric acid and about 3% as oxygen.

In the absence of reducing agents, about 61% of the oxidizing power appears as ClO_2, 36% as chloric acid and about 3% as oxygen.

A number of reactions are involved in the generation of ClO_2 by the acidification route and are reported to proceed simultaneously (19).

Most industrial water treatment facilities generate their ClO_2 from sodium chlorite using the "chlorine water" from an existing chlorinator or from sodium hypochlorite and hydrochloric acid (20). The ClO_2 is produced by the following reactions:

From Cl₂

The overall reaction is:

$$2NaClO_2 + Cl_2 + H_2O \rightarrow 2ClO_2 + 2NaCl + H_2O$$

The stepwise reactions are:

$$Cl_2 + H_2O \rightarrow HOCl + HCl$$

$$HOCl + HCl + 2NaClO_2 \rightarrow 2ClO_2 + 2NaCl + H_2O$$

From Hypochlorite

The overall reaction may be summarized as:

$$2NaClO_2 + NaOCl + 2HCl \rightarrow 2ClO_2 + 3NaCl + H_2O$$

The stepwise reactions are:

$$NaOCl + HCl \rightarrow NaCl + HOCl$$

$$HCl + HOCl + 2NaClO_2 \longrightarrow 2ClO_2 + 2NaCl + H_2O$$

These routes are simple, require little capital investment and easy to install, operate and maintain. The chlorine dioxide produced is much more costly than that produced by the large scale commercial routes.

Typical flow diagrams for the two routes are detailed in Figures 8 and 9.

In the chlorine route, 1.68 pounds of sodium chlorite (80%) are required to react with 0.5 pound of chlorine to yield 1.0 pound of chlorine dioxide. In practice, the chlorine solution from a solution feed type gas chlorinator is mixed with a solution of chlorite and fed into the base of a packed glass reaction tower. The required contact time to form ClO_2 is about one minute. A good visual indicator for this operation is the development of the characteristic greenish-yellow color of ClO_2.

The chlorinator is capable of producing solutions ranging from 500 to 5,000 ppm, thus providing flexibility of operation.

The chlorine to chlorite ratio 1:1 is recommended so that the excess of Cl_2 will increase the reaction rate and insure complete activation of the chlorite to give the best possible yields of ClO_2. The feed rates, pH about 3, and residence time are controlled for maximum efficiency of the generator.

Hypochlorite feed systems may also be adapted to Cl_2-chlorite generation processes. In this approach, both the sodium chlorite and sodium hypochlorite solutions are acidified with sulfuric or hydrochloric acid.

In practice, the three solutions may be fed via multiple head pulsafeeder type pump to the base of the mixing tower where the residence time is controlled to yield the desired ClO_2 concentration.

A typical installation using a 1% NaOCl solution requires a 2.42% $NaClO_2$ solution for mixing with a 2.0% H_2SO_4 solution. The H_2SO_4 is used in excess of stoichiometric in order to give the desired pH range, 3-4.

Uses Of Chlorine Dioxide

Early research indicated that ClO_2 possessed some rather unique and distinct properties. These include:

1. ClO_2 gas could not be safely concentrated or liquified as is chlorine.

2. ClO_2 could however be generated for use on-site.

3. ClO_2 was an effective pulp bleaching agent that expanded the brightness range with little or no cellulose degradation.

4. ClO_2 possessed bactericidal and viricidal properties.

5. ClO_2 was a unique oxidizing agent that reacted with a broad spectrum of organic and inorganic compounds.

Some of these properties have been developed into commercial applications. For example, the use of ClO_2 in pulp bleaching has become the standard of the industry. Some 500+ tons per day are produced in about 60 U.S. mills. Newer developments in pulp bleaching, pioneered by Hooker and others, will accelerate this current use. ClO_2 for this application is produced on-site from sodium chlorate at a level of 3 tons—20 tons per day. Installations of thirty to forty tons per day capacity are now being built.

The second largest application for ClO_2 is in the textile industry where it is used in bleaching and dye stripping.

The bleaching properties of ClO_2 are also used in the processing of flour, fats, oils and waxes.

The biological properties of ClO_2 have been used for the disinfection of potable water with improved taste and odor control. This application area is under intensive study and the use of ClO_2 in water and wastewater treatment is expected to expand in the near future.

The lower levels of ClO_2 required for some of these applications have in the past, been produced from sodium chlorite.

Other potential uses for ClO_2 come to mind when one studies the chemical reactions of ClO_2. Some of these are illustrated in the following sections.

Reactions Of ClO_2

Chlorine dioxide reacts with a number of organic and inorganic compounds to produce a variety of products. Typical reactions are illustrated in the following equations:

Inorganic Reactions (6,9)

- Oxidation of iron

$$ClO_2 + FeO + NaOH + H_2O \longrightarrow Fe(OH)_3 + NaClO_2$$

- Oxidation of manganese

$$2ClO_2 + MnSO_4 + 4NaOH \longrightarrow MnO_2 + 2NaClO_2 + Na_2SO_4 + 2H_2O$$

- Oxidation of sodium sulfide

$$2ClO_2 + Na_2S \longrightarrow NaCl + Na_2SO_4 + S$$

- With sodium iodide

$$2ClO_2 + 2NaI \longrightarrow I_2 + 2NaClO_2$$

- With NaOH

$$2ClO_2 + 2NaOH \longrightarrow NaClO_2 + NaClO_3 + H_2O$$

- With hydrogen peroxide

$$2ClO_2 + H_2O_2 + 2NaOH \longrightarrow 2NaClO_2 + 2H_2O + O_2$$

- With ozone to yield chlorine hexoxide

$$ClO_2 + O_3 \longrightarrow ClO_3 + O_2$$

$$ClO_3 + ClO_3 \longrightarrow Cl_2O_6$$

The reaction of ClO_2 with iron and manganese is the basis for their removal from reservoir and well water sources.

The liberation of iodine by the reaction of ClO_2 with sodium iodide is the basis for its analytical determination.

A mixture of $NaClO_2$ and $NaClO_3$ formed by reaction of ClO_2 with NaOH is called R-I salt and may be used in some ClO_2 generation processes.

ClO_2 reacts with hydrogen peroxide in the presence of NaOH to produce sodium chlorite. This is the basis for the commercial production of chlorite.

Organic Reactions (5,7)

In the organic reactions of ClO_2 the products formed and the yields obtained vary depending on the reaction conditions. Therefore, the following represents an over-simplification of the reactions involved and are meant to be only illustrative.

- With lignin model compounds (pulp bleaching)

Vanillin + ClO_2 ⟶ β-Formyl Muconic Acid Monomethylester

- With saturated hydrocarbons ⟶ acids
- With ethylenic double bonds ⟶ ketones
  ⟶ epoxides
  ⟶ alcohols
- With saturated acids ⟶ no reaction
- With anhydrides ⟶ no reaction but catalyzes hydrolysis
- With aldehydes ⟶ acids
- With ketones ⟶ alcohols
- With amines - primary aliphatic ⟶ slow reaction
  - secondary aliphatic ⟶ slow reaction
  - tertiary aliphatic ⟶ rupture of CN bond no N-Oxides formed
- With phenols ⟶ dicarboxylic acids

Biological Properties

Chlorine dioxide was found to possess biological properties in the early studies of Ridenour et al. (20,21,22). They reported the poliomyelitis virus was inactivated by chlorine dioxide.

They also demonstrated the bactericidal properties of chlorine dioxide with E. coli and artificially polluted solutions of peptone in water.

In addition, they evaluated the following group of test organisms:

Common Water Pathogens

    Eberthella Typhosa
    Shigella Dysenteriae
    Salmonella Parathyphi B

Relatively Disinfection-Resistant Organisms

> Pseudomonas Aeruginosa
> Staphylococcus Aureus

Sanitary Test Organisms

> Escherichia Coli
> Aerobacter Aerogenes

For all of the organisms studied, bactericidal efficiency of ClO_2 increased with an increase in pH.

The sporicidal properties of chlorine dioxide were determined using the spore family bacteria.

> B. subtilis
> B. megatherium
> B. mesentericus

ClO_2 was found to be an effective sporicide.

Benarde et al. (23,24) questioned the validity of some of these earlier studies because the physicochemical characteristics and kinetics of ClO_2 were unavailable to these investigators. The biocidal effect per se was not questioned, but the efficiency was.

Benarde noted that the germicidal activity of Cl_2 results from its hydrolysis in aqueous solution to form hypochlorous acid which is the disinfectant constituent.

The optimum activity occurs in acidic solution where HOCl exists in an undissociated form. Hypochlorous acid ionizes with an increase in pH and the resulting hypochlorite ion, OCl^-, has little disinfectant value.

In contrast, Benarde reported that ClO_2 does not hydrolyze in aqueous solutions. The spectral analysis of ClO_2 solutions at pH at 4.0, 6.45, and 8.42 showed the concentration of ClO_2 was unaltered and therefore, the ClO_2 molecule appeared to be the bactericidal compound.

Using very sophisticated techniques, Benarde et al. studied:

1. Effect of pH on kill,

2. Effect of contact time on organism survival -- E. coli was the test organism,

3. Effect of contact time on disinfectant utilization.

Their results showed a decided effect of pH on maximum time for 99+% kill.

At a pH of 6.5, Cl_2 was somewhat more efficient than ClO_2. ClO_2 was dramatically more effective than Cl_2 at pH 8.5. These data are illustrated in the following:

pH 6.5

Effect Of pH On Maximum Time For 99+% Kill

Initial Disinfection Con. (mg/l)	Maximum Time For 99+% Kill (sec.)	
	ClO_2	Cl_2
0.25	60	30
0.50	30	15
0.75	15	15

pH 8.5

Initial Disinfection Con. (mg/l)	Maximum Time For 99+% Kill (sec.)	
	ClO_2	Cl_2
0.25	15	300
0.50	15	60
0.75	15	15

In order to study the comparative bactericidal efficiencies in simulated plant conditions, a known cell density of E. coli was added to a sterile, unchlorinated sewage effluent. ClO_2 and Cl_2 residuals were determined by the OT test. The combination of organic matter and pH was found to sharply reduce their disinfectant activity. This is shown in the table on the following page.

Effect Of Contact Time On Organism Survival
Disinfection Of E. Coli In Sewage Effluent
ClO_2 vs. Cl_2 - pH 8.5 - 15,000 Cells/ml - 24°C

Dosage, mg/l	Organism Kill, %	
	ClO_2	Cl_2
0.25	31	17
0.50	50	20
0.75	75	25
2.0	100	--
5.0	--	95

The effect of contact time on disinfectant utilization was also studied with the same organism system as in the above. Residual values for ClO_2 and Cl_2 were determined. When an initial dose of 2.0 ppm ClO_2 was applied, complete organism removal was achieved in 30 seconds. This actually required 0.9 ppm of ClO_2. In the case of Cl_2, 5.0 ppm were applied to yield 90% removal in 5 minutes, which used 2.25 ppm of ClO_2. These data indicate that the major reduction of organisms occurred in the first minute of contact, even though residuals may remain after 5 minutes.

This then suggests that residuals are poor measures of disinfectant value. These data also demonstrate the greater bactericidal activity of ClO_2 versus Cl_2.

The kinetics and mechanism of bacterial disinfection with ClO_2 were studies by Benarde et al. Their data indicated that the mechanism of chlorine dioxide kill occurred via the disruption of protein synthesis and not enzyme inactivation as had been postulated by others.

Analysis

The higher concentrations of chlorine dioxide free of chlorine, or containing only trace amounts, may be readily analyzed by titrating the liberated iodine from an acidified KI solution with standard sodium thiosulfate solution.

If however, the chlorine concentration is not negligible, both the Cl_2 and ClO_2 must be determined. This is done by first conducting a neutral titration, since in neutral solution ClO_2 liberates one equivalent of iodine. So too does the free chlorine. The solution is then acidified, and the ClO_2 liberates four additional equivalents of iodine which may then be titrated. With the appropriate calculations, both concentrations of ClO_2 and Cl_2 may be determined.

Very dilute solutions of ClO_2 are measured colorimetrically and several methods are available. Since the conventional OTA (ortho-tolidine arsenite) test does not distinguish between free chlorine and chlorine dioxide, a modification is required. This modification called OTO (ortho tolidine oxalic acid) is simple and fast. It is based on the fact that chlorine will react with oxalic acid, but ClO_2 will not. The standard OTA procedure is carried out with the addition of another step wherein the sample is added to a saturated oxalic acid solution followed by the OT solution and arsenite reagent. The necessary corrections and calculations may then be made.

The DPD method, diethyl-phenylenediamine, has been reported to be well suited for biocontrol applications. This is also a colorimetric procedure.

Microbiological analysis may also be used to control the ClO_2 dosage rate.

Summary

In summation we have seen that:

ClO_2 is a greenish-yellow gas, soluble in water.

ClO_2 reacts with a wide range of organic and inorganic compounds.

ClO_2 is not amenable to bulk shipment.

ClO_2 is, therefore, generated on-site from sodium chlorate or chlorite, with the chlorate being the least expensive route.

ClO_2 via sodium chlorate, is used in tonnage quantities for wood pulp bleaching.

ClO_2 has been used in potable water treatment for taste and odor control.

ClO_2 is effective against a broad spectrum of organisms including viruses, bacteria, and spores.

ClO_2 possesses particular appeal as a Cl_2 replacement in water disinfection, since under most conditions, it does not produce chlorinated organics.

LITERATURE CITED

(1) Davy, H., Phil. Trans. 101, 155-162 (1811).

(2) Watt, C. and Burgess, H., U.S. Pat. 11,343 (1854).

(3) Schmidt, E., Ber. 54 B: 3111-3114 (1921).

(4) Synan, J.F., MacMohon, J.D., and Vincent, G.P., Water Works and Service 91:423 (1944).

(5) Rapson, W.H., TAPPI Monograph Series No. 27, 130-131 (1963).

(6) Kirk-Othmer-Encyclopedia of Chemical Technology, Vol. 5, 27-47 (1964).

(7) Masschelein, W., Monographics Dunod, "Chlorine Oxides and Sodium Chlorite," 16-57 (1969).

(8) Haller, J.F., and Northgraves, W.W., TAPPI, Vol. 38, No. 4, 199-202 (1955).

(9) White, G.C., Handbook of Chlorination, Van Nostrand Reinhold Company (1972).

(10) Woodside, V. and MacLoed, K.S., Paper Trade J., 137, No. 8:26-31 (August 21, 1953).

(11) Northgraves, W.W., Nicolaisen, B.H., Dexter, T.H., and Jaszka, D.J., Paper Mill News, 79, 10-13 (1956).

(12) Hampel, C.A., Soule, E.C., Canadian Patent 434,213 (1946).

(13) Partridge, H.D. and Rapson, W.H., TAPPI 44, No. 10: 698-702 (1961).

(14) Partridge, H.D., Schoepfle, B.O., Schulz, A.C. and Rosen, H.J., U.S. Patent 3,563,702 (1971).

(15) Atkinson, E.S. and Simonette, R., Pulp and Paper, No. 4:32-36 (1968).

(16) Rapson, W.A., and Wayman, M., U.S. Patent 2,481,240 (1949).

(17) Jaszka, D.J., U.S. Patent 3,950,500 (1976).

(18) SUPCA - Technical Bulletin, 23 bis Rue Balzor Paris.

(19) Synan, J.F. and Malley, H.A., Paper presented to Engineering Panel of the Campden Food Preservation Research Associations, Chipping Campden, Glos., England (1975).

(20) Ridenour, G.M. and Ingols, R.S., J. Public Health, Vol. 36, No. 6 (1946).

(21) Ridenour, G.M. and Armbruster, E.H., J. AWWA, 537-550 (1949).

(22) Ridenour, G.M., Ingols, R.S. and Armbruster, E.H., Water and Sewage Works, Vol. 96, No. 8 (1949).

(23) Benarde, M.A., Israel, B.M., Oliveri, V.P. and Granstrom, M.L., J. Applied Microbiology, Vol. 13, No. 5 (1955).

(24) Benarde, M.A., Snow, W.B., Oliveri, V.P. and Davidson, B., J. Applied Microbiology, Vol. 15, No. 2 (1967).

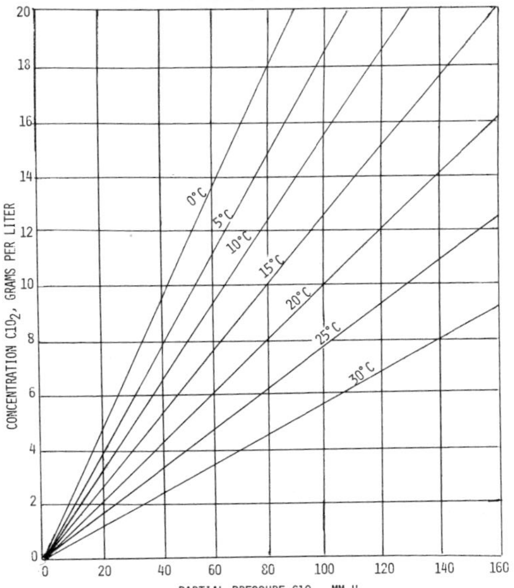

FIGURE 2
MATHIESON PROCESS FLOWSHEET

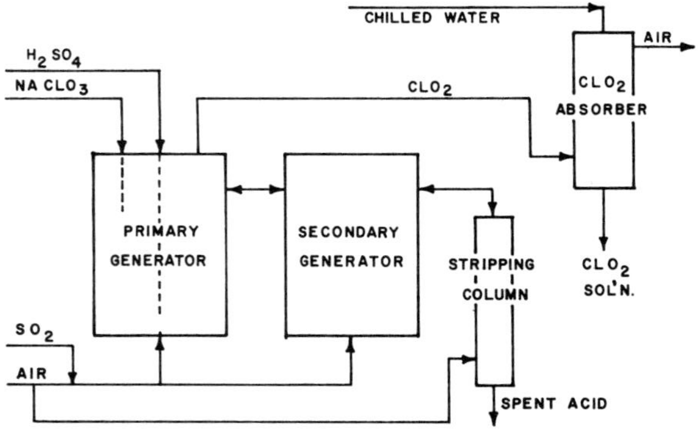

FIGURE 3
SOLVAY PROCESS FLOWSHEET

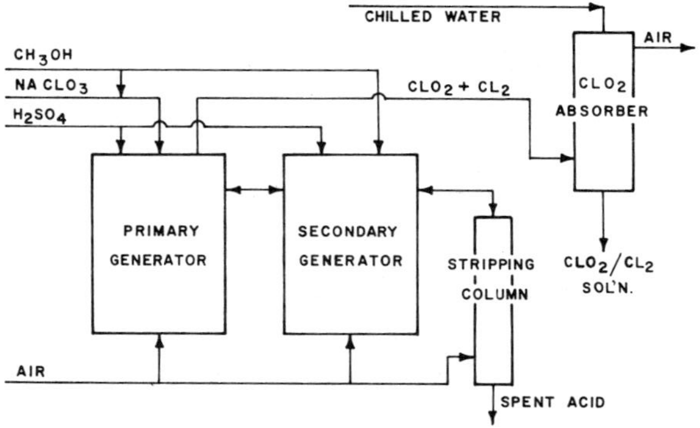

FIGURE 4
HOOKER R-2 PROCESS FLOWSHEET

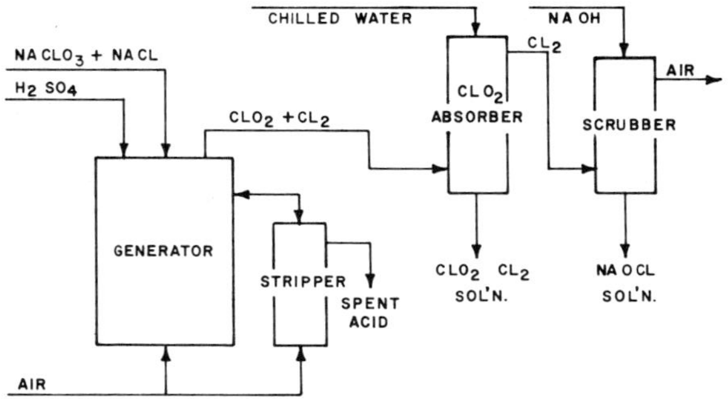

FIG. 1 - SOLUBILITY OF CHLORINE DIOXIDE IN WATER

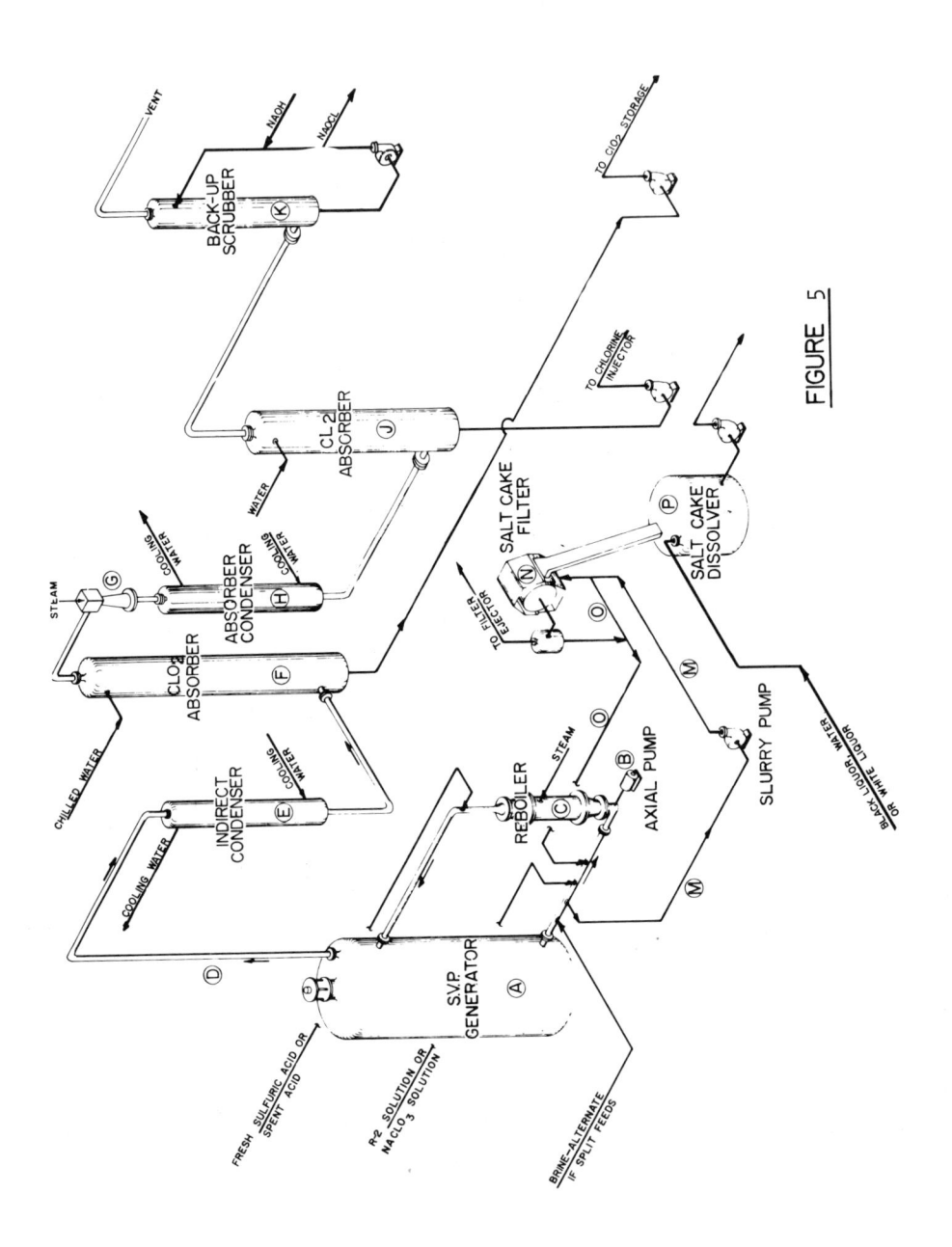

FIGURE 5

FIGURE 6
CIP PROCESS

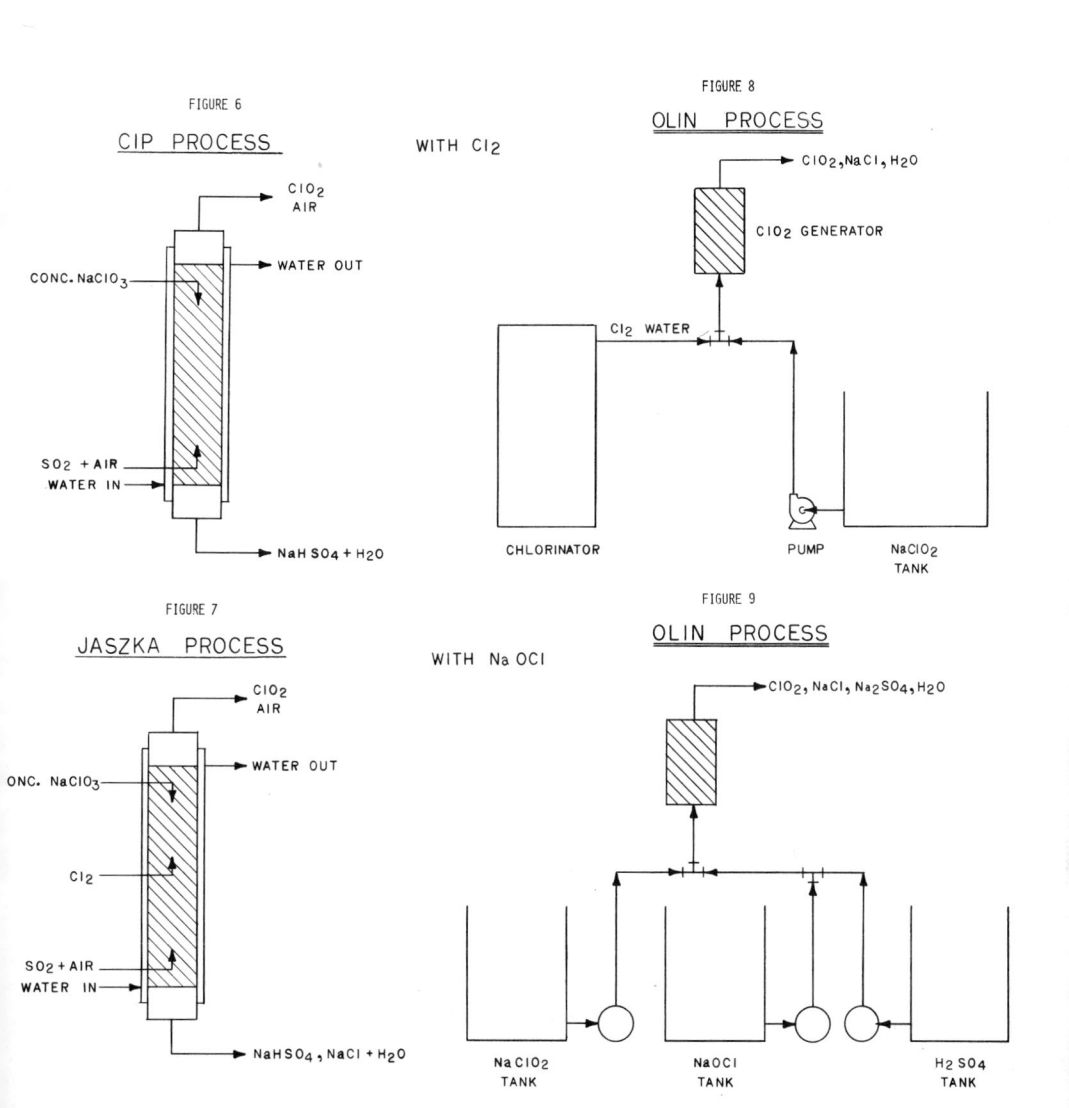

FIGURE 8
OLIN PROCESS
WITH Cl$_2$

FIGURE 7
JASZKA PROCESS

FIGURE 9
OLIN PROCESS
WITH NaOCl

TABLE 1

RAW MATERIAL REQUIREMENTS

LB/LB ClO$_2$

REACTANT	MATHIESON	PROCESS SOLVAY	R-2	SVP
NaClO$_3$	1.80	1.81	1.66	1.66
NaCl	0.10	--	0.98	0.95
CH$_3$OH	--	0.21	--	--
SO$_2$	0.65	--	--	--
H$_2$SO$_4$	1.55	2.90	4.85	1.70

TABLE 2

TYPICAL BY-PRODUCTS OF ClO$_2$ PROCESSES

PROCESSES LBS

BY-PRODUCT	MATHIESON	SOLVAY	R-2	SVP
Cl$_2$	0	0.1	0.58	0.58
Na$_2$SO$_4$, Or Equiv.	1.3	1.3	2.3	2.3
H$_2$SO$_4$	1.7	1.7	3.2	-
Total Sulfur, Na$_2$SO$_4$ Basis	3.66	4.08	6.94	2.3

TABLE 3

ClO$_2$ - TYPICAL CHEMICAL COSTS

(FROM CHLORATE)

¢/LB

	AT FULL BY-PRODUCT CREDIT	PARTIAL SPENT ACID CREDIT
MATHIESON	22 - 24	31 - 33
R-2	16 - 18	30 - 32
SVP	16 - 18	---

PRODUCTS OF CHLORINE DIOXIDE TREATMENT OF ORGANIC MATERIALS IN WATER

Alan A. Stevens
Dennis R. Seeger and Clois J. Slocum

Environmental Research Center
Municipal Environmental Research Laboratory
Water Supply Reseach Division
26 West St. Clair Street
Cincinnati, Ohio 45268
U.S.A.

ABSTRACT

Concern has arisen over the presence of trihalomethanes and other chlorinated organic chemical by-products from chlorine disinfection of drinking water supplies. Treatment options for preventing formation of or for removal of these by-products include the use of chlorine dioxide as an alternate disinfectant. If this treatment option is selected, a new and different array of organic by-products may be produced. Although little information exists describing the reactions of chlorine dioxide with organic compounds under water treatment conditions, the available literature indicates that the products of these reactions might be predominantly aldehydes, carboxylic acids, ketones, and quinones. Trihalomethanes are not formed, but certain other chlorinated products may be. Studies on this subject are just beginning at our laboratory.

INTRODUCTION

During the last two years, interest in the study of organic compounds in drinking water has increased. This trend in drinking water research started with the disclosure of the results of a 1974 study of New Orleans drinking water (1) where source water contaminants were still found to be in the treated drinking water. About the same time a special problem was described in two papers -- trihalomethanes (chloroform, bromodichloromethane, dibromochloromethane and bromoform) were found in finished drinking waters, but were not present at detectable concentrations in the source water (2,3). Indications were that the trihalomethanes were formed during the chlorination/disinfection part of the treatment process.

Because of the concern as to the significance of the formation of chlorinated organic compounds during the water treatment process, the U.S. EPA included the measurement of these four trihalomethanes in its National Organics Reconnaissance Survey (NORS) of 80 selected cities (4).

Source and finished waters were analyzed, and as a result, the occurrence of trihalomethanes in finished drinking water was demonstrated to be widespread and a direct result of the chlorination practice. During the survey, the median concentration of chloroform found was 21 µg/l in finished waters. The other three trihalomethanes were generally found in lower concentrations, although this was not always the case (4).

The organic precursors to trihalomethane formation have since been demonstrated to be the natural humic materials (5,6) present in virtually all source waters.

Because of recent findings concerning the carcinogenicity of chloroform and the confirmation of the ubiquity of trihalomethanes in chlorinated drinking water, the USEPA has established a policy initiating a Voluntary Nationwide Chloroform Reduction Program (7) in which utilities will use available knowledge and technology to adjust or change treatment processes to attempt to reduce concentrations of trihalomethanes reaching the consumer.

Three treatment approaches which can be considered for preventing trihalomethanes from reaching the consumer are: (a) Remove precursor (humic material) before chlorination, (b) remove trihalomethanes after they are formed, and (c) change disinfectant. This paper is being presented with consideration of the third option only.

Two oxidants sometimes used and now often suggested as disinfectant alternatives to chlorine are ozone and chlorine dioxide. Ozone as an alternate was discussed by others at this meeting, therefore only chlorine dioxide will be discussed here.

Some results of work by Miltner (8) at our laboratory to monitor trihalomethane production with time after chlorine dioxide, chlorine, and chlorine dioxide with chlorine were separately applied to pilot plant settled water are shown in Figure 1. The upper curve (Figure 1) represents the action of chlorine alone; the curve coincident with the abcissa represents the application of chlorine dioxide alone; and the curve between these two represents the action of chlorine to form trihalomethanes in the presence of chlorine dioxide. These curves demonstrate two important points: (a) chlorine dioxide does not cause the formation of trihalomethanes and (b) chlorine dioxide plus excess chlorine (as is often the case in water treatment) causes formation of lower concentrations of trihalomethanes than the same amount of chlorine alone. Therefore, with consideration only to minimizing trihalomethane formation, chlorine dioxide is a viable alternative to chlorine as a disinfectant in water treatment. Although this is the case, consideration still must be given to other by-products that may be formed by chlorine dioxide treatment.

ORGANIC REACTIONS OF CHLORINE DIOXIDE

There is considerable evidence that chlorine dioxide reacts with organic material during water treatment and therefore is likely to produce organic by-products:

1. Because chlorine dioxide is a good disinfectant, some reaction is taking place between the cell components of the organism and the chlorine dioxide.

2. Surface waters exhibit a chlorine dioxide demand.

3. At applied chlorine dioxide concentrations higher than those encountered in drinking water treatment, identifiable by-products have been isolated.

4. Chlorine dioxide destroys phenolic compounds when the oxidant is used for taste and odor control in water supplies.

5. Most importantly, as was shown earlier, the presence of chlorine dioxide reduces the formation of trihalomethanes by chlorine. This and other evidence obtained by Miltner (8) indicates that chlorine dioxide reacts with natural humic acid. This is not at all surprising because chlorine dioxide is used in the paper industry for delignification of wood pulp and is somewhat effective for reducing color in drinking water supplies (9).

Before an informed decision can be made as to the relative safety of the two oxidants with regard to the formation of organic by-products, a detailed investigation and identification of the products formed during disinfection with both chlorine and chlorine dioxide is necessary. The purpose of this paper is to briefly consider the possible organic by-products arising from the use of chlorine dioxide as a disinfectant in drinking water treatment. In this regard, it is most informative to present the reactions of chlorine dioxide in perspective with the reactions of chlorine when both are applied under the same circumstances.

Although the literature describing the organic reactions of chlorine dioxide is brief, numerous products of oxidation and chlorination by chlorine dioxide are described. However, the majority of this literature describes chlorinated and non-chlorinated derivatives, including acids, epoxides, quinones, aldehydes, disulfides, and sulfonic acids, that are products of reactions carried out under conditions vastly different from those experienced at water treatment plants. This paper is not a complete review of organic reactions of chlorine dioxide and discusses only a few reaction types which are known or suspected to be important in water treatment practice on the basis of some experimental evidence. A more complete review of chlorine dioxide chemistry is available elsewhere (10). The possible reactions with saturated aliphatic hydrocarbons, olefins, amines, and aromatic compounds (especially phenols) are in that group. In all examples presented for chlorine dioxide reactions the authors claimed that the applied chlorine dioxide was free of chlorine or hypochlorite.

Reactions with Saturated Aliphatic Hydrocarbons

No evidence exists that chlorine dioxide or chlorine undergoes reactions with saturated aliphatic hydrocarbons under water treatment conditions. The question of whether or not free chlorine reacts with saturated aliphatic hydrocarbons in aqueous systems to form chlorinated derivatives frequently arises because of the presence of methane in many ground waters, and the root name "methane" in "trihalomethane" implies that such reactions take place. Such reactions of free halogen with aliphatic hydrocarbons require a free radical mechanism, however, and are improbable in aqueous systems where chlorinated derivatives are formed. The trihalomethane formation can be more readily explained by the classical haloform reaction with methyl ketones. However, the classical haloform reaction is not the complete explanation because certain phenol derivatives, notably resorcinol, have been demonstrated to react quite readily with aqueous chlorine to produce chloroform (2).

Reactions with Olefins (See Fig. 2)

Chlorine can react with olefins by addition

across the double bond to produce the saturated dichloro derivative. A more likely course in aqueous systems, however, is to produce the chlorohydrin by reaction with hypochlorous acid (11).

The reactions of chlorine dioxide with olefins are apparently very complex, producing a host of chlorinated and nonchlorinated products. Methyl oleate is reported to react at the site of the double bond to produce aldehydes, the epoxide, chlorohydrin, a dichloro derivative, and α-chloro and α-unsaturated ketones. The aldehyde formation seems to be subject to argument: Leopold and Mutton (12) reported aldehyde formations by chlorine dioxide cleavage of the double bond in triolein (the triglyceride of oleic acid), and Lindgren and Svahn (13) did not find the aldehydes after chlorine dioxide reaction with methyl oleate.

Cyclohexene in aqueous mixture with pure chlorine dioxide has been shown by Lindgren and Svahn (14) to produce a similar complex mixture: cyclohexene-3-one, 3-chlorocylohexene, trans-2-chlorocyclohexanol, 2-chlorocyclohexanone, trans-1,2-dichlorocyclohexane, 1,2-epoxy-cyclohexane, and adipic acid. Note that some of these products are the same as would be expected from reaction of cyclohexene with chlorine.

Reactions with amines (See Fig. 3)

Aqueous chlorine and chlorine dioxide react very differently with amines. Chlorine reacts with ammonia, primary, and secondary amines to produce the well recognized chloramines by replacement of hydrogen. Tertiary amines are a special case producing a chloramine and an aldehyde (11). Chlorine dioxide, however, does not react with ammonia and reacts only slowly with primary amines. In general, amines produce the respective aldehydes upon reaction with chlorine dioxide in the following order of reactivity: tertiary>secondary>primary. Chlorine dioxide does not react to form chloramines.

Two examples (amines and possibly olefins) have now been presented where aldehydes might be the end products of reaction of organic materials with chlorine dioxide. If aldehydes are formed, they will react further at low pH with the chlorite also formed by reduction of chlorine dioxide and produce the respective acids. This reaction is slow at treatment pH, however.

These observations suggest that if certain organic substrates are present in a source water, chlorine dioxide treatment should cause an increase in the concentrations of the respective aldehydes. Preliminary results of some of our in-house work indicate that this may be occurring.

Two gas chromatograph/mass spectrometer (GC/MS) total ion current profiles (TICP) for compounds purged from samples of Ohio River water, one sample treated with chlorine dioxide and the other untreated, are shown in Figure 4. Some compounds appear to be higher in concentration in the treated sample, and some new compounds appear. Many of these were identified as aldehydes by their respective mass spectra. Because the chlorinated compounds are present in such large concentrations compared to the aldehydes, apparent changes in aldehyde concentrations can be observed more clearly after additional data manipulation. A fragment ion common to the mass spectra of low molecular weight aldehydes but not present in the mass spectra of the chlorinated compounds was selected. The extracted ion current profile (EICP) for this ion was then plotted.

Figure 5 shows the EICP for m/e = 29 of the Figure 4 data. In Figure 5 the peaks corresponding to some aldehydes are labeled. Because the computerized output from the GC/MS is normalized on the largest peak in the profile after setting it to 100% of scale, the Figure 5 plots were adjusted in the vertical scale to equalize the peak heights of diethyl ether which also has an m/e = 29 fragment and was assumed not to change in concentration with treatment. The halogenated compounds are not represented in these plots, and the comparatively higher aldehyde peaks in the treated sample are clearly evident.

A similar set of chromatograms was obtained for an untreated source water (Ohio River) and a corresponding chlorinated tap water. The increase in chloroform concentration is obvious from the TICP in Figure 6. The m/e=29 EICP does not show the obvious differences in aldehyde concentrations caused by chlorine dioxide treatment (Figure 7).

Although the general conclusions above are valid as qualified in the text, some care must be taken in interpreting the quantitative aspects of these data. These data were extracted from some of our early screening studies with chlorine dioxide in which we were looking for gross changes upon treatment. If aldehyde formation is considered important, considerably more work with analytical methodology is required in order to assure a good quantitative estimation of these compounds to adequately assess the magnitude of the observed changes.

Reactions with Aromatic Compounds

The best known reactions of aqueous chlorine with aromatic compounds in the water treatment field are those that occur with phenols. Chlorine reacts rapidly with phenol to form mono-, di-, and tri-chloro derivatives as shown in Figure 8. These compounds are highly odorous and are slowly decomposed by excess chlorine. Other phenolics and substituted aromatics can also be chlorinated (11). The formation of chlorophenols by chlorine treatment is one of the chief reasons that chlorine dioxide has been used in drinking water disinfection applications.

Chlorine dioxide does not seem to cause formation of odorous compounds with phenol but, through a complex mechanism, forms the quinones and chloroquinones, and when in excess, oxalic and maleic acids (Figure 9, Reference 10). Note that chlorine substitution in the products, as with olefins, is not entirely absent. Chlorine dioxide treatment of phenols can cause chlorine substitution or ring cleavage or both depending on the phenol reacted and the conditions of the reaction. The chlorinated products will generally be of different structure and are either less odorous or are formed in much lower yield than those formed when chlorine is applied. Additional examples of chlorine dioxide reactions with phenol derivatives are shown in Figure 10.

Vanillin reacts at pH 4 with chlorine dioxide to give the non-chlorinated β-formylmuconic acid monomethyl ester. Vanillyl alcohol reacts at low pH to produce both a chlorinated quinone and a nonchlorinated product of ring cleavage. The lactone structure is shown in Figure 10. Veratryl alcohol produces 4,5-dichloroveratrole (10,15). This is one of the products expected from chlorination of veratryl alcohol (16).

Some of the more detailed investigations of the reactions of chlorine dioxide with phenols have been accomplished by Glabisz and by Paluch at the Polytechnic University, Szczecin, Poland. A generalization for the reaction type (ring cleavage or ring retention with or without chlorine substitution) has been given by Glabisz (17). Glabisz states that, at least at concentrations of one mg/ℓ and above, the character of the reaction products of phenols falls into two groups. The first is the group in which the ring structure is retained and the end products are quinones. This group is made up of para-dihydric and monohydric phenols that are not para-substituted. The second group is characterized by those phenols which undergo ring cleavage to give carboxylic acids as end products. Examples include para-alkyl phenols and ortho or meta-dihydric phenols.

In general, monohydric phenols reacting with chlorine dioxide undergo chlorination along with oxidation, and those of the first group form the chloroquinones as well as chlorophenols. Glabisz considers this to be somewhat similar to the reaction of chlorine with these phenols. Although chlorine dioxide tends to favor oxidation over chlorination, the relative amount of chlorinated versus oxidized products depends on the relative amounts of both chlorine dioxide used and phenols present. Excess chlorine dioxide favors oxidation.

The two most significant points about reaction products from chlorine dioxide disinfection are:

(a) Chlorinated products. Although the final extent of chlorination may be less when chlorine dioxide reacts with phenols than when chlorine reacts, chlorinated products do not seem to be absent except when chlorine dioxide is applied in large excess. Thus, the chlorinated by-products of chlorine dioxide treatment may be present when chlorine dioxide has been used as a disinfectant at concentrations exceeding the short term demand by only a small margin. This may also be true for the products of chlorine dioxide reaction with olefins mentioned earlier. In the experiments cited, the olefin concentrations were held in large excess to maximize yields of reaction intermediates. The results of applying chlorine dioxide in small excess are not known.

(b) <u>Oxidation products.</u> The quinones seem to be likely by-products of chlorine dioxide treatment of water containing phenols when chlorine dioxide is not applied in large excess. These species are of largely unknown toxicological significance. In addition there is possible epoxide formation from olefins, and epoxides are suspected of being metabolic intermediates in carcinogenesis caused by organic compounds. The questions of whether or not we would simply be exchanging one problem for another by changing disinfectants in an effort to eliminate trihalomethane formation must be answered through some thorough toxicological testing of the products formed.

Whether disinfection levels of chlorine dioxide will produce significant concentrations of these potentially undesirable compounds in the water sample or whether doses of chlorine dioxide much higher than anticipated for disinfection alone will be required to obtain more complete oxidation is not known.

Finally, the probability that these potential problems are not limited to waters contaminated with industrial wastes must be emphasized. The significance of the reactivity of phenols goes far beyond that. The potential by-products of phenol reactions may be the most significant in treatment of any natural water.

The core structure of natural aquatic humic materials as described by Christman and Ghassemi (18) is shown in Figure 11. Recall that this is the same material that is responsible for trihalomethane formation when chlorine is used as a disinfectant and that this material probably makes up the bulk of organic substances in natural waters. This is a poly-phenolic material that could produce a variety of undesirable chlorine dioxide reaction products.

One unforeseen result has been noted on reaction of chlorine with humic material; that is the formation of trihalomethanes. It may be safely assumed that other chlorinated materials are formed. Work is progressing at this laboratory (EPA, Cincinnati) to determine what the results of using chlorine dioxide to disinfect water containing humic materials will be. Some early results were presented above. Solvent extraction and low temperature evaporation, both with and without derivatization before gas chromatography, are representative of our present approach for detecting organic by-products of chlorine dioxide treatment.

SUMMARY

An inspection of the available literature reveals that a detailed investigation of the aqueous organic chemistry of chlorine dioxide and systematic identification of products formed during water disinfection has not been performed. This must be done before an informed assessment as to the relative safety of using chlorine dioxide as a disinfectant alternative to chlorine can be made.

Although trihalomethanes are not formed by the action of chlorine dioxide, the products of chlorine dioxide treatment of organic materials are oxidized species, some of which contain chlorine. The relative amounts of species types are likely to depend on the amount of chlorine dioxide residual maintained and the concentration and nature of the organic materials present in the source water. Studies on this subject are just beginning at the Water Supply Research Division, U.S. Environmental Protection Agency Laboratory.

Some of the nonchlorinated products of chlorine dioxide oxidation reactions, specifically quinones and epoxides are of questionable safety, but still of largely unknown toxicological significance.

LITERATURE CITED

(1) Lower Mississippi River Facility, USEPA, New Orleans Area Water Supply Study (Draft Analytical Report), Slidell, Louisiana (1974).

(2) J.J. Rook, "Formation of Haloforms During Chlorination of Natural Waters." *Water Treatment and Examination*, 23, (Part 2):234 (1974).

(3) T.A. Bellar, J.J. Lichtenberg and R. C. Kroner, "The Occurrence of Organohalides in Chlorinated Drinking Water." *Jour. AWWA*, 66:703 (1974).

(4) J.M. Symons, T.A. Bellar, J.K. Carswell, J.J. DeMarco, K.L. Kropp, G.G. Robeck, D.R. Seeger, C.J. Slocum, B.L. Smith and A.A. Stevens, "National Organics Reconnaissance Survey for Halogenated Organics." *Jour. AWWA*, 67:634 (1975).

(5) A.A. Stevens, C.J. Slocum, D.R. Seeger and G.G. Robeck, "Chlorination of Organics in Drinking Water." Proceedings of the Conference on the Environmental Impact of Water Chlorination, Oak Ridge National Laboratory, Oak Ridge, Tennessee, October 22-24, 1975. Also: *Jour. AWWA*, 68:615 (1976).

(6) J.J. Rook, "Haloforms in Drinking Water," *Jour. AWWA*, 68:168 (1976).

(7) Train, R.E., USEPA News Release, March 29, 1976.

(8) R.J. Miltner, "The Effect of Chlorine Dioxide on Trihalomethanes in Drinking Water," M.S. Thesis, University of Cincinnati (1976).

(9) A.P. Black and R.F. Christman, "Chemical Characteristics of Fulvic Acids." *Jour. AWWA*, 55:897 (1963).

(10) G. Gordon, R.G. Kieffer, and D.H. Rosenblatt, "The Chemistry of Chlorine Dioxide," in *Progress in Inorganic Chemistry*, Vol. 15, S.J. Lippard, Editor, Wiley - Interscience, New York (1972) p. 201.

(11) J.C. Morris, "Formation of Halogenated Organics by Chlorination of Water Supplies," EPA-600/1-75-002, U.S. Environmental Protection Agency, Washington, D.C. (1975).

(12) B. Leopold and D.B. Mutton, "The Effect of Chlorinating and Oxidizing Agents on Derivatives of Oleic Acid." Tappi, 42:218 (1959).

(13) B.O. Lindgren and C.M. Svahn, "Reactions of Chlorine Dioxide with Unsaturated Compounds, II, Methyl Oleate." Acta Chem. Scand., 20:211 (1966).

(14) B.O. Lindgren and C.M. Svahn, "Chlorine Dioxide Oxidation of Cyclohexene." Acta Chem. Scand., 19:7 (1965).

(15) C.W. Dence, M.K. Gupta, and R.V. Sarkanen, "Studies on Oxidative Delignification Mechanisms, Part II. Reactions of Vanillyl Alcohol with Chlorine Dioxide and Sodium Chlorite." Tappi 45:29 (1962).

(16) C.W. Dence and K.V. Sarkanen, "A Proposed Mechanism for the Acidic Chlorination of Softwood Lignin." Tappi 43:87 (1960).

(17) U. Glabisz, "The Reactions of Chlorine Dioxide with Components of Phenolic Wastewaters - Summary. Monograph 44, Polytechnic University, Szezecin, Poland (1968).

(18) R.F. Christman and M. Ghassemi, "Chemical Nature of Organic Color in Water," Jour. AWWA, 58:723 (1966).

DISCUSSION OF ALL CHLORINE DIOXIDE PAPERS

Cotruvo: With respect to the use of chlorine dioxide, you are going to end up with inorganic reaction products. What is the toxicological situation on those? Chlorites and chlorate seem to be the possibilities.

Rosenblatt: As far as the inorganic products are concerned, we ought to look at them. I don't think we can simply gloss over them at this point. Our organization has just done a pretty complete literature review on the toxicology of chlorates, and if anybody is interested in getting that review, let him write to me and I will send him a copy.

As far as chlorite is concerned, less is known. I think I heard somebody saying something about it interfering with hemoglobin. I would suspect that it might be one of the relatively small number of toxic materials in which the important toxicity is the acute toxicity rather than chronic toxicity. Things like carbon monoxide and well, not nitrite, because nitrite has other properties, but things that form methemoglobin. I'd rather not say anything more than that because really, this has to be done by a toxicologist, and I'm not one of them.

George C. White: In talking with some analytical people in France, they believe, they are not sure, but they think that chlorine dioxide followed by ozone reverts any of the chlorite back to chlorine dioxide. This ought to be investigated.

Sherman T. Mayne, Duke Power & Light: I'd like to know the cost of treating the waste that you generate after you get your chlorine dioxide. No one has talked about the cost of that.

Sussman: I think in large measure that depends upon what process and what application you are referring to. Starting with sodium chlorite and using chlorine, the waste product, if you will, is in your drinking water. The other system, such as large scale chlorine dioxide generation, those wastes, the acid wastes from some processes, have to be neutralized, and that's one of the reasons why the SVP process was developed, because there is no acid waste there. You have sodium sulfate as the by-product. In the small scale that we had referred to, the Jaska process, or CIP, we haven't investigated that area.

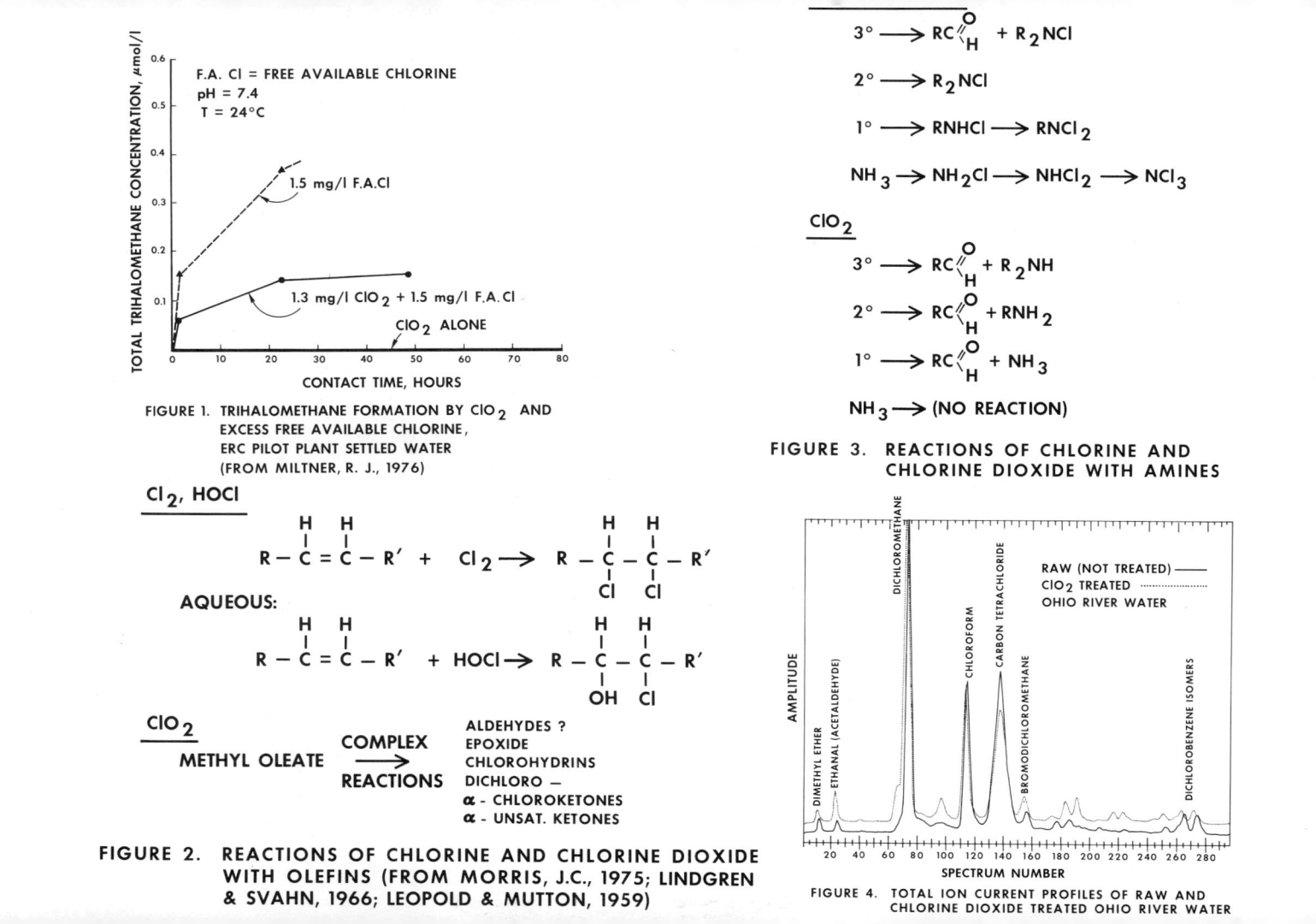

FIGURE 3. REACTIONS OF CHLORINE AND CHLORINE DIOXIDE WITH AMINES

$3° \rightarrow RC\overset{O}{\underset{H}{\diagdown}} + R_2NCl$

$2° \rightarrow R_2NCl$

$1° \rightarrow RNHCl \rightarrow RNCl_2$

$NH_3 \rightarrow NH_2Cl \rightarrow NHCl_2 \rightarrow NCl_3$

ClO_2

$3° \rightarrow RC\overset{O}{\underset{H}{\diagdown}} + R_2NH$

$2° \rightarrow RC\overset{O}{\underset{H}{\diagdown}} + RNH_2$

$1° \rightarrow RC\overset{O}{\underset{H}{\diagdown}} + NH_3$

$NH_3 \rightarrow$ (NO REACTION)

FIGURE 4. TOTAL ION CURRENT PROFILES OF RAW AND CHLORINE DIOXIDE TREATED OHIO RIVER WATER

FIGURE 1. TRIHALOMETHANE FORMATION BY ClO_2 AND EXCESS FREE AVAILABLE CHLORINE, ERC PILOT PLANT SETTLED WATER (FROM MILTNER, R. J., 1976)

Cl_2, HOCl

AQUEOUS:

$R-C\overset{H}{\underset{H}{\mid}}=C-R' + Cl_2 \rightarrow R-\overset{H}{\underset{Cl}{\overset{\mid}{C}}}-\overset{H}{\underset{Cl}{\overset{\mid}{C}}}-R'$

$R-C\overset{H}{\underset{H}{\mid}}=C-R' + HOCl \rightarrow R-\overset{H}{\underset{OH}{\overset{\mid}{C}}}-\overset{H}{\underset{Cl}{\overset{\mid}{C}}}-R'$

ClO_2

METHYL OLEATE $\xrightarrow{\text{COMPLEX REACTIONS}}$ ALDEHYDES ?
EPOXIDE
CHLOROHYDRINS
DICHLORO –
α - CHLOROKETONES
α - UNSAT. KETONES

FIGURE 2. REACTIONS OF CHLORINE AND CHLORINE DIOXIDE WITH OLEFINS (FROM MORRIS, J.C., 1975; LINDGREN & SVAHN, 1966; LEOPOLD & MUTTON, 1959)

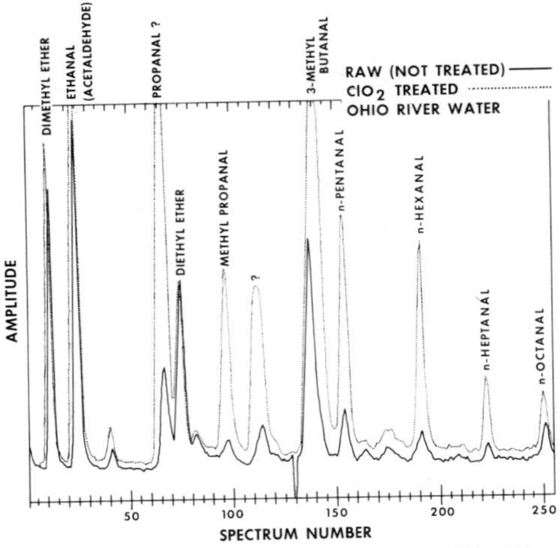

FIGURE 5. EXTRACTED ION CURRENT PROFILES (M/E = 29) OF RAW AND CHLORINE DIOXIDE TREATED OHIO RIVER WATER

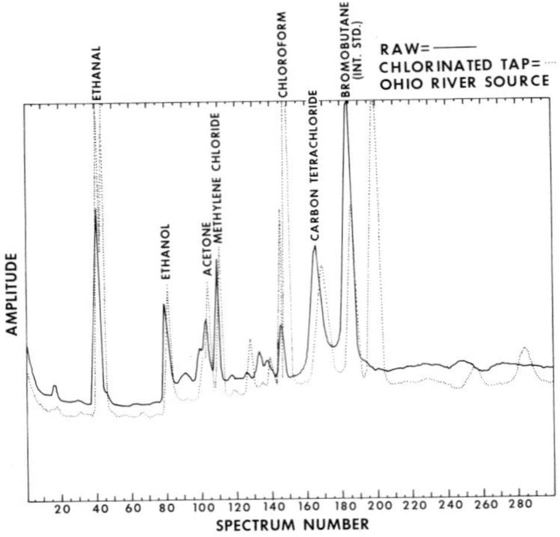

FIGURE 6. TOTAL ION CURRENT PROFILES OF RAW AND CHLORINATED TAP WATER FROM AN OHIO RIVER SOURCE

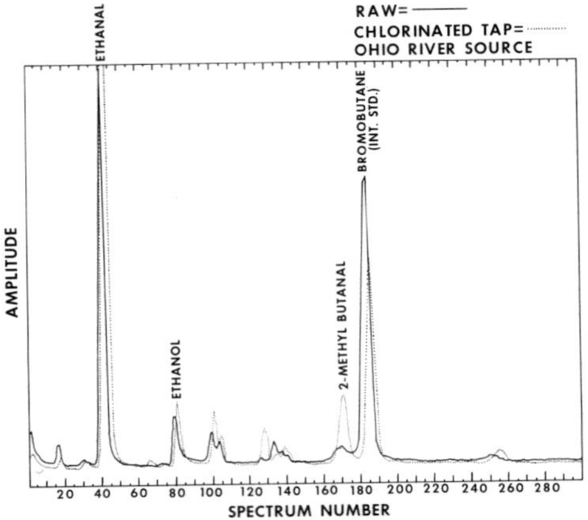

FIGURE 7. EXTRACTED ION CURRENT PROFILES (M/E = 29) OF RAW AND CHLORINATED TAP WATER FROM AN OHIO RIVER SOURCE

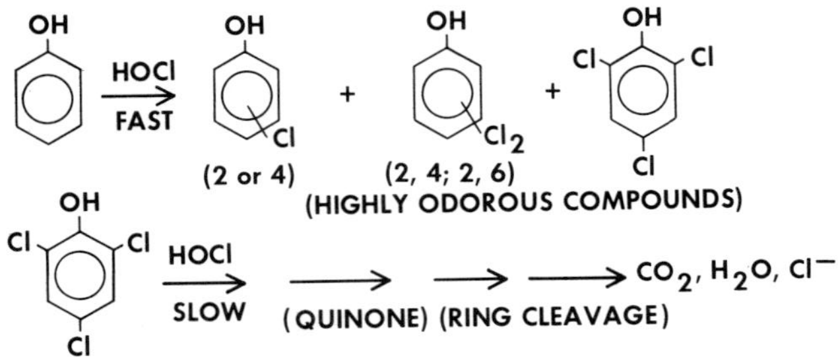

FIGURE 8. REACTIONS OF CHLORINE WITH PHENOL (FROM MORRIS, J. C., 1975)

FIGURE 9. REACTIONS OF CHLORINE DIOXIDE WITH PHENOL (FROM GORDON, KIEFFER, & ROSENBLATT, 1972)

FIGURE 10. REACTIONS OF CHLORINE DIOXIDE WITH PHENOLIC DERIVATIVES (FROM GORDON, KIEFFER, & ROSENBLATT, 1972)

FIGURE 11. A PROPOSED HUMIC STRUCTURE (FROM CHRISTMAN & GHASSEMI, 1966)

COMPETITIVE OXIDATION AND HALOGENATION REACTIONS

IN THE DISINFECTION OF WASTEWATER

Jack F. Mills
The Dow Chemical Company, U.S.A.
Midland, Michigan 48640

INTRODUCTION

What are the chemicals produced and the hazards generated during the process of disinfecting municipal wastewaters? The answer remains unknown, to a large extent, because we lack the chemical and toxicological data from the many competitive reactions which take place during disinfection processes. The chemical reactions between halogen species and various constituents of wastewater are best characterized by their numerous, competitive-reaction pathways. Recent studies using ^{36}Cl tracer (1), high resolution chromatography (2), and gas chromatography mass spectroscopy techniques (3) have helped identify and quantify some of the chemical products resulting from disinfection. However, many of the products from disinfection reactions are found at levels below one part per billion, which is very low to be determined accurately.

The toxicity of "suspected" hazardous products at very low concentrations becomes extremely difficult to assess. The carcinogenicity and mutagenicity data presently available was obtained at concentrations several orders of magnitude greater than typical use levels which most certainly clouds the interpretation of these data.

Somewhat more concern may exist where a suspected carcinogen bioaccumulates in the food chain by several orders of magnitude. Even in this case, the current evidence is, to a large degree, presumptive or circumstantial without additional data.

An important consideration in evaluating the potential hazards in disinfecting wastewaters is the stability of the resulting by-products in our environment. The products from and the extent of biodegradation and hydrolytic or photochemical decomposition will determine the real hazards of these materials downstream from their source. For example, 5-chlorouracil, a known mutagenic compound, has been identified as a by-product from chlorination of wastewater (4). Yet this chemical is relatively unstable in the environment and there is no evidence of bioaccumulation or public health hazard resulting from its formation in wastewater.

This paper will describe some of the chemistry and factors influencing the reaction products from disinfection of wastewater by both chlorination and chlorobromination processes. In particular, the competitive oxidation and halogenation reactions in wastewater will be discussed with emphasis on the difficulties involved in predicting the products from these reactions.

WASTEWATER VARIABLES

Predicting the products of disinfection wastewater with halogens is extremely difficult because the actual products which result, along with their concentrations, depend on the highly variable chemical composition of wastewater effluents, presence of ammonia compounds, halogen concentration, pH, temperature, rate of mixing, and reaction duration.

In general, the reactions of halogens with organic compounds in wastewater may be by addition, substitution, or oxidation. Reactions of halogens with inorganic waste compounds are almost exclusively oxidation. By far the predominant reaction in disinfecting wastewater using either chlorine or bromine chloride is oxidation.

To draw some insights into the competitive reactions going on during disinfection, one can look at individual reactions and the data comparing one reaction pathway against another alternative pathway. In citing the published literature, it is often important to distinguish between chemical reactions in aqueous solution versus reactions in non-aqueous solvents. The reaction mechanisms, the rates, and the end products in dilute aqueous halogen solutions can be very different from those in concentrated non-aqueous solutions, in which most organic reactions have been studied in the laboratory and published.

In aqueous solutions, both halogens (chlorine and bromine chloride) hydrolyze rapidly to form hypochlorous and hypobromous acids, respectively.

$$Cl_2 + H_2O \longrightarrow HOCl + HCl$$

$$BrCl + H_2O \longrightarrow HOBr + HCl$$

Both of these hypohalous acids are relatively strong oxidants, undergoing a host of redox reactions with both inorganic and organic compounds found in wastewater.

COMPETITIVE REACTIONS OF BrCl

Compared to chlorine, the results of competitive reactions of bromine chloride with water, ammonia, and various reducing agents (including organics) are more complementary to disinfection (5). Bromine chloride appears to hydrolyze exclusively to hypobromous acid, an active disinfectant which is more active at higher pH than hypochlorous acid, the product of chlorine hydrolysis. The hydrolysis rate for bromine chloride is estimated at less than 0.35 milliseconds at these concentrations, which is faster than the hydrolysis of either bromine or chlorine.

In wastewater, bromine chloride and its hydrolysis products react rapidly with ammonia to form bromamines. The major products are mono- and dibromoamine. Both are more chemically active, and less stable, than chloramines (5).

$$3BrCl + 2NH_3 \longrightarrow NH_2Br + NHBr_2 + 3HCl$$

Bromamines are much less stable in sewage or its receiving water, breaking down into harmless elements in a relatively short time without producing the hazards to marine life which are typical of the more stable and toxic chloramines.

Bromamines are far superior to chloramines in bactericidal and virucidal activities, almost equal to those of free bromine at the pH range of wastewater (6). A recent EPA study on alternative disinfectants concluded that bromine chloride is an effective disinfectant for secondary effluent with less toxic effects to aquatic life than chlorine (7).

The dominant reactions involving BrCl in wastewater are reduction reactions producing innoxious halide salts. Like chlorine, the major portion of the available halogen during the disinfection of wastewater effluents was consumed in oxidation reactions. The total halogenation yield was estimated at less than 1% of the halogen dosage applied.

In addition to inorganic reducing agents, such as sulfides and nitrites, there are many organic reducing agents. Organic alcohols, aldehydes, amines and mercaptans are oxidized by bromine chloride reaction species resulting in harmless chloride and bromide salts. Typical reactions between bromine chloride and organics in aqueous solution can be illustrated by the following examples:

$$BrCl + H_2O \longrightarrow HOBr + HCl$$

Oxidation

$$HOBr + CH_3SH \longrightarrow CH_3SO_3H + HBr$$

Bromination

$$HOBr + CH_2=CH_2 \longrightarrow CH_2BrCH_2OH$$

The oxygen-consuming characteristics of reducing substances have an analogy in their halogen demand. These reducing agents markedly increase bromine chloride requirements and thus compete with and inhibit the desired disinfection reactions.

COMPETITIVE OXIDATION REACTIONS

Bromine species are much better oxidizing agents than analogous chlorine species, whereas the latter are generally more active halogenating species. For example, the oxidation of cellulose by hypobromous acid is much faster than by hypochlorous acid (8). Also, the oxidation of glucose to gluconic acid using hypobromite is 1360 times faster than with hypochlorite (9).

However, the relative oxidizing strengths of these halogen species as measured by their oxidation potentials are not true indicators of their chemical reactivity.

Species	$E°$ (Volts) (10)
O_3	2.07
HOCl	1.50
Cl_2	1.40
HOBr	1.35
BrCl	1.30
ClO_2	0.95
OCl^-	0.90
OBr^-	0.75
NH_2Cl	0.75
NH_2Br	0.74

As shown above, the "effective oxidation potentials" of the bromine systems are lower than those of the analogous chlorine systems. Since the bromine systems are better oxidizing agents, the oxidation potential cannot be the decisive factor in determining the reaction velocity, although the disinfectant species must have a certain oxidation potential in order to oxidize. The much higher rate of oxidation by the bromine systems than by the chlorine systems indicates that different factors, such as reaction mechanisms, bond strengths, and steric factors influence the reaction rates much more than the oxidation potentials.

Ozone, with its high oxidation potential, can be used to generate hypobromous acid from bromide salts (11).

$$O_3 + Br^- \longrightarrow HOBr$$

Similarly, chlorine, with its relatively high oxidation potential, can oxidize bromide and iodide ions to hypobromous acid and hypoiodous acid, respectively (12). Thus, the presence of halide ions in both fresh water and particularly seawater (67 ppm Br^-) can influence the competitive oxidation pathways in these waters.

The results of chemical reactions in dilute aqueous solutions can be far different from those reactions carried out under typical laboratory procedures. For example, the bromination of toluene in dilute aqueous solution requires 22 times as long to attain 10% completion as chlorine under the same conditions:

Dilute in Water	Time to Attain 10% Yield
Br_2 + toluene	11 hours
Cl_2 + toluene	30 minutes

Concentrated in CCl_4	Time to Attain 98% Yield
Br_2 + toluene	<2 minutes

This is in contrast to the uncatalyzed bromination of toluene in carbon tetrachloride which requires less than two minutes to give 98% yield of bromotoluene.

FORMATION OF HALOGENATED ORGANICS

It has been reported recently that chlorination of water and wastewater results in the formation of halogenated organic compounds, some of which are suspected of being carcinogenic to man. The formation of chloroform and other similar volatile chlorinated organics identified in potable water samples from 80 major cities has been attributed mainly to the chlorination of naturally occurring humic and fulvic acids (13). However, in the disinfection of wastewater, Bellar and his coworkers (14) have reported much lower levels of volatile halogenated organics than observed in drinking water surveys. The formation of mixed haloforms containing chlorine, bromine, and even iodine can be accounted for by the oxidation of their respective halide ions by chlorine as mentioned before.

$$\text{Organics (humates)} \xrightarrow{HOCl} CHCl_3, CHBrCl_2, \text{etc.}$$

$$HOCl \xrightarrow{Br^-} HOBr \xrightarrow{Humates} CHBr_3$$

The lower yields of haloforms produced during chlorination of wastewater compared to drinking water may be explained in part by the higher levels of ammonia and different pH conditions which are generally found in wastewater.

The presence of ammonia can be an important factor influencing the course of the haloform reaction. Monochloramine appears not to undergo the haloform reaction nor to oxidize bromide ions. Therefore, it is significant that the reaction between hypochlorous acid and ammonia predominates even in the presence of bromide ions.

$$HOCl + NH_3 \xrightarrow{K_1} NH_2Cl + H_2O$$
$$HOCl + Br^- \xrightarrow{K_2} HOBr + Cl^-$$

The ratio of the above reaction rate constants $K_1/K_2 = 1700$ (at 25°C) (15) shows the dominating effect of ammonia on this reaction. Only 1 ppm of ammonia is needed to convert 0.5 ppm of chlorine into the inactive monochloramine.

The chlorination of municipal wastewater results in the formation of numerous other non-volatile, chlorinated organic compounds (1,2). However, the yield of these chlorinated organic compounds comprised in total less than 1% of the chlorine dose and, furthermore, nearly all the compounds identified contained only a single chlorine atom. In general, such chlorinated compounds are much more readily degraded biochemically and photochemically than are the polychlorinated materials that have been found to be so persistent in the environment.

With some exceptions, brominated derivatives are more readily degraded than their chlorinated analogs (16). Activated aromatic compounds which are more readily halogenated were also found to be more susceptible to photochemical degradation by sunlight and by ultraviolet light. Also, experiments with bromine additions to seawater showed that the bromine derivatives were less stable than the similar chlorine derivatives (17).

Great care must be exercised in predicting the properties of structurally analoguous compounds. In contrast to hexachlorobenzene, hexabromobenzene did not accumulate in juvenile Atlantic salmon (Salmo salar) either from water or from food (18). The carbon-bromine bond is less stable than the carbon-chlorine bond, which may be part of the explanation for this difference. Also, there is a possibility that, in addition to metabolism by hydroxylation, well established for chlorobiphenyls, reductive debromination may be a degradation pathway of hexabromobenzene.

Although hexahalobenzenes are unlikely products of the disinfection of wastewater, the foregoing discussion merely exemplifies the problem of speculation based on structurally analogous compounds. Toxicity, even more than chemical reactivity, is extremely difficult to predict on the basis of chemical structure. On the basis of their low concentrations and ease of degradation, it appears unlikely that any of these disinfectant products will persist in receiving waters long enough to cause major problems at down-stream water treatment plants (19) or that they will biologically accumulate in organisms to any great extent.

SUMMATION

1. Several factors contribute to the sometimes surprising course of competitive reactions in wastewater disinfection.

2. The major reaction products from halogenation of wastewaters are oxidation products.

3. Bromine species are much better oxidizing agents than analogous chlorine species; whereas, the latter are more active halogenating species.

4. Oxidation potentials are only partial indicators of chemical reactivity.

5. In contrast to potable water, wastewater chlorination and bromochlorination results in insignificant haloform production.

6. Known organic oxygen and halogen compounds produced in wastewater disinfection are relatively unstable in the environment and are most likely non-hazardous.

LITERATURE CITED

1. Jolley, R. L., J. Water Poll. Control Fed. 47(3):601-618 (1975).

2. Jolley, R. L., Environ. Letter, I:321 (1974).

3. Jolley, R. L., Katz, S., and Mrochek, J. E., Chemtech, 310-318, May, 1975.

4. Proceedings of the conference on "The Environmental Impact of Water Chlorination," Oak Ridge National Laboratory, Oak Ridge, Tennessee, October, 1975, Conf. 751096, pages 115, 247.

5. Johnson, J. D., Ed., "Disinfection - Water and Wastewater, Chapter 8, Ann Arbor Science Publishers, Ann Arbor, Michigan, 1975.

6. Johannesson, J. D., Am. J. Public Health, 50:1731-1736 (1960).

7. "Disinfection Efficiency and Residual Toxicity of Several Wastewater Disinfectants - Volume I - Grandville, Michigan," EPA Report No. 600/2-76-156, October, 1976.

8. Giertz, H. W., Tappi, 34:209 (1951).

9. Lewin, M., Ph.D. Thesis, Hebrew University, Jerusalem, 1947.

10. Stande, Phys. Chem. Taschenb II:1572 (1949).

11. Marks, H.C., and Strandskov, F. B., U. S. Patent 2,580,809, Jan. 1, 1952.

12. Marks, H. C., and Strandskov, F. B., U. S. Patent, 2,443,429, June 15, 1948.

13. Rook, J. J., J. Water Treatment and Examination, 23(2):234 (1974).

14. Bellar, T. A., Lichtenberg, J. J., and Kroner, R. C., "The Occurrence of Organohalides in Chlorinated Drinking Waters," EPA Report No. 670/4-76-008, November, 1974.

15. Weil, I., and Morris, J. C., J. Am. Chem. Soc. 71:1664-1671 (1949).

16. Plonka, J., The Dow Chemical Company, Midland, Michigan, 1972, unpublished data.

17. Duursma, E. K., and Parsi, P., Netherlands J. of Sea Research 10(2):192-214 (1976).

18. Zitko, V., and Hutzinger, O., Bull. Environ. Contam. & Toxicology, 16(6):665-673 (1976).

19. Morris, J. C., "Formation of Halogenated Organics by Chlorination of Water Supplies," EPA Report No. 600/1-75-002 (March, 1975).

IRON (VI) FERRATE AS A GENERAL OXIDANT FOR WATER AND WASTEWATER TREATMENT

By

Thomas D. Waite and Marsha Gilbert
Dept. of Civil Engineering
Northwestern University
Evanston, IL 60201

Ferrate Chemistry

Iron in its familiar forms exists in the +2 and +3 oxidation states, however, in a strong oxidizing environment it is possible to obtain higher oxidation states. Compounds of iron (IV), (V) and (VI) have been isolated as the metal salts of ferric acid (1,2); however, it is the hexavalent form of iron that is of interest in this paper.

For the past three years, considerable effort has been put into investigating the action of ferrate (VI) ion on the constituents in water and wastewater. This seemingly exotic form of iron has been of interest to analytical chemists since 1841, when Fremy first synthesized potassium ferrate (3). By 1925 a wide variety of metallic iron (VI) salts had been synthesized. It was not until 1948 however, that procedures were developed whereby a stable, crystalline solid of high purity could by synthesized, and analyzed for its iron (VI) content (4,5,6,7). As a result of work by Schreyer, physical chemists and kineticists have been able to establish the structure for ferrate (VI) ion and find evidence to support its existence (8,9). Although ferrate chemistry is in a state of infancy, several U.S. patents are currently held that relate to the use of ferrate in aqueous solutions. Three of these patents include: removal of color from industrial electrolytic baths (10), use in making catalysts for the Fischer-Tropsch process (11,12) and purification of hemicellulose (13).

2-

Ferrate (VI) ion has the molecular formula FeO_4 and is a powerful oxidizing agent through the entire pH range. Wood (9) has reported the redox potential of ferrate to vary from -2.2V to -0.7V in acid and base, respectively. The standard electromotive force for the half reaction is:

$$Fe^{3+} + 4H_2O \rightarrow FeO_4^{2-} + 8H^+ \; 3e^-; \; E_o = -2.2 \pm .03V$$

Latimer (14) gives as a calculated estimate for the reaction:

$$Fe(OH)_3 + 5OH^- \rightarrow FeO_4^{2-} + 4H_2O + 3e^-$$

$E_o = 0.72 \pm 0.03V$.

Nearly thirty metallic salts of ferric acid ($MFeO_4$) have been prepared, but only a few of these compounds are found to yield a highly pure and stable product. As a matter of interest, heavy metal ferrate compounds containing Ag, Al, Zn, Cr, Cu, Co, Pb, Mn, Ni, Hg, or Tl have been synthesized by double decomposition of $BaFeO_4$ and the corresponding metal nitrate in aqueous solution (15). For example:

$$BaFeO_4 + 2\;Al(NO_3) \rightarrow Ba(NO_3)_2 + Al_2FeO_4$$

It is difficult to isolate many of these compounds from solution as they are subject to decomposition at 30°C and react rapidly with CO_2 while being dried in air.

Of more practical interest are the metal ferrates which form either stable solutions or stable crystalline solids. These compounds vary widely in their aqueous solubilities. Lithium, sodium, calcium and magnesium ferrate are reported to be extremely soluble and can be synthesized by double decomposition with alkali metal perchlorate ($MClO_4$) and potassium ferrate (15).

Murmann (16) has reported on an electrochemical procedure for generating concentrated solutions of Na_2FeO_4. Using a 35-40% NaOH solution, 10-15 cm² electrodes with 2 cm separation, and an initial 2-5 ohms resistance, scrap iron can be converted to the FeO_4^{2-} ion with an expected efficiency of 40%. Both H_2 and O_2 gas are produced as by-products.

A third procedure developed by Schreyer, et al. (5), employs wet chemical oxidation of Fe(III) by hypochlorite followed by chemical precipitation of FeO_4^{2-} with KOH to form K_2FeO_4. Re-crystallization results in a high purity crystalline solid. This method was utilized in generating the potassium ferrate for all of the experiments performed in this laboratory.

Aqueous solutions of ferrate ion have a characteristic violet color much like that of permanganate in solution. Spectroscopic analyses of visible spectra of aqueous ferrate solutions show one maximum peak at 505 nm and two minima at 390 nm and 670 nm. The molar extinction coefficient as determined by Wood (9) is 1070 ± 30 in 10^4M NaOH (moles/l) and 4 m NaOH (moles/kg at 505 nm.

Potassium ferrate decomposes in aqueous solution generating hydroxide ion and molecular oxygen. The overall decomposition of ferrate (VI) ion in aqueous medium is described below:

$$2FeO_4^{2-} + 3H_2O \rightarrow 2FeO(OH) + 3/2\ O_2 + 4OH^-$$

The decomposition rate is highly dependent on pH, initial ferrate concentration and temperature; it is also influenced to some extent by the surface character of the hydrous iron oxide formed upon decomposition. Ferrate is most stable in strong base with two regions of maximum stability, one at pH 10-11 and the other in solutions greater than 3M in base (9), although this is highly dependent on the initial Fe(III) concentration which apparently accelerates decomposition (17).

Studies on the stability of ferrate in aqueous solution have shown that dilute solutions of ferrate are more stable than concentrated solutions (18). The decay curves presented in Figure 1 show the effect of initial concentration on decomposition rates at 26 ± 0.5°C. and K_2FeO_4 concentrations varying from 0.020 molal to 0.085 molal. It can be seen that a rapid initial decay occurs in the first few minutes followed by a period of stability. Analyses made after sixty minutes reveal all of the ferrate to have decomposed for solutions between 0.030 molal and 0.085 molal. In the two least concentrated solutions, i.e., 0.020 and 0.025, nearly 89% of ferrate remains. An inconsistancy occurs at 0.03 molal ferrate, where decomposition with time becomes erratic. No explanation was given by the authors, but the effect may be a result of the stability of an iron (IV) intermediate which is thought to be generated under alkaline conditions (19). In another study, Wagner, et al. (20), found 1.9×10^{-3}M ferrate solutions to be only 37.4% decomposed after three hours and fifty minutes at 25°C. Ferrate decomposition rate has also been found to decline markedly in the presence of phosphate, and at low temperatures (9,18,20).

Kinetic studies on ferrate decomposition in aqueous systems also show the decay rate to be a complicated function of initial ferrate concentration, pH and the nature of the resulting Fe(III) solid phase. Decomposition rates of 1×10^{-3}M K_2FeO_4 solutions are second order in ferrate for pH 7.7 through 9.5 with a rate constant of 1.97×10^{10} $l^2ml^{-2}min^{-1}$ (21). Less concentrated solutions (10^{-4}M and 10^{-5}M) for the same pH range are first order in ferrate with approximate constants of 0.02 to 0.2 min^{-1}.

Ferrates exhibit a variety of responses to inorganic impurities. Under alkaline conditions, solutions of potassium ferrate are stabilized by the presence of phosphate ion. Examination of the absorbance spectra, and evaluation of the extinction coefficient for alkaline ferrate-phosphate systems (0.05M PO_4), show that phosphate most likely does not influence the structure of the aqueous ion. However, where concentrations of iron (III) and phosphate are sufficient, precipitation occurs. It has been observed that a solution containing 10^{-4}M K_2FeO_4 buffered with 0.05M PO_4^{3-} will precipitate an iron phosphate compound at pH 5 as evidenced by loss of buffering capacity and formation of a bulky white precipitate. At pH 8.5 phosphate is evidently complexed with iron (III), thus enhancing stability of FeO_4^{2-}.

Nitrate has been shown to have little effect on ferrate stability, while borate is reported to stimulate decay. Contact with hydrous ferric oxide causes rapid decomposition (18). Figure 2 illustrates the relative stability effects of several salts and ferric hydroxide on ferrate. Other solute domains probably exist in the presence of SO_4^{2-}, F^- and dissolved or colloidal organic matter, and these groups can form stable complexes with Fe(III) which also alter the decomposition rate of K_2FeO_4.

Ferrate reacts rapidly with most inorganic reducing agents under both acid and basic conditions. Reactions involving inorganic ammonia have been studied in detail, and oxidation of ammonia appears to have an optimum conversion in the pH range 9.5 to 11.2, although losses to the gaseous phase might be suspected at the higher pH. Strong (22) reported the degree of conversion of ammonia to increase as the molar ratio of ferrate to ammonia became greater and as temperature was increased. Murmann (16) reports a pseudo first order rate constant for ferrate oxidation of NH_3 to products to be 7.0×10^{-3} sec^{-1} at pH 10.6 and 2.5×10^{-2} sec^{-1} at pH 9.0. The proposed sequence of the reaction was given as:

$$NH_3 > NH_2OH > N_2O > N_2^- > NO_3^-$$

The possibility of converting aqueous ammonia to N_2 gas by ferrate oxidation was also investigated (23), but conditions to be met for large conversions appear to be too rigorous to have any industrial application.

Iron (II) and iron (III) enter into a wide variety of reactions with organic compounds which can include complexation, chelation, precipitation and oxidation-reduction. The extent of organo-iron (VI) interactions is not known; however, some work has been completed which reflects upon these reactions, providing some insight on mechanisms. The degree of oxidation of amino acids by FeO_4^{2-} varies with initial ferrate concentration (16) and cystine and glycine react completely with excess ferrate to form CO_2 and N_2. When the amino acid is in excess, a variety of oxidation products is generated. Some sugars and glycols are slowly oxidized to organic acids. Murmann indicated that ferrate oxidation of alkenes was not by the same mechanism as OsO_4. (Osmium tetroxide adds to the double bond as a molecule and not as an ion).

Several postulates have been made regarding the nature of possible intermediates formed during the course of decay from iron (VI) to iron (III). However, mechanisms for decomposition are expected to be different with changes in pH, and variation in solute species present. Possible intermediates are: (1) complexes involving other oxidation states of iron, (2) oxygen-bearing radicals, and (3) hydroxo-metal complexes. Several researchers have determined that iron (VI) does not convert directly to iron (III), but passes through its +5 and +4 oxidation states (16,19,21). Magee (21) suggests the formation of peroxo-complexes of iron (V) and iron (IV).

Mechanistic studies have revealed that all four bound oxygens of FeO_4^{2-} are equivalent. Goff and Murmann (24) have investigated O^{18} exchange rates between FeO_4^{2-} and solvent water, and they determined that no predominant chemical changes occur, and that the coordination number of ferrate does not vary from four in 5×10^{-3}M solutions at pH 9.6 - 14. They also concluded that the formula of the aqueous ion is most likely FeO_4^{2-} with the coordination sphere of four oxygens arranged tetrahedrally. The reduction of ferrate was shown to be first order in $[FeO_4^{2-}]$ and independent of $[OH^-]$. The mechanism for oxygen exchange was not determined and changes in ionic strength had little effect on overall rates. It is interesting to note that exchange rates were found to be unaffected by the presence of the Fe(III) phase. Murmann and Goff also determined the activation energy, $E_a = 15.4 \pm 1.0$ Kcal/mol, to be similar to those of MnO_4^- and ReO_4^-, and the authors suggest similar activation states for these isomorphically related species.

Tracer studies using O^{18} labeled water at near neutral conditions, show at least half of the oxygen evolved from the FeO_4^{2-} - H_2O interaction comes from ferrate, while in 1M acid the O^{18} concentration of the solvent water is in equilibrium with O^{18} liberated during reaction. According to Goff and Murmann, exchange becomes competitive with water oxidation in the acidic region. Additionally, it was shown that interactions between ferrate and sulfite may likely proceed through an oxygen-transferring intermediate:

$$\begin{bmatrix} & O & & O & \\ & | & & | & \\ O- & Fe & -O- & S & -O \\ & | & & | & \\ & O & & O & \end{bmatrix}^{4-}$$

The reaction rate is first order in $[FeO_4^{2-}]$ and second order in $[SO_3^{2-}]$ when neither reagent is in excess. The rate equation governing FeO_4^{2-} -- H_2O_2 interaction has also been determined and is given by:

$$-[FeO_4^{2-}]/dt = K [FeO_4^{2-}][H_2O_2]$$

Ferrate as a disinfectant

Inactivation of E.coli by ferrate was studied in relatively simple, pH-buffered systems. Die-off in these systems showed typical initial deviations from Chick's Law, followed by first order, rapid inactivation of bacteria.

Results from the inactivation study show a ferrate concentration of 6 mg/l as FeO_4^{2-} to effect 99% kill of bacteria in 8.5 minutes, 7.2 minutes and 6.4 minutes at pH 8.0, 8.2 and 8.5 respectively. It was observed in all experiments using the borate buffer, that bacterial cells had a tendency to clump together. Aggregation of cells did not appear to retard die-off, but difficulties in obtaining an even distribution of cells on growth plates made it necessary to abandon the use of borate buffer. A 0.01M phosphate buffer had sufficient capacity to hold pH at the ferrate concentrations used for testing.

The kinetics of disinfection were too rapid to study at pH 7.5 for the concentration range selected. Since concentrations of K_2FeO_4 less than $1 \times 10^{-5}M$ are difficult to measure, no further experiments were run at pH 7.5. It was observed, however, that ferrate exhibits a remarkable increase in disinfection capacity as pH shifts below 8.0 to the more acidic region. Die-off is essentially instantaneous even at a low concentration of $1 \times 10^{-5}M$ K_2FeO_4. The sharp rise in germicidal effectiveness at pH 7.5 is consistent with the higher reactivity of FeO_4^{2-} in increasingly acidic water.

The rate constants calculated for the die-off curves are presented in Table 1, assuming disinfection follows a first order relationship. Very little difference is noted in rates as a function of pH at $1 \times 10^{-5}M$ ferrate. Rate constants are observed to have the expected concentration dependency. There is essentially no difference in rates as a function of pH at 8.2 and 8.5, however, the value of rate constants calculated at pH 8.0 are apparently more dependent on pH than on concentration. Values for the rate constant at pH 8.0 and $5 \times 10^{-5}M$ ferrate are five times lower than the value at pH 8.5 for the same ferrate concentration. The depressed rates of die-off at pH 8.0, again reflect some transitional chemical state for iron at this pH. Although the rate constant as shown in Table 1 for tests run at pH 8.2 with borate buffer ($5 \times 10^{-5}M$ K_2FeO_4) is higher than rates calculated for either of the phosphate buffered systems for the same conditions, no conclusions have been drawn concerning effects of borate on the rate of E.coli die-off.

Two experiments on the disinfection of wastewater by ferrate were conducted on secondary effluent taken from an activated sludge treatment plant. The initial pH of the effluent was 7.5 with about 10 mg/l PO_4^{3-} as P, 75 mg/l BOD and 10^7 E.coli per ml. One hundred milliliters of effluent were disinfected with 1 x 10^{-4} M and 5 x 10^{-5}M K_2FeO_4. The sample was agitated for a few minutes and permitted to settle. Rapid floc formation was observed, followed by sedimentation, which left an apparently clear supernatant. No mixing of the supernatant was provided.

Results from the study are shown in Figure 3. An arithmetic plot of the reciprocal mean number of surviving E.coli against contact time in minutes, demonstrates that, in this investigation, inactivation does not follow Chick's Law. In addition to the reasons provided by Collin (25), Popp (26) and Gard (27) for the nth order inactivation kinetics in wastewaters, consideration must be made for coagulation in ferrated wastewaters. It is most likely that although a variety of events are taking place, removal of bacteria by sedimentation may be the controlling mechanism. The sludge obtained during this procedure was not assayed for viable bacteria.

Results presented in Figure 3 show 99% removal of E.coli in eight minutes with 1 x 10^{-4}M K_2FeO_4 (12 mg/l as FeO_4^{2-}). This reflects a five-fold increase in the ferrate molar concentration over that required for the same degree of disinfection at pH 8.0 in buffered distilled water. The increase is due to a demand on ferrate by constituents in the wastewater. Under similar conditions, and to provide the same degree of germicidal efficiency, however, chlorine requires as much as a thirty-fold increase in molar concentration as a result of a chlorine demand exerted by the wastewater.

Observations of a number of qualitative experiments where ferrate was added to wastewater, indicate that FeO_4^{2-} may be less sensitive to parameters such as pH and mixing, with respect to coagulation, than are the iron (III) salts. Removal of particulate organic matter and color was observed to occur nearly instantaneously with 10^{-4}M and 10^{-5}M K_2FeO_4 in all samples tested over a pH range from 6.0 to 11.0. Wastewater which had been treated with ferrate was of excellent apparent clarity and was not observably colored. When coagulated wastewater solutions were agitated to break up the floc, the suspension settled again, leaving a clear supernatant.

Activity of ferrates against soluble organic substances

In order to further investigate the potential of ferrate (VI) ion as a treatment chemical, a study was initiated to evaluate its capabilities as a general oxidant in water. The interest in dissolved organics has been especially stimulated by recent reports on the detection of potentially toxic organic halides in drinking waters, which may be products of the chlorination process. Bellar, et al. (28) indicate ethanol as the suspected precursor to chloroform formation during the chlorination of drinking water. A mechanism was presented by which ethanol might be chlorinated.

$$CH_3-CH_2OH \rightarrow CH_3\overset{O}{\overset{\|}{C}}H \rightarrow Cl_3-C\overset{O}{\overset{\|}{C}}H \rightarrow Cl_3C-\overset{H}{\underset{|}{C}}(OH)_2 \rightarrow CHCl_3$$

   Ethanol   Aldehyde   Chloraldehyde   Chloralhydrate   Chloroform

Rook (29) has recently indicated that chlorination of alcohols and aldehydes appears to occur too slowly to implicate these compounds as precursors, but Rook's findings do not discount this possibility under conditions experienced in treatment plant operation.

In conjunction with these problems which originate from the organic fraction in water and in wastewater, it was felt that a study on the effects of ferrate on soluble organic substances would yield valuable information regarding its use as a broad spectrum oxidant of soluble compounds in aqueous solution. This study was, therefore, designed to:

1. Postulate when ferrate reacts directly with organic substrates; and when, if at all, ferrate acts through more reactive intermediates

2. Establish which parameters are important in maximizing oxidations and to optimize these parameters

3. Demonstrate the selectivity of ferrate for specific functional groups

4. Determine the extent of oxidation of selected substrates by ferrate for the purpose of establishing the effectiveness of ferrate as a general oxidant

5. Observe whether synergism occurs between two substrates in aqueous systems containing ferrate.

Three observable trends for ferrate decomposition were established in the substrate-ferrate systems: (1) substrates were found to have no effect on normal decomposition at each pH value tested; (2) substrates caused acceleration in the rate of ferrate decomposition; (3) substrates retarded the rate of ferrate decomposition. Allylbenzene, chlorobenzene, nitrobenzene and 1-hexene-4-ol did not alter ferrate decomposition over a pH range of 7.0 to 10.5. However, it was determined by gas chromatographic studies made under similar conditions, that allylbenzene, chlorobenzene and 1-hexene-4-ol were oxidized by ferrate. (No G-C analysis was conducted using nitrobenzene). Therefore, although these substrates did not stimulate ferrate decay, they did enter into a reaction, and were to some extent converted to products by ferrate. Interpretation of this slow but positive response may be considered from two mechanistic viewpoints: (1) ferrate reacts directly with the substrate; (2) ferrate acts through a 'stable' intermediate such as Fe(X) where (X) is a lower oxidation state of iron. In either case, allylbenzene, benzene, chlorobenzene and 1-hexene-4-ol apparently do not alter the mechanism of ferrate decay through the pH range of 7 to 10.5. Figures 4 to 6 compare ferrate decay in the presence of allylbenzene, chlorobenzene, and nitrobenzene to normal ferrate decay, under conditions of similar pH and concentration. Each set of curves was presented to show that regardless of initial ferrate concentration, and thus independent of the actual rate of ferrate decay, interaction between substrate and ferrate proceeded by some mechanism involving a slowly reacting species.

Figure 7 and 8 show the effects of aniline and phenol on ferrate decomposition from pH 7 to 11. It was not surprising that phenol and aniline stimulated ferrate decomposition, owing to the high reactivity of these substrates. Although the rapid decomposition of ferrate in the presence of phenol (Figure 8), suggests the presence of a fast-reacting intermediate, it is apparent that iron is somehow directly utilized in the reaction. Some irregularity is noted in Figure 8, for curves representing reactions above pH 9. This may be due to the predominance of phenolate ion, $C_6H_5O^-$ which occurs above pH 9.

Benzene was found to retard decomposition as illustrated in Figure 9. This protective effect could include involvement between Fe (III) and benzene.

Some reaction was taking place, however, since some benzene was converted to products in the system. Reaction between benzene and ferrate, thus, takes place slowly, with the ultimate precipitation of iron (III)-hydroxide.

Substrate oxidation by ferrate was evaluated by monitoring the disappearance of substrate using gas chromatography. Tests were conducted over a pH range of 2 to 10.5 at 20°C. A wide variety of substrate to ferrate molar ratios was used, and all samples contained 0.05M phosphate buffer. The following substrates were selected on the basis of compatibility with the gas chromatographic column: benzene, allylbenzene, chlorobenzene and 1-hexene-4-ol. The data have been evaluated in terms of pH dependency, effect of substrate-ferrate molar ratio, and synergistic effects in two-substrate systems.

Ferrate was found to significantly reduce the concentrations of allylbenzene and chlorobenzene, while benzene and 1-hexene-4-ol were converted to about 50% products. The oxidations were dependent on s:Fe (VI) molar ratios, where an excess of ferrate was shown to be most effective in reducing substrate concentrations. Molar ratios of s:Fe (VI) greater than 1:3 did not significantly enhance conversions, which points to the formation of products like organic acids, rather than to complete oxidation to CO_2.

○ ○ ○ ○ ○ ● ● ● ● ○ ● ● ● ● ●

The above data were collected from preliminary experiments utilizing potassium ferrate. There is a tremendous amount of exploratory work remaining to further quantify the reaction of ferrate in water and wastewater systems. Many questions are still unanswered, but initial findings are very promising. Ferrate has been shown to be an adequate biocide, and good oxidant of organic material. The ability of ferrate to act as a coagulant, as well as nutrient scavenger has yet to be investigated, but it is anticipated that positive results will be obtained. Therefore, ferrate has the potential of becoming an efficient, multi-purpose water and wastewater treatment chemical. It is important that new chemical regimes such as ferrate, receive serious national attention for possible application to water and wastewater treatment. Many red flags have been raised with regard to continued use of chlorine as a general oxidant. It is time we seriously consider some alternatives.

REFERENCES

(1) Erchak, M.J., et al. (1946). Jour. Amer. Chem. Soc. 68:2085,

(2) Scholder, R., H.V. Bensen, F. Kindervater and W. Zeiss (1955). Ztschr. anorg. allgem. Chem. 282:268.

(3) Fremy, E.F. (1841). Compt. rend. 12:23.

(4) Schreyer, J.M., L.T. Ockerman and G.W. Thomson (1950). Anal. Chem. 22:691.

(5) Inorganic Synthesis (1953). 4:164.

(6) Schreyer, J.M., et al. (1951). Anal. Chem. 22:1426.

(7) Schreyer, J.M., et al. (1951). Jour. Amer. Chem. Soc. 73:1379.

(8) Hrostowski, H.J. and A.B. Scott (1950). Journ. Chem. Phys. 18:105.

(9) Wood, R.H. (1958). The heat, free energy and entrophy of the ferrate (VI) ion. Journ. Amer. Chem. Soc. 80:2038.

(10) Schreyer, J.M. (1951). USP 2536703.

(11) Mosesman, M.A. (1949). USP 2470784.

(12) Mosesman, M.A. (1948). USP 2455696.

(13) Harrison, J.B. (1965). USP 2728695.

(14) Latimer, W.M. (1952). Oxidation Potentials. Prentice Hall, N.Y.

(15) Lozana, L. (1925). Acido ferrico e ferrati (VI). Gazz. Chim. Ital. 55:468.

(16) Murmann, R.K. (1974). The preparation and oxidative properties of ferrate (FeO_4^{2-}). NTIS Publication PB-238-057.

(17) Haire, R.G. (1965). A study of the decomposition of potassium ferrate (VI) in aqueous solution. D. Abstr.

(18) Schreyer, J.M. and L.T. Ockerman (1950). Stability of ferrate (VI) ion in aqueous solution. Anal. Chem. 24:1498.

(19) Jezowska-Trzebiatowska, B. and M. Wronska (1957). Congr. Intern. Chim. Pure et Appl. 16e, Paris.

(20) Wagner, W.F., J.R. Gump and E.N. Hart (1952). Factors affecting the stability of aqueous potassium ferrate (VI) solutions. Anal. Chem. $\underline{24}$:1397.

(21) Magee, J.S. (1961). The kinetics of the decomposition of potassium tetraoxoferrate (VI) in aqueous solution. D. Abstr.

(22) Strong, A.W. (1973). An exploratory work on the oxidation of ammonia by potassium ferrate (VI). NTIS Publication PB 231873.

(23) Murmann, R.K. Personal Communication (1973).

(24) Goff, H. and R.K. Murmann (1971). Studies on the mechanism of isotopic oxygen exchange and reduction of ferrate (VI) ion (FeO_4^{2-}). Jour. Amer. Chem. Soc. $\underline{93}$:6058.

(25) Collins, H.F., R.E. Selleck and G.C. White (1971). Problems in obtaining adequate sewage disinfection. JASCE, SA5 $\underline{97}$:549.

(26) Popp, L. (1954). Bacteriological Studies of the effect of chlorine for the disinfection of water. Gas and Wasserfach. $\underline{95}$.

(27) Grad, S. (1957). Chemical inactivation of viruses. CIBA Foundation Symposium, Nature of Viruses. p 123.

(28) Bellar, T.A., J.J. Lichtenberg and R.C. Kroner (1974). Journ. Amer. Water Works Assn. $\underline{66}$:703.

(29) Rook, J.J. Formation of haloforms during chlorination of natural waters. Water Treat. Exam. $\underline{23}$:234. (1974).

Table 1 - Rate constants for disinfection of E.coli in buffered water by ferrate (VI) ion at 27°C.

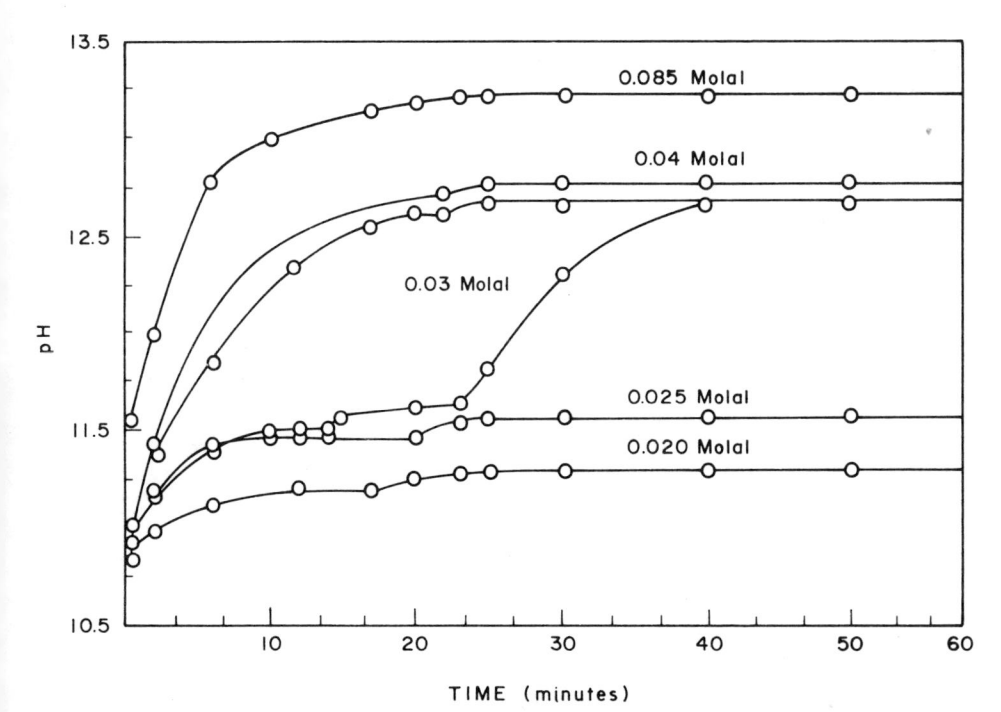

pH	Ferrate M/l	Rate Constant, K min^{-1}
8.0	1 x 10^{-5}	0.156
	2 x 10^{-5}	0.259
	5 x 10^{-5}	0.346
8.2	1 x 10^{-5}	0.147
	2 x 10^{-5}	0.397
	5 x 10^{-5}	1.739
8.5	1 x 10^{-5}	0.172
	2 x 10^{-5}	0.387
	5 x 10^{-5}	1.435

Figure 1- Effect of initial concentration of potassium ferrate upon decomposition in aqueous solution (from Schreyer (4)).

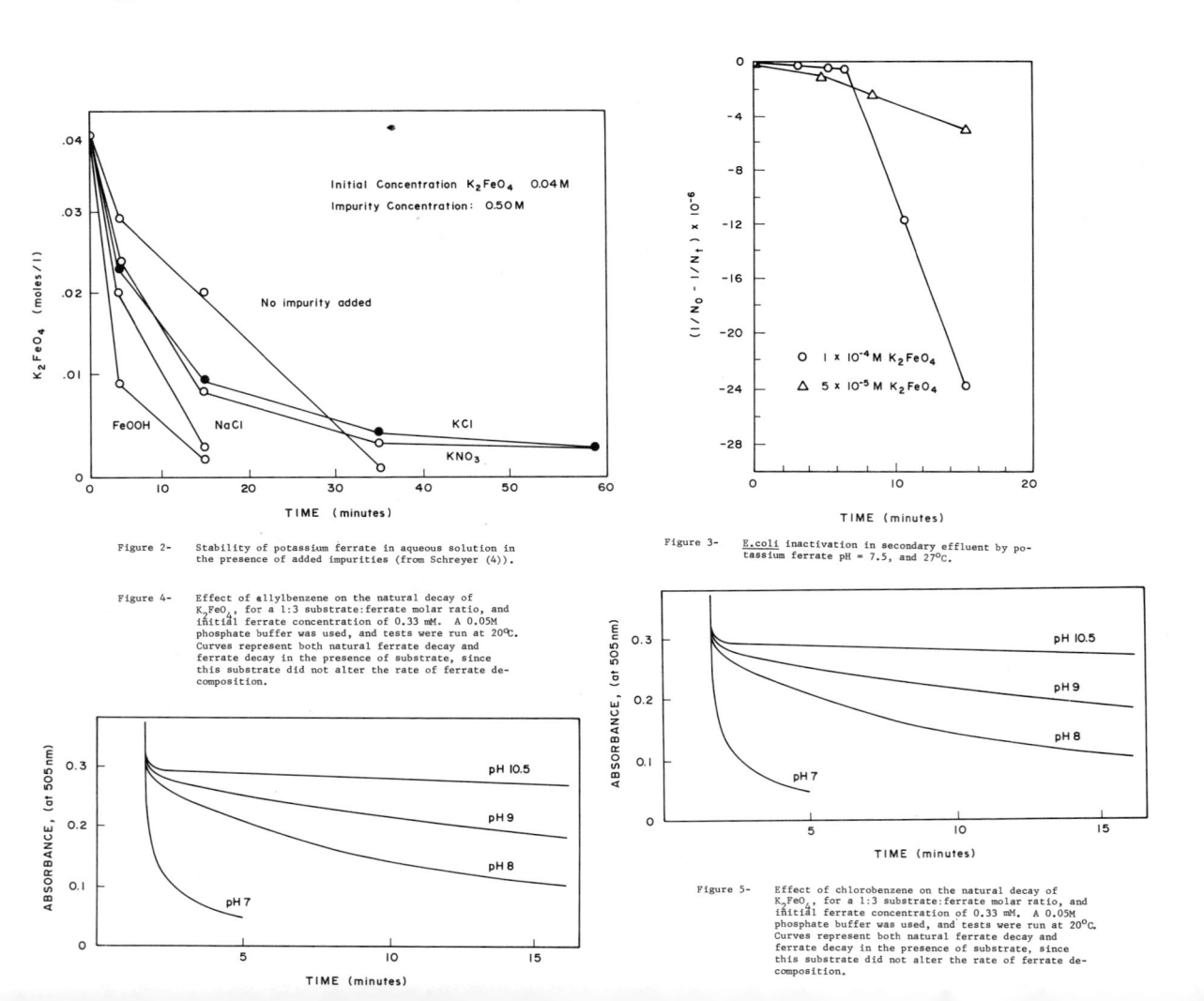

Figure 2- Stability of potassium ferrate in aqueous solution in the presence of added impurities (from Schreyer (4)).

Figure 3- E.coli inactivation in secondary effluent by potassium ferrate pH = 7.5, and 27°C.

Figure 4- Effect of ethylbenzene on the natural decay of K$_2$FeO$_4$, for a 1:3 substrate:ferrate molar ratio, and initial ferrate concentration of 0.33 mM. A 0.05M phosphate buffer was used, and tests were run at 20°C. Curves represent both natural ferrate decay and ferrate decay in the presence of substrate, since this substrate did not alter the rate of ferrate decomposition.

Figure 5- Effect of chlorobenzene on the natural decay of K$_2$FeO$_4$, for a 1:3 substrate:ferrate molar ratio, and initial ferrate concentration of 0.33 mM. A 0.05M phosphate buffer was used, and tests were run at 20°C. Curves represent both natural ferrate decay and ferrate decay in the presence of substrate, since this substrate did not alter the rate of ferrate decomposition.

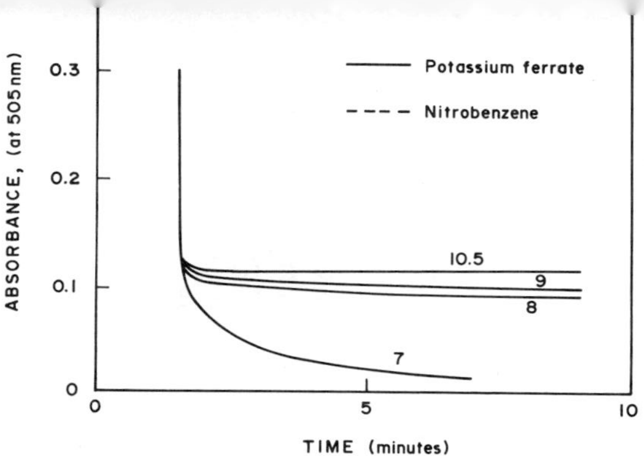

Figure 6— Effect of nitrobenzene on natural decay rate of potassium ferrate, for a 2:1 substrate:ferrate molar ratio, and initial ferrate concentration of 0.15 mM. A 0.05 M phosphate buffer was used, and tests were run at 20°C.

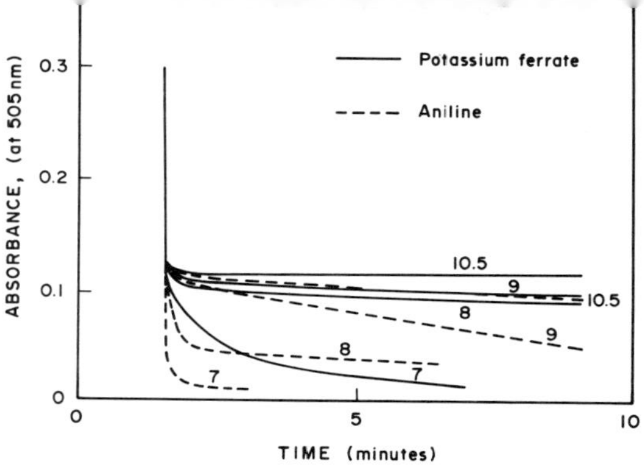

Figure 7— Effect of aniline on natural decay rate of potassium ferrate, for a 2:1 substrate:ferrate molar ratio, and initial ferrate concentration of 0.15 mM. A 0.05 M phosphate buffer was used, and tests were run at 20°C.

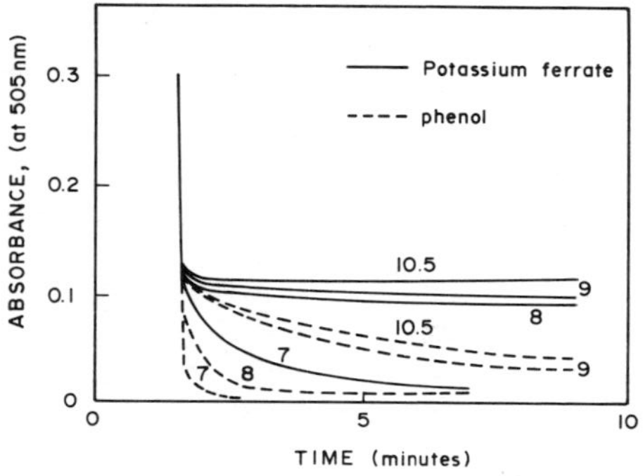

Figure 8— Effect of phenol on natural decay rate of potassium ferrate, for a 2:1 substrate:ferrate molar ratio, and initial ferrate concentration of 0.15 mM. A 0.05 M phosphate buffer was used, and tests were run at 20°C.

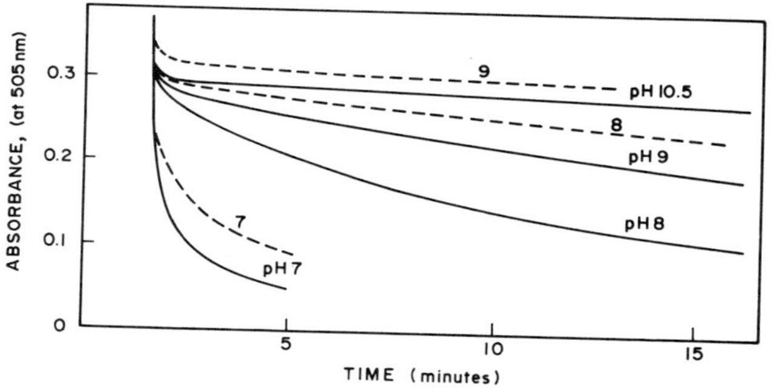

Figure 9— Effect of benzene on the natural decay rate of K_2FeO_4 for a 1:3 substrate:ferrate molar ratio, and initial ferrate concentration of 0.33mM. A 0.05M phosphate buffer was used, and tests were run at 20°C. Curves represent both natural ferrate decay and ferrate decay in the presence of substrate, since this substrate did not alter the rate of ferrate decomposition.

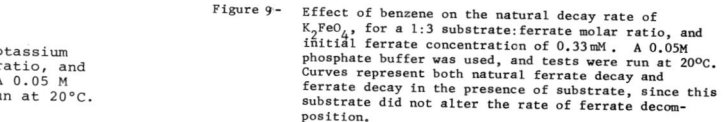

-425-

USE OF OZONE AND CHLORINE IN WATER WORKS IN THE
FEDERAL REPUBLIC OF GERMANY

Kühn, W., Sontheimer, H. and Kurz, R.

Engler-Bunte-Institut
der Universität Karlsruhe
Bereich Wasserchemie
Postfach 6380
D-7500 Karlsruhe 1
Federal Republic of Germany

Introduction

One cannot write about experiences in German drinking water works without mentioning the essential criteria that had a definite influence on the improvement in our country. These criteria of drinking water philosophy are the reasons why the formation of haloforms during chlorination occurred in a considerably smaller scale than, for example, in the United States of America. Summarizing, one perhaps can make the following statements:

1) The historical improvement of drinking water supply has led to the fact that ground water was, and today still is, the most important raw material for drinking water. Such a ground water source was always preferred, and depending on the surrounding area of the wells and the conditions in the soil, is chemically and bacteriologically unobjectionable without any treatment and without chlorination. Nowadays there are still hundreds of waterworks which

provide water without any chlorination and which have no problems with bacteria in the distribution system. Among these works are big cities such as Karlsruhe and Munich.

2) In the overcrowded regions along the river Rhine and the river Ruhr there was no way to avoid the use of surface water as a source of drinking water. Because of the good experiences with ground water one always strived to use treatment procedures that included passage through the ground, by sand bank filtration, or by an artificial percolation, reducing the total concentration of organic material below 2 mg TOC/l. With the additional occurrence of chemical ground water pollution, in addition to these natural procedures, physicochemical treatment became necessary. So 40 years ago ozone and activated carbon filters were applied for the first time. The goal was always to reach a water quality identical to that of pure ground water. This was the reason for limiting the overall amount of organic impurities to 2 mg TOC/l or less and for limiting the chlorine dosage for disinfection purposes to 0.5 mg/l or less.

These improvement trends have to be kept in mind when experiences and research results gained in German water works are reported. It has to be pointed out that the problem of chlorinated organics in our country arose as a consequence of increasing pollution of the surface waters, and not because of drinking water chlorination.

Analytical Methods

The observations obtained were reason enough to develop an analytical method for the determination of the overall amount of these toxic and hygienically important compounds in the raw water as well as within the different treatment steps. The most important questions are: (1) How are the chloroorganics adsorbed in an activated carbon filter? (2) How to develop a suitable analytical control method to determine the breakthrough behavior of activated carbon filters, and to determine the necessity of reactivation?

It was obvious, therefore, that it was necessary to develop an analytical method which is preceeded by an adsorptive enrichment onto activated carbon.

Consequently we not only obtain the concentration of these compounds in water (DOCl - Dissolved Organic Chlorine) but also the loading of the activated carbon, and by this technique, indirectly, the specific activated carbon quality.

Figure 1 shows the method of pyrohydrolysis (1,2) developed for this purpose. About 1 gram of sample in a quartz boat is placed within the tube furnace I. While this furnace is heated up to 600°C, the quartz tube is purged with oxygen and steam. In furnace II, where the reaction zone is, the outgoing material is mineralized at 1000°C. This means the organic bonded chlorine is changed to inorganic chloride. This chloride is dissolved in the water that condenses in the cooler and drops into a beaker. Afterwards the chloride is measured with an ion selective electrode.

Table 1 shows the concentrations in some rivers of organic chloro compounds. The results were obtained after concentration in an activated carbon minisampler (3).

The data for AOS (Adsorbable Organic Substances measured in the activated carbon extract), DOCl (Dissolved Organic Chlorine - measured by pyrohydrolysis) and DOClN (Dissolved Organic Chlorine Nonpolar--measured in the extract by a microcoulometric technique) show how surface water pollution is increasing especially along the river Rhine. This group parameter, dissolved organic chlorine, gives extensive evidence about the condition of a water source with respect to chemical pollution. Also, it is a technique that is easy to handle and to perform for the routine control of activated carbon filters.

Nevertheless, in special circumstances, such as legal cases, it is necessary to investigate the presence and additional concentration of single organic compounds as well as group parameters. For this purpose the Grob method (4) is used. This is described schematically in Figure 2.

Similar techniques have been used in the USA by Bellar and Lichtenberg (5). In the Grob method the purgable organics are stripped out of the water sample by recirculating the headspace gas. After concentration on a small activated carbon filter trap, an elution with CS_2 follows and the eluate is then injected into a combined GC/MS system.

In this way many single compounds were determined. Figure 3 shows typical GC/MS diagrams for a sample from the river Rhine and its sand bank filtrate. Besides the disturbing number of peaks, one can see qualitatively the treatment efficiency of the sand bank filtration step.

Table 2 and also the following examples show a good agreement between single substance measurements and group parameters (6). One can recognize certain points of pollution by looking at the single compound data, which are related to wastewater discharges. The high concentration of chlorobutadiene is clearly seen at Basle, and those of chloroaromatic compounds at Duisburg. The polluters could be located by the single compound analysis; measuring group parameters only would not have identified these polluters. On the other hand, it is obvious that only around 10% of the overall chloroorganics can be measured by single substance analysis. Therefore, to obtain a complete overview concerning pollution, we cannot ignore the measurement of group parameters.

Influence of Chlorination

In Table 3 the behavior of the concentration of these compounds within the different treatment steps is shown for a water utility located on the lower Rhine. The influence of chlorination, which is practiced at this utility, is especially important. Raw Rhine river water is dosed with 2 mg/l of chlorine and takes about 3 hours to reach the treatment plant. The consequence of this chlorination is a water with a DOCl concentration 7 µg/l higher than the raw water, 78 µg/l. The value for GOCl (Gas Chromatographic Detectable Organic Chlorine) has also nearly doubled.

Besides the formation of aliphatic compounds, like tetrachloroethylene, chloroform and the other well-known trihalomethanes, an increase in the concentration of chlorobenzenes, dichlorobenzenes and chlorotoluenes occurred. Beyond this, other chloroorganics, not detectable by GC, were formed, which are probably polar and of high molecular weight. Table 3 shows a 10 times higher result of the pyrohydrolysis compared with the sum of GC/MS detectable compounds (GOCl).

Remarkable also is the formation of bromoorganics, $CHBr_2Cl$ and $CHBr_3$, that are detected in concentrations around 2 µg/l. However, besides the formation of these well-known bromoorganics, other brominated hydrocarbons are formed. In the example here it could be shown that a number of bromoaromatics like bromobenzenes and bromotoluenes were produced in relatively high concentration. Altogether 4.1 ug/l of organically bonded bromine was identified by GC/MS. Laboratory tests have shown that the bromoform reaction is possibly not only related to the content of bromide in the raw water, but also to contamination of the chlorine gas with bromine. It is technically and practically difficult to manufacture chlorine without any bromine content.

Influence of Ozonation

Table 4 gives data of a water utility also located on the lower Rhine, but using ozonation instead of chlorination. In contrast to Table 3, there is no increase of the DOCl concentration after ozonation. After ozonation and manganese removal it is surprising to see an increase of DOClN as well as of GOCl. As the single substance data show, mostly very volatile compounds like $CHCl_3$, CCl_4, C_2HCl_3, C_2Cl_4, tetra- and hexachlorobutadiene are found in a higher concentration. The explanation of this effect with our present knowledge is hypothetical and has to be proven by experiments. It is suggested that ozone splits off small fragments from high molecular weight organic chlorocompounds. These smaller units are then detectable by GC/MS.

Independent of the use of either ozone or chlorine, the data show the high treatment efficiency of an activated carbon filtration step. This is once again shown in Table 5, where the percentages of single compounds remaining are listed for the different treatment steps. Note that sand bank filtration treatment efficiencies decrease as the chlorine content of chloroaromatic compounds increases. This effect is generally for chlorinated hydrocarbons. It is also of interest that the concentrations of trichlorobenzenes are fairly well reduced during ozonation.

The most important fact, however, is that even in the nearly exhausted activated carbon filters which were used here, most of the compounds are removed to around 1 percent of their influent concentration. It is certain that the biological activity of the carbon filter is an important factor here. The formation of some aliphatic hydrocarbons points out that there is biodegradation within the activated carbon filter, which in this case is positively assisted by the ozonation.

Use of Ozone

Because many utilities in the Federal Republic of Germany use ozone, it is worthwhile to try to summarize their collective experiences.

With few exceptions ozone is not used for disinfection alone. This means that ozonation is not used as the last treatment step. This is related to the observation that at many places oxidation with ozone leads to a transformation of the organic matter, which makes many non-degradable compounds biodegradable. This effect possibly causes an increase in bacterial growth in the distribution system, if there is not additional chlorination after the ozonation. Detailed practical experiences in this field were gained about 10 years ago in the utilities of Kreuzlingen and Zurich.

Figure 4 gives typical results of recent laboratory tests with refiltered Lake of Constance water. The curves show that the growth of bacteria is much faster in the ozone treated water, which was seeded after ozonation, than in the raw water. This effect is greater the higher the ozone dosage.

This kind of effect, the regrowth of bacteria, is undesirable in the water distribution system. If one wants to utilize this phenomenon positively it would be efficient to have these biological processes take place in a filter. An activated carbon filter is well qualified for this purpose (7). This statement is confirmed by the experiences in the Rhine Waterworks where, for other reasons and with another motivation, from the very beginning ozonation is conducted before the activated carbon treatment. Though there are about 10^9 bacteria per gram of activated carbon within a filter, the biodegradation and the parallel nitrification are completed at the outlet of the filter, and the number of bacteria in the filtered water are very low, e.g., under 10/ml in the Düsseldorf utilities. Such a biological post-treatment could also be carried out in sand filters if the ozone consumption is satisfied before the filtration.

Such a combination of ozonation and adsorption, carried out in most of the German waterworks using ozone, leads on one hand to the nearly complete removal of bacteria and viruses and, on the other hand, improves the removal of dissolved organic compounds by adsorption as discussed below. The biological activity in the granular activated carbon filters makes it possible to lengthen the operating time of the filter, because it causes a type of biological reactivation and subsequent lowering of the carbon loading (8).

Another complicating factor that one has to consider is that during ozonation of organics in water more polar compounds are formed, which also changes their adsorbability. This is illustrated in Figure 5, which shows the adsorption, as measured by the percent removal of DOC, of ozonated and non-ozonated activated sludge effluent on alumina and activated carbon. The change of the organics by ozonation is seen by a decrease of activated carbon adsorption and an increase in the adsorption onto the more polar alumina. How this is influenced by the ozone dose is shown in Figure 6.

In Figure 6 are presented on log-log paper the Freundlich isotherms, that means UV-loading as shown by UV-removal per gram of activated carbon versus equilibrium UV-extinction. It is obvious in this case, in which we analyzed Lake of Constance water with a TOC concentration of about 1.1 mg/l, that even a small amount of ozone is enough to shift the linearized Freundlich adsorption isotherms to the right, which means a poorer adsorption onto activated carbon.

These results caused us to investigate whether or not these more polar humic acids which form during ozonation could be removed with flocculation, because they are well adsorbed onto aluminum hydroxide, as Figure 7 shows. Just the contrary effect occurs in the case of activated carbon adsorption. Very small amounts of ozone gave a much higher adsorption, which cannot be enhanced by a higher ozone dosage. In addition to this, using ozone, a precipitation of humic material and a higher turbidity removal can be attained (9). This so-called "microflocculation" occurred especially at the Lake of Constance Water Works, where it is utilized (10).

Figure 8 points out that this effect is caused by an oxidation of humic acids in the raw water. In these studies a kaolin suspension was mixed with different amounts of humics, which were extracted from Lake of Constance water by ion exchange. The lower half of the figure shows the residual turbidity of such a humic-containing suspension after dosage with different amounts of ozone. After a preliminary increase of turbidity one can see a dramatic decrease. This can be related to the ozonated humic acid studies in which these materials act like synthetic polymers which trap the kaolin particles within a forming floc.

As has already been mentioned, this effect is aided by an additional application of flocculant like aluminum sulfate, whereby a low dosage of 0.5-1 mg of Al^{3+}/l is sufficient. This secondary flocculation, after ozonation and before filtration, has proven effective in several waterworks, and gives residual turbidities as low as those in the best ground waters.

Finally, it should be mentioned that the ozonation of organics in water leads also to a better adsorption onto calcium carbonate ($CaCO_3$). This is shown in Figure 9. It is again contrary to what occurred with activated carbon adsorption (Figure 7) and shows the same trend that took place in the case of alumina as an adsorbent (Figure 8), but here we have a stronger influence of the ozone concentration. Other tests in our laboratories have shown that this is of importance for practice in the field. The adsorbability onto $CaCO_3$ is a simple method for evaluation of the inhibitor efficiency of organic compounds against corrosion. This desirable effect of organics in water is therefore intensified by ozonation.

Conclusions

The following statements can be made from the results discussed above and the experiences in German water works with the oxidizing agent ozone:

1) Ozone treatment of waters makes humic materials more biodegradable and, therefore, increases their treatability by natural biological processes. Because of this, there should be a filter after ozonation, in which the biodegradation can take place.

2) With such a filter, the floc formed during ozonation can be also removed, if there is some flocculant added. By this technique, very low concentrations of residual turbidity can be achieved.

3) This type of ozone treatment gives such a remarkable improvement in water quality and, simultaneously, such complete inactivation of bacteria and viruses, that one can use very small amounts of chlorine in the final chlorination step. In most of these utilities a chlorine dosage of 0.1 - 0.3 mg/l is sufficient. Also, chlorine dioxide is being used more and more, because of the good experiences observed in doing this.

These conclusions make it clear why there is an organic chlorine problem only in water utilities of the Federal Republic of Germany where breakpoint chlorination is used--because of high ammonia concentrations, or where there are already high concentrations of chlorinated hydrocarbons in the raw water because of industrial pollution. In the latter case, for control, one sees that the pollution decreases and in addition to this, that the treatment, especially in the granular activated carbon filters, is carefully controlled and often reactivated.

In the case of ammonia removal, one strives more and more for biological processes. Initial successful tests were made along the Rhine, where better conditions for biological nitrification are attained by ozonation. By doing this, the total amount of ammonia can be oxidized in the activated carbon filter following ozonation without breakpoint chlorination.

It should be emphasized also that there is no general treatment scheme that is optimal for every raw water. One has to consider that there is an optimum efficiency for each treatment step, as shown in Table 6.

There are possible disadvantageous effects in most treatment processes in cases of too little or too much treatment, especially when using chlorine or ozone. An optimum in water treatment can be reached only if the individual treatment steps supplement each other. This type of procedure will produce an unobjectionable, high quality finished water, which, in our opinion, should be equivalent to a natural, pure drinking water derived from ground water.

Literature Cited

(1) W. Kühn and H. Sontheimer, "Einige Untersuchungen von organischen Chlorverbindungen auf Aktivkohlen." Vom Wasser, 41:65-79 (1973).

(2) W. Kühn and H. Sontheimer, "Zur analytischen Erfassung organischer Chlorverbindungen mit der temperatur-programmierten Pyrohydrolyses." Vom Wasser, 43:327-341 (1974).

(3) W. Kühn and F. Fuchs, "Untersuchungen zur Bedeutung der organischen Chlorverbindungen und ihrer Adsorbierbarkeit." Vom Wasser, 45:217-232 (1975).

(4) K. Grob and F. Zürcher, "Stripping of trace organic substances from water--equipment and procedure." J. Chromatography, 117:285-294 (1976).

(5) T.A. Bellar, J.J. Lichtenberg and R.C. Kroner, "The occurrence of organohalides in chlorinated drinking water." Jour. AWWA, 66:703 (1974).

(6) L. Stieglitz, W. Kühn and W. Leger, "Verhalten von Organohalogenverbindungen bei der Trinkwasseraufbereitung." Vom Wasser, 47: in prearation.

(7) W. Hopf, "Zur Wasseraufbereitung mit Ozon und Aktivkohle." GWF - Wasser/Abwasser, Bd. 111, Heft 2, S. 83-92 (1970).

(8) M. Eberhardt, S. Madsen, and H. Sontheimer, "Untersuchungen zur Verwendung biologisch arbeitender Aktivkohlefilter bei der Trinkwasseraufbereitung." Veröffentlichungen des Bereichs und des Lehrstuhls für Wasserchemie am ENGLER BUNTE INSTITUT der Universität Karlsruhe, Heft 7, (1974).

(9) E. Wurster and G. Werner, "Die Leipheimer Versuche zur Aufbereitung von Donauwasser." GWF, Wasser/Abwasser, Bd. 112, Heft 2, S. 89-91 (1974).

(10) D. Maus, "Wirking von Ozon auf die gelösten Substanzen im Bodenseewasser." Vom Wasser, Bd. 43, S. 127-160 (1974).

DISCUSSION

George White, Consulting Engineer, San Francisco: Dr. you confuse me a little bit by interchanging the terms "residual" and "dosage" when you were mentioning chlorine in the distribution system. I wonder if you would go over that again. I think you meant in both cases that it was residual and wasn't dosage.

Kühn: I talked about residual, mentioning the 0.1 mg/l at the tap after the distribution system, which is allowed by government regulations. The dosage at the outlet of the water works is sometimes a little higher, for example 0.2 mg/l. It shouldn't be much higher because there shouldn't be much chlorine demand in the distribution system. That means that if you produce a good water it will have no demand for chlorine.

White: What would the dosage at the plant have to be to produce the 0.1 mg/l at the tap?

Kühn: This would be 0.2 or 0.1 mg/l. There is nearly no chlorine demand in the distribution system. For example, in the city of Pulse or in the city of Munich, they add no chlorine at all. We made some tests on these waters at the University of Karlsruhe. We added bacteria to the distribution system, and there was no chance for the bacteria to grow because there was just no food for them. And to be very honest we had the part of Karlsruhe where there is chlorination. We have an American village there and they make chlorination, high chlorination with our safe water.

White: That's what you get for having the Americans over there.

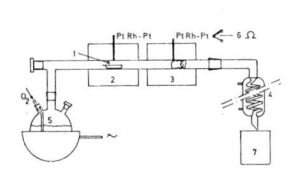

1) Sample
2) Furnace 1
3) Furnace 2
4) Condenser
5) Water
6) Quartz wool plug
7) Beaker

PYROHYDROLYSIS APPARATUS

FIGURE 1

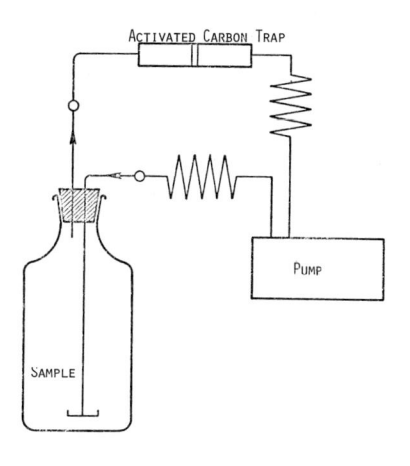

A PURGING AND ADSORBING UNIT FOR PURGABLE ORGANICS

FIGURE 2

Sampling Point	AOS (g/m^3)	DOCl (mg/m^3)	DOCl N (mg/m^3)
Lake of Constance	1.5	6	3
Rhine above Basle	3.1	93	27
Rhine at Cologne	5.8	192	37
Rhine at Duisburg	9.0	228	55
Danube at Leipheim	~2.6	~47	~6
Neckar at Ludwigsburg	6.2	48	16
Ruhr at Mülheim	7.5	~160	24
Fulda at Kassel	5.0	35	7
Weser at Bremen	7.0	~200	16

ADSORBABLE COMPOUNDS AND ORGANIC CHLORO COMPOUNDS IN SOME GERMAN SURFACE WATERS (AVERAGE VALUES FOR 1974)

TABLE 1

Substance	Basle	Cologne	Duisburg
Chloroform	1.1	0.3	1.1
Carbon Tetrachloride	0.3	2.4	3.3
Chloroethane	2.5	0.2	0.6
Trichloroethylene	0.9	0.8	0.6
Tetrachloroethylene	0.3	0.3	1.5
Chlorobutadiene	4.4	0.4	0.3
Chlorobenzene	0.9	0.6	5.3
Chlorotoluene	-	0.5	1.3
Benzene	0.2	0.3	0.8
Toluene	0.8	0.7	1.9
0 - C$_2$	0.2	0.7	1.4
0 - C$_3$	0.1	0.4	1.0
0 - C$_4$	-	0.1	0.5
DOCl	55	78	100
DOClN	10	5	15
DOC	2300	4600	5000

DATA FOR 3 LOCATIONS ALONG THE RIVER RHINE (mg/m^3)

TABLE 2

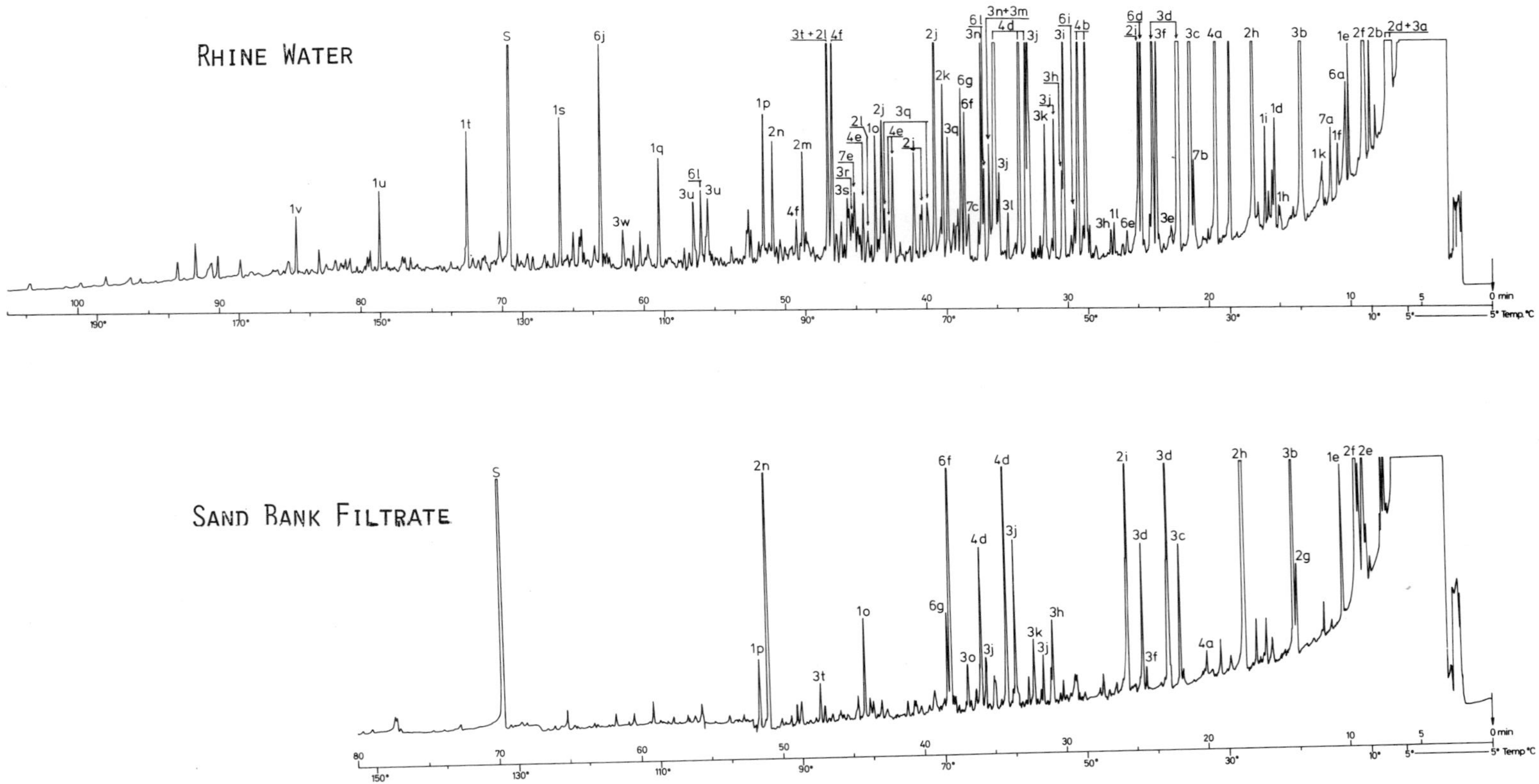

GC/MS CHROMATOGRAMS OF RHINE WATER AND SAND BANK FILTRATE

FIGURE 3

-438-

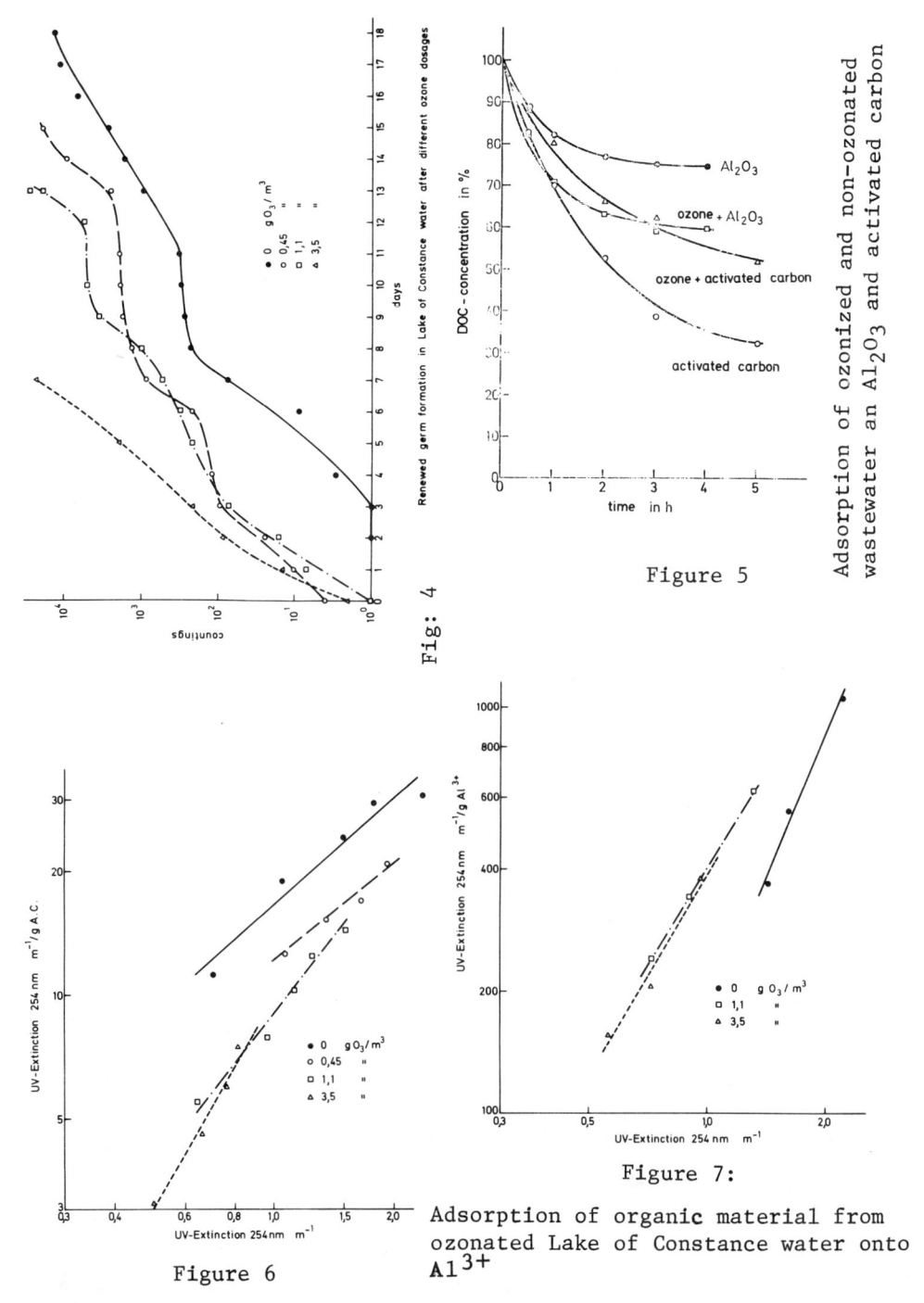

Figure 4: Renewed germ formation in Lake of Constance water after different ozone dosages

Figure 5: Adsorption of ozonized and non-ozonated wastewater an Al$_2$O$_3$ and activated carbon

Figure 6: Adsorption of organic material from ozonated Lake of Constance water onto Activated Carbon

Figure 7: Adsorption of organic material from ozonated Lake of Constance water onto Al^{3+}

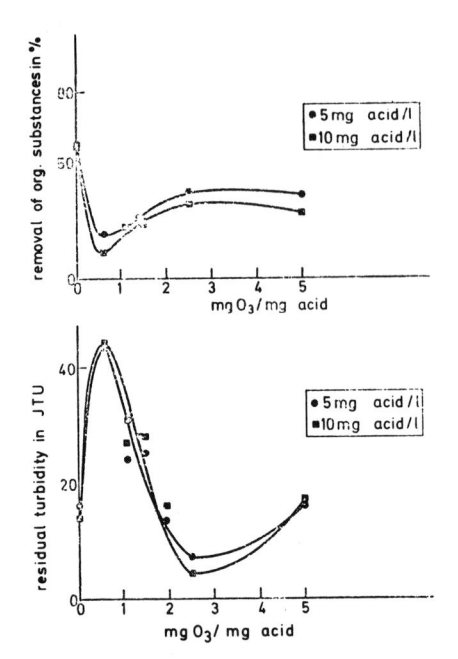

Figure 8.

Residual turbidity and removal of organic material at various ozone dosages (100 mg kaolin/l and 5 mmol $CaCl_2/l$)

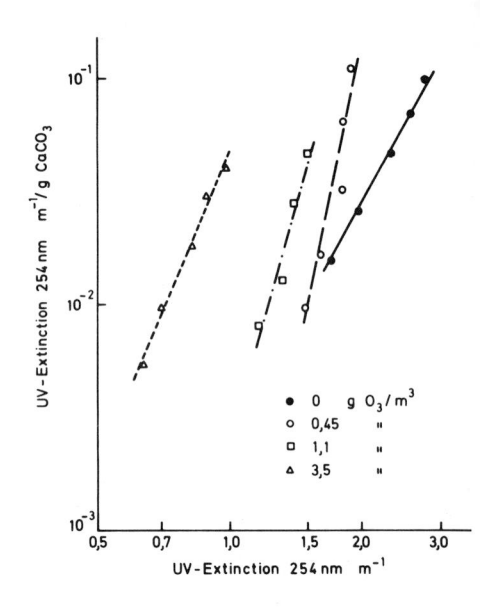

Figure 9.

Adsorption of organic material from ozonated Lake of Constance water onto $CaCO_3$

ELEMENT	C_L			C		Br
PARAMETER (µG/L) SAMPLE	DOCL	DOCLN	GOCL	DOC	GOC	GOBR
Rhine Water	73	3	4.6	4600	4.1	0
Sand Bank Filtrate	17	2	2	900	0.9	0
Chlorinated Sand Bank Filtrate	85	7	3.7	1000	1.5	4.1
Activated Carbon Filtrate	2	1	4	3700*	2.2	0

* Contaminated

MEASURED TOTAL, GROUP, AND SINGLE COMPOUND DATA WITHIN DIFFERENT TREATMENT STEPS
(RHINE WATER UTILITY USING CHLORINATION)

Table 3

Element	CL			C	
Parameter (μg/L) Sample	DOCL	DOCLN	GOCL	DOC	GOC
Rhine Water	99	15	9.5	5000	11
Sand Bank Filtrate	61	4	4.8	1600	2.4
Ozonated Sand Bank Filtrate	45	10	8.3	1600	3.5
Activated Carbon Filtrate	27	4	6	1200	4

MEASURED TOTAL, GROUP, AND SINGLE COMPOUND DATA WITHIN DIFFERENT TREATMENT STEPS
(RHINE WATER UTILITY USING OZONATION)

Table 4

Substance	Sand Bank Filtrate	Ozonated and Filtered	Activated Carbon Filtered
Chlorobenzene	4	7	<1
Dichlorobenzene (o,m,p)	38	42	5
Trichlorobenzene (3 isomers)	82	<1	<1
Chlorotoluene	6	9	<1
Dichlorotoluene	38	38	<1

PERCENT REMAINING AFTER SEVERAL TREATMENT STEPS FOR SELECTED COMPOUNDS
IN THE RHINE RIVER

Table 5

PROCESS	POSSIBLE DISADVANTAGEOUS EFFECTS OF THE PROCESSES IN CASE OF	
	TOO LITTLE TREATMENT	TOO MUCH TREATMENT
Chlorination	Unsatisfactory Disinfection	Production of Organic Chlorocompounds
Ozonation	Not Sufficient Efficiency	Bacteria in the Distribution System
Adsorption	Too Little Organic Removal	Corrosion Problems
Flocculation	Unsatisfactory Removal of Colloids	High Concentrations of Neutral Salts
CO_2 Removal	Trouble with $CaCO_3$ Precipitate	Corrosion Problems, Increase of Heart Attacks

Table 6: Water Treatment Processes

COMPARISON OF PRACTICAL ALTERNATIVE TREATMENT SCHEMES

FOR REDUCTION OF TRIHALOMETHANES IN DRINKING WATER

by

James M. Symons, O. Thomas Love, Jr., and Keith Carswell

U.S. Environmental Protection Agency
Municipal Environmental Research Laboratory
Water Supply Research Division
Cincinnati, Ohio 45268

INTRODUCTION

It is an interesting position to be the last speaker in a two and a half day conference that is essentially on the same subject. On the one hand, everything you have to say has been said before, but on the other hand, the audience is well prepared for your information. Therefore I hope that when I share with you some of our research on ozonation, chlorination and the use of chlorine dioxide over the last two years it will be meaningful for you, based upon the excellent presentations we have had previous to this talk.

ORGANICS REMOVAL

Two years ago we began looking at ozone as a water treatment scheme for the removal of organic compounds. We started by ozonating Cincinnati, Ohio tap water with various dosages in an attempt to look at the percent reduction of the general organic content of the water. We were using two parameters in those days. One was CCE-m, the mini-sampler carbon-chloroform extract procedure, and the other was TOC, at least what we now call the non-purgable portion of the total organic carbon. The point can be made from Table I that in order to obtain significant reductions of either parameter, it takes a rather large applied dose of ozone (in Cincinnati tap water). About the time we were beginning this experimentation, the problem of trihalomethanes surfaced and we focused our attention on that as one of the major organic problems in drinking waters in the United States.

OTHER METHODS FOR REMOVAL OF TRIHALOMETHANES

To briefly summarize, we have looked at aeration and adsorption for removing trihalomethanes from drinking water. Aeration does strip chloroform and its related by-products (the other three trihalomethanes) from water, but we have found that rather large volumes of gas are needed relative to the volumes of water. Because the purging and trapping procedure for chloroform analysis uses a stripping procedure, we know that these compounds can be stripped from water; it is just a matter of economics in terms of how much gas you are willing to use to remove them. Furthermore, because precursors are not removed by aeration, if you have post-chlorination you will continue to have some chloroform formed in the distribution system; so that must be taken into account.

Adsorption by both powdered and granular activated carbon is effective for trihalomethane removal. The only problem is one of capacity. Chloroform is not a particularly adsorbable compound, and rather high doses (50 to 100 mg/l) of powdered activated carbon are needed to effect significant removals of trihalomethanes. This results in additional sludge that must be disposed of at the water treatment plant.

Granular activated carbon in columns operated at a conventional flow rate and bed depth, which in this country is in the two to three gallons per minute per square foot range and a bed depth of two to three feet, was effective for trihalomethane removal, but this effectiveness did not last very long. The first breakthrough for Cincinnati, Ohio tap water, which contains anywhere from about 5-100 micrograms per liter (ug/l of chloroform depending on the time of year, occurred in our tests within three or four weeks, and complete exhaustion could be measured in 8-10 weeks.

OZONE FOR REMOVING TRIHALOMETHANES

Now focusing on the theme of this conference, we considered the two oxidants, ozone and chlorine dioxide, as possible techniques for removing trihalomethanes from drinking water. We started with a procedure in which we merely ozonated Cincinnati, Ohio drinking water. Like some of the other speakers, we had a high-speed mixer in our contactor in an attempt to improve ozone mass transfer. This test apparatus had about a five or six minute contact time and the ozone dose was approximately 20-25 mg/l.

Table II shows that ozone at these dosages and these contact times was ineffective for removing the trihalomethanes. The other tests were made to determine whether or not we were getting any purging from the water because of the gas flow, and there is little, if any, difference between those four values. It should be noted that the trihalomethane concentrations are expressed as ΣTHM, which is simply an arithmetic sum of the concentrations of four commonly found trihalomethanes: chloroform, bromodichloromethane, dibromochloromethane, and bromoform.

CHLORINE DIOXIDE FOR REMOVING TRIHALOMETHANES

Chlorine dioxide was considered in another experiment, again working with Cincinnati, Ohio tap water. In this experiment the chlorine dioxide dose was 7-10 mg/l, a rather high dose, and contact time in that system was one and two days. Table III shows that chlorine dioxide was ineffective under these conditions for removing trihalomethanes.

TRIHALOMETHANE PRECURSOR REMOVAL

Having completed this phase of our experimentation, we went on to study the influence of the various disinfectants on trihalomethane precursor. We took undisinfected Ohio River water, coagulated and settled it with alum or iron salts as they do in conventional water treatment, and passed it in parallel either through a dual media filter or a granular activated carbon (GAC) filter-adsorber, following both with post-disinfection.

First, we wanted to see what dosages of ozone and chlorine dioxide were needed to obtain good disinfection of the pilot plant effluent. Even though we were using undisinfected Ohio River water as the source water, we found that few, if any, coliforms were coming through the dual media filter, so Figures 1 and 2 are related to standard plate counts in organisms per one ml. You will notice that 0.1 mg/l of ozone virtually eliminated the standard plate count organisms in the GAC effluent, but in the dual-media effluent, it took about 0.5 mg/l.

The GAC filter was about eight weeks old at this time, and in spite of the fact that some TOC was coming through, the influence of treatment and the reduction of disinfectant demand is clearly in evidence. This supports what Wolfgang Kühn said just a few minutes ago, that disinfection of a high quality water does not take much disinfectant.

The situation with chlorine dioxide, Figure 2, is similar. Here, 0.1 mg/l of chlorine dioxide removed the standard plate count organisms in the GAC effluent. During these experiments the carbon was 24 weeks old, which is old as related to life expectancy for TOC removal. The dual media effluent required 0.3 mg/l to accomplish the same kill.

Table IV shows the comparison of chloroform formation, either with chlorine alone, or with ozone followed by chlorine in the dual-media effluent. We think that a residual disinfectant at the tap is essential, so we always think of following ozonation with chlorination. Chlorine alone produced about 10 μg/l of chloroform; however 0.7 mg/l of ozone followed by chlorine formed approximately the same amount of chloroform. In another experiment, 18.6 mg/l of ozone had no effect on chloroform precursors, as monitored by subsequent THM formation.

In these studies we used chlorine. We stored these samples for six days prior to analysis, and we wanted to make sure that during the storage period we would not run out of chlorine and stop the trihalomethane formation reaction. We dosed the chlorinated samples with enough chlorine to make sure that the chloroform reaction would not be limited by the chlorine concentration. I was wondering when the speaker from Toronto yesterday mentioned about ozonation having an influence on precusor, which we do not see, whether or not he was looking at instantaneous chloroform values right after application of chlorine or whether he actually stored his samples for some time to simulate passage through the distribution system.

The final test relates trihalomethane production from 8 mg/l of chlorine to 227 mg/l dose of ozone. In this case we did show that under these conditions oxidation by very high dosages of ozone can reduce chloroform precursors, see Table IV.

I might also add that in our studies ozonation does not produce chloroform. As you recall, Wolfgang Kühn noted previously that in their experience, they did get some increases in certain low molecular weight chlorinated species upon ozonation, which he postulated was due to the breaking up of large, chlorinated molecules into small ones that were then amenable to the particular analytical procedure that he was using. I hope you noticed from his slide, however, that the total organic chlorine concentration was going down, while the non-polar organic chlorine concentration (as measured by solvent extraction) was going up. So the analytical procedure has some influence on that.

Another possibility for the production of low molecular weight chlorinated compounds during ozonation is the oxidation of chloride to chlorine, which we get several inquiries about. But at least in our waters, we have not seen the production of chloroform upon ozonation, provided that you do not use chlorine at all.

CHLORINE DIOXIDE FOR PRECURSOR REMOVAL

Figure 3 shows a comparison to the ozone situation just described. Chlorine dioxide alone does not produce chloroform. Chlorine alone produces a trihalomethane growth curve, as shown by the upper curve. Chlorine dioxide apparently has some influence on precursors, because in the presence of chlorine dioxide, even though you have excess chlorine, you have a lower production of chloroform, as shown by the middle curve.

END-PRODUCTS AND BY-PRODUCTS

Figure 4 illustrates the problem of by-product formation. We have a number of by-products from ozone and we probably should have included also the possibility of an end-product of hydrogen peroxide. We have chlorine dioxide undergoing oxidation reactions. Also, as Al Stevens pointed out yesterday, chlorine dioxide undergoes chlorination reactions. We have ClO_2 end products of chloride, chlorite, chlorate, and chlorine itself.

I think our problem with any disinfectant is one of looking at the complete picture of by-products and end-products. I think that focusing, for example, on the toxicity of chlorite by itself, which was mentioned yesterday and is something we are concerned about and are looking into, is good information to have, but we must also look at the <u>total</u> picture. I mean, for example that the toxicity of ClO_2 by-products and end products should be compared to the toxicity of chlorination by-products. This should be done for any disinfecting agent.

Figure 5 summarizes briefly the tests that we are doing in terms of trying to evaluate the toxicity of disinfectant by-products. One is the limulus lysate test for detecting endotoxins. Another is the bacteria mutagenicity screening test mentioned yesterday by the Stanford Research Institute speaker (Ron Spanggord). We are just beginning a program in which we will be evaluating the cytotoxicity of disinfection end products and by-products. That is the test that Riley Kinman mentioned on Monday. What we are going to do here is to add various disinfectants (separately of course) chlorine, chlorine dioxide, ozone, and chloramine, and then we will examine the finished water by these various tests of biological activity.

COST

A big problem with any disinfectant along with effectiveness and by-products is cost. We have prepared Table V, a desk top analysis of the estimated cost of disinfection with three candidate disinfectants. This table shows four treatment plants from 1 to 150 million gallons per day (MGD) all operating at 70% of design capacity. We have compared, arbitrarily,

2 mg/l of chlorine (30 minute contact time), 1 mg/l of ozone, and 1 mg/l of chlorine dioxide generated from chlorite and chlorine. We have chosen lower dosages for ozone and chlorine dioxide because we assumed that on a weight to volume basis they would probably be more effective disinfectants than chlorine. The total cost is the amortized cost of the feeding equipment, contact chambers, plus the labor cost, maintenance, operation, and so forth.

FUTURE RESEARCH

Briefly, what are the research needs that we feel are important? The International Ozone Institute has developed a list of research needs, Table VI. To us the first three are most important, and we are looking into all three of these ourselves.

In terms of monitoring techniques, measuring the oxidants and generating and handling the oxidants, we are relying on the equipment manufacturers and oxidant producers for information.

Table VII is a list of our extramural studies on the use of ozone and chlorine dioxide. Public Technology Incorporated, in connection with IOI, is looking at the status of ozone and chlorine dioxide use in water treatment both here and abroad. Dr. Glaze talked about his work sponsored by EPA on UV (ultraviolet) catalyzed oxidation with ozone at North Texas State University. The comparison of ozone, UV by itself, and chlorine as disinfectants for small systems, is the subject of a contract with the Health Department of the State of Vermont. The effect of particulates on disinfection by ozone and chlorine dioxide as alternative disinfectants is being studied at the University of Missouri using a raw water fairly heavily contaminated with viruses to look into the virucidal aspects of alternative disinfectants. The use of ozone and powdered activated carbon for removal of organic materials in raw water in New Orleans, and chlorine dioxide and granular activated carbon as treatment schemes in Evansville, Indiana, are the last two projects.

So, where does this leave us in the final analysis? We are seriously concerned with various disinfectants: chlorine, ozone, chlorine dioxide, and others. By what criteria are we judging them? We feel that the disinfectant must have an easy measurement technique, it must produce a residual at the tap, and be a good disinfectant. On the other hand there are the problems of by-product toxicity, cost, and maybe other unknown factors. The EPA is at the crossroads now in terms of disinfection. We have to investigate any new disinfectant and measure it against the one we are presently using. I hope that soon we will have these data and the Agency can make effective judgements.

DISCUSSION

Manfred Noack, Olin, New Haven Research Center: Among the products of disinfection with chlorine dioxide, you mentioned chlorate, and on the basis of the chemistry of chlorine dioxide, you would expect chlorate as a possible by-product of disinfection. Chlorate at very low levels in the presence of chlorite and chlorine dioxide is very difficult to detect. Is your mentioning of chlorate based on actual observation of chlorate, and if so, which method did you use to identify and quantify the chlorate concentration?

Symons: I would like to preface the answer to your question by stating what I probably should have stated at the beginning. We have a team working on this overall project, and I am here just representing the team. You met some of the chemists yesterday. The two engineers on the team are Tom Love and Keith Carswell. I meant to identify them all.

I say all this because another of our team members has worked on the analytical schemes for these materials in attempts to have a balance of adding chlorine dioxide to water, and then measuring the chlorine dioxide residual, chlorite, chlorate, and chloride that are produced during the reactions in order to get a material balance. He is giving a paper on this methodology which is a modification of the Palin DPD method, as I understand it, at the San Diego Conference of the American Water Works Association, and if you would like a copy of his printed paper give me your name at the end of the meeting, I'll be glad to send you one. He has worked out these procedures, and he feels he can account for the materials and make a mass balance.

Howard Huang, University of Missouri at Rolla: In your cost estimates using ozone, a contact time of twenty minutes was estimated. Do you really think that twenty minutes of contact is necessary, or could a shorter time be used?

Symons: It's possible that a shorter contact time could be used. European experience is to use contact times in that range, and so we decided to use that, although in our pilot studies we were using five minutes and, as you can see, getting good disinfection. So it would mean that the size of the basin then would be a little bit smaller, and this might affect the cost somewhat. But the contact chamber didn't turn out to be a big percentage of the cost anyway.

Huang: In using chlorine, as far as the operation and control is concerned, the dosing rate can be easily adjusted with the flow. But can this be done conveniently with ozonation? Also, to determine whether the dosing rate is right or not, in chlorination the operator can measure the residual at the effluent, or after a certain time of contact. In ozonation I wonder whether this can be easily done or not, and whether, in this respect, it will affect the overall operating costs?

Symons: You noticed in the list of research needs there were needs related to monitoring, control, generation, and so forth. People are doing it, so it can be done. I suspect that it could be done better, and I know that people are working on that, the manufacturers themselves, and people in the area of analysis. I tend to agree with you. I think that this is a slight drawback for these alternates in terms of their measurements.

There was mention in terms of chlorine dioxide yesterday that an experienced operator could judge the color of the gas and decide if his dosage was accurate, but they would prefer some sort of real time monitor. So I think this is a drawback, but I don't think it's insurmountable, even at the present time, and we're hopeful that it will improve in the future.

TABLE I

EFFECT OF OZONATION ON GENERAL ORGANIC PARAMETERS

Applied Ozone Dose** mg/l	Ozone Utilitzed*** mg/l	Contact Time Minutes	CCE-m Reduction %	Fluorescence* Reduction %	NPTOC Reduction %
Air Only	-	-	0	3	0
Oxygen Only	-	-	25	3	0
8.9	8.8	19	33	11	18
19.2	18.0	38	67	30	24
26.8	17.7	95	73	11	43
38.2	19.9	19	60	25	-
46.7	39.7	38	90	8	50
56.5	26.5	95	75	19	50
71.0	29.0	19	83	29	-
140.0	62.5	38	80	29	75

* The details of this analytic determination are contained in the National Organics Reconnaissance Survey. J. Am Water Works Assoc. 67:634-647 (1975).

**Applied Dose, continuous flow studies, mg/l =

$$\frac{mg\ ozone}{std\ liter\ of\ gas\ (O_3 + O_2)} \times std\ l\ gas(O_3 + O_2)/min \times \frac{min}{liters,\ water}$$

This parameter may not be directly related to the actual oxidation of organic compounds because of unaccounted for variations in mass transfer and/or chemical reaction rates.

*** Applied ozone dose minus ozone escaping the top of the contactor.

Table II

Removing Trihalomethanes From Cincinnati, Ohio Drinking Water by Ozonation

APPLIED OZONE DOSE - 25 mg/l

CONTACT TIME - 4-5 min

	ΣTHM µg/l
TAP WATER	26
MIXER	29
MIXER + OXYGEN	27
MIXER + OZONE	27

TABLE III

Removing Trihalomethanes From Cincinnati, Ohio Drinking Water with Chlorine Dioxide

ClO_2 DOSE, mg/l	CONTACT TIME, days	ΣTHM µg/l
0	0	52
7	1	58
7	2	51
0	0	75
10	2	78

TABLE VI

WSRD RESEARCH NEEDS IN OXIDANTS RESEARCH

- Identification of End Products and By-Products
- Determine Toxicity of These Identified Compounds
- Effects of Combined Use of Oxidants and Adsorbents
- Rapid and Reliable Techniques of Monitoring Oxidants
- Optimize Generation, Handling and Application of Oxidants

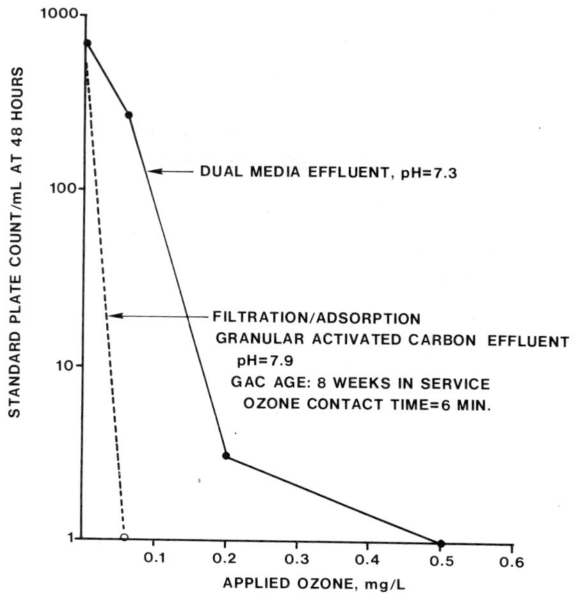

Fig. 1. POST DISINFECTION WITH OZONE

(Graph: Standard Plate Count/mL at 48 Hours vs. Applied Ozone, mg/L)
- DUAL MEDIA EFFLUENT, pH=7.3
- FILTRATION/ADSORPTION GRANULAR ACTIVATED CARBON EFFLUENT pH=7.9 GAC AGE: 8 WEEKS IN SERVICE OZONE CONTACT TIME=6 MIN.

TABLE IV

Effect Of Ozonation of Dual Media Effluent

Contact Time - 5-6 min

APPLIED O_3 DOSE mg/l	CHLORINE DOSE mg/l	SUMMATION TRIHALOMETHANE CONC. AFTER 6 DAYS, μg/l
0.7	0	<0.2
0	8	20
0.7	8	23
18.6	0	<0.2
0	8	23
18.6	8	30
0	8	123
227	8	70

TABLE V

Estimated Cost of Disinfection

(All Costs In Cents/1000 Gallons)

DESIGN CAPACITY	1 mgd	10 mgd	100 mgd	150 mgd
AVERAGE DAILY FLOW	0.7 mgd	7 mgd	70 mgd	105 mgd
2 mg/l CHLORINATION, 30-MIN. CONTACT TIME				
CHLORINE 15¢/lb	4	1	0.7	0.6
1 mg/l OZONE, 20-MIN. CONTACT TIME				
O_3 GENERATED BY AIR	6	2	0.9	0.8
O_3 GENERATED BY OXYGEN	8	2	1.0	0.8
1 mg/l ClO_2 GENERATED FROM Cl_2 and $NaClO_2$				
CHLORINE 15¢/lb	4	2	1	1

-453-

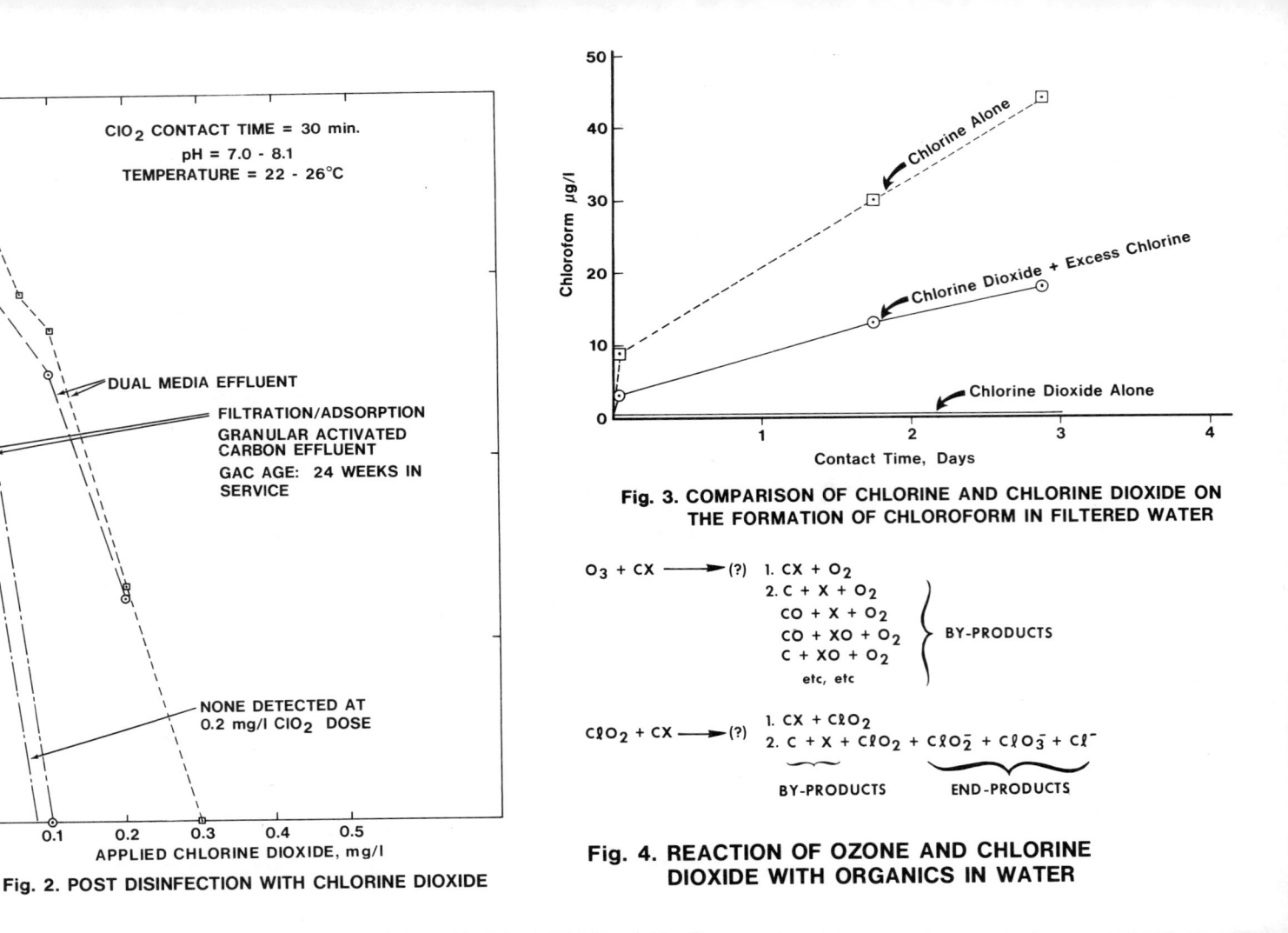

Fig. 3. COMPARISON OF CHLORINE AND CHLORINE DIOXIDE ON THE FORMATION OF CHLOROFORM IN FILTERED WATER

$$O_3 + CX \xrightarrow{(?)} \begin{array}{l} 1.\ CX + O_2 \\ 2.\ C + X + O_2 \\ CO + X + O_2 \\ CO + XO + O_2 \\ C + XO + O_2 \\ \text{etc, etc} \end{array} \Big\} \text{BY-PRODUCTS}$$

$$ClO_2 + CX \xrightarrow{(?)} \begin{array}{l} 1.\ CX + ClO_2 \\ 2.\ C + X + ClO_2 + ClO_2^- + ClO_3^- + Cl^- \end{array}$$

⎱ BY-PRODUCTS
⎰ END-PRODUCTS

Fig. 4. REACTION OF OZONE AND CHLORINE DIOXIDE WITH ORGANICS IN WATER

Fig. 2. POST DISINFECTION WITH CHLORINE DIOXIDE

ClO_2 CONTACT TIME = 30 min.
pH = 7.0 – 8.1
TEMPERATURE = 22 – 26°C

FILTRATION/ADSORPTION
GRANULAR ACTIVATED
CARBON EFFLUENT
GAC AGE: 24 WEEKS IN SERVICE

NONE DETECTED AT 0.2 mg/l ClO_2 DOSE

DUAL MEDIA EFFLUENT

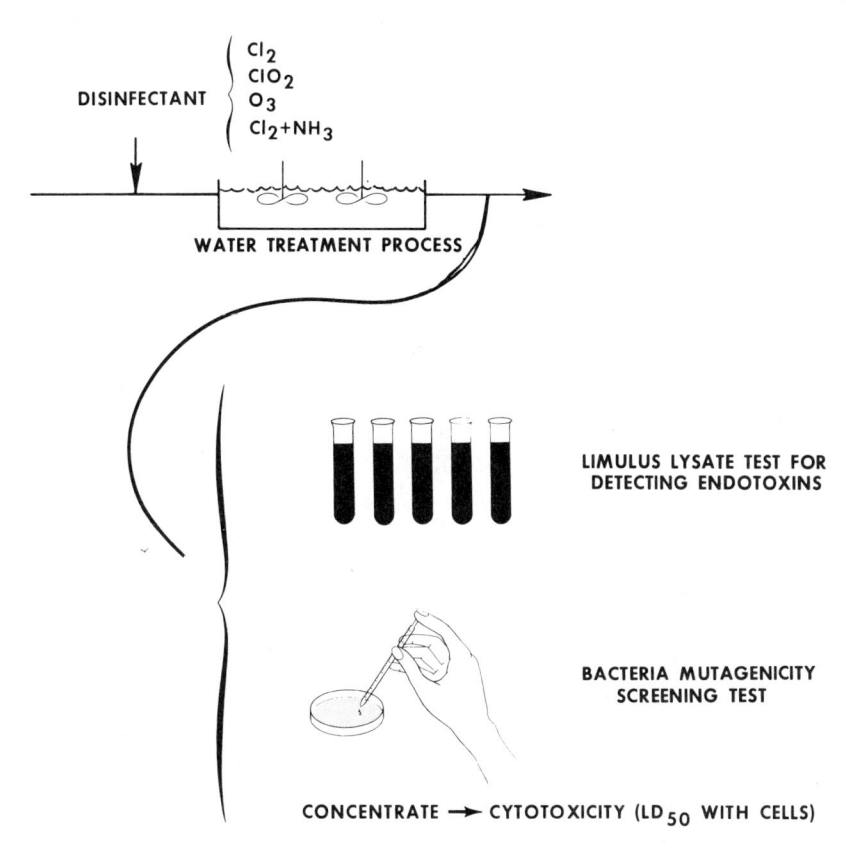

Fig. 5. PROMISING TECHNIQUES FOR MONITORING TOXICITY

(Diagram labels: DISINFECTANT — Cl_2, ClO_2, O_3, Cl_2+NH_3 — WATER TREATMENT PROCESS; LIMULUS LYSATE TEST FOR DETECTING ENDOTOXINS; BACTERIA MUTAGENICITY SCREENING TEST; CONCENTRATE → CYTOTOXICITY (LD_{50} WITH CELLS))

TABLE VII

WSRD Extramural Studies on Ozone and Chlorine Dioxide

"Status of Ozone and Chlorine Dioxide In Water Treatment	Public Technology, Inc.	1976-1977
"UV Catalyzed Ozone"	North Texas State Univ./ Houston Research	1976-1979
"Comparison of Disinfectants In Small Systems	State of Vermont	1975-1977
"Effect of Particulates on Disinfection By:		
Ozone"	Univ. of Maine	1976-1978
Chlorine Dioxide"	Univ of Cincinnati	1976-1978
"Alternative Disinfectants"	Univ. of Missouri	1976-1978
"Ozone + Powdered Act. Carbon"	City of New Orleans	1976-1978
"Chlorine Dioxide + Granular Act. Carbon"	City of Evansville, Ind.	1976-1978

ROUND TABLE WORKSHOP DISCUSSION

          Chairmen: Dr. Rip G. Rice
                     Dr. Joseph A. Cotruvo

Rip Rice: Before we get into questions and answers, we will first have summaries of three areas of fundamental importance to this meeting: Regulatory Factors, Organic Chemistry and Toxicology. To introduce the summation of Regulatory Factors, I present again your Workshop Co-Chairman, Dr. Joseph A. Cotruvo of the U. S. Environmental Protection Agency.

Joe Cotruvo: The subject of the meeting has been oxidation, but it has developed into a discussion of very diverse subject matter. We have talked about many different disinfectants: bromine chloride, chlorine dioxide, ozone, ferrates, and ultraviolet. We have talked about recycling water, wastewater and drinking water. We have had some discussions of economics by Jim Symons which had a lot of heads bobbing out in the audience; there is always concern about costs. We have had some extremely interesting discussion by Dr. Kühn on the practical application, particularly of ozone and granular carbon and other adsorbents, in Germany. The topic that has been left out so far is, what is the bottom line on all of this? The bottom line, as far as we at EPA are concerned at least, is the regulatory decision, in terms of the applicability of one or more of these processes to water treatment in the United States.

The man who will discuss that for you briefly is Mr. Victor Kimm, who is Deputy Assistant Administrator for Water Supply at EPA. Vic is the fellow who has to sign on the bottom line eventually, and usually sooner than he would like to, as is always the case when you have a moving target, with new information being developed on a daily basis. Mr. Kimm has been with EPA for several years, and in Water Supply for two. He is a sanitary engineer, and has also had very diverse experience in public policy and economics areas. I give you my boss, Victor Kimm.

Victor Kimm: Thanks Joe. Normally scientific and technical meetings like this are interesting and challenging, but far removed from decisions that anybody in the real world has to make. However, that's not the case here. The Safe Drinking Water Act included a requirement to publish the Interim Regulations immediately, a requirement for a major study by the National Academy of Sciences, which will be available at the end of this year, and then a revised set of regulations with a set of instructions on being protective of public health, and especially concerned about organic materials. That makes this an issue which we must address.

As a practical matter, the Agency is already under suit for not having dealt with the organics in the Interim Regulations, and if you recall, at that point all that we did with the Interim Regs was to conclude that the existing information was not sufficient, and we began a round of some additional monitoring.

We have issued an Advance Notice of Proposed Rule-Making, in which we laid out the technical nature of the organics problem as we see it, and we are now collecting all of that information leading up to our regulatory decision.

At any rate, that's just exactly where we are. That is, many of the things that were in these revisions in our research strategy as Jim Symons and Gordon Robeck were alluding to earlier, are things that we've worked our way into as we move into this rule-making process.

I expect that we will have to deal with some type of organics standard; actually we have to deal with it in the revised regulations, as my attorneys who are trying to defend me in the suit on interim regs read the statute, and we hope to do something sooner than that. The details of that are exactly what we're all about at the moment. I don't want to create the image that we really have the answers at this time.

So often when we talk to people at the agency, they assume that we are sitting there with a big stack of cards, and we are turning them over one at a time and seeing how people react. Because of your meeting, and all of the scientific uncertainty here, it is clear that all of us would like to say, "gee, if we could just give Jim Symons and Company a little bit of money and let them go off for a year or two, they'd come back with all the answers for us." This is one of those situations in which that's not going to happen. The decisions are being put together and packaged right now, and the Agency is going to come down on this one very shortly.

Thank you very much for an interesting meeting. We are pleased to have participated with IOI in its presentation.

Rip Rice: Thank you Vic. Our next summation is of the Organic Chemistry that has been discussed, and this is a monumental task. I had not figured out just how to do this, until I had breakfast this morning with a truly international gentleman that I believe has admirable competence. This is a man who was born in Czechoslovakia of Hungarian descent, is schooled in Hungary and Romania, has a European Chemical Engineering degree, a Ph.D. in Organic Chemistry from England, was a university professor in Canada for 10 years, and is currently with the Ontario Ministry of the Environment of the Government of Canada. Ladies and gentlemen, Dr. Otto Meresz.

Otto Meresz: Ladies and gentlemen, until this morning, I was an innocent observer at this workshop, but this morning, Rip Rice bought me breakfast, and now I'm having to work for it. He was looking for a man who could summarize all the chemistry, and especially the organic chemistry that was presented at this conference. I didn't claim to be able to do that, but what I can do is give you the impressions of an organic chemist who has heard most of the papers presented here, and who is interested in environmental organic chemistry.

What I have seen is that we have the foundations for future work in this field. We have seen first the classical ozone chemistry presented in connection with organic materials, especially the aspects which are used in synthetic and industrial chemistry, that was presented by Dr. Oehlschlager from Emery Industries. This chemistry concerned ozone and the chemistry of the carbon-carbon unsaturated bond. A whole branch of the synthetic organic chemicals industry has been built up on these reactions.

Then we had aqueous ozone chemistry which initially was directed completely towards disinfection. And then the problem arose with organics in drinking water, and all the work that was previously carried out in order to get good yields has been turned around completely. For organics in drinking water we _don't_ want fine chemicals and large yields of organic compounds. We want to destroy _all_ the organic compounds, so we actually use a different kind of ozone chemistry.

We have seen from the papers that there are extremely complicated collections of chemical reactions and chemical aspects. We have 1,3-dipolar cyclo-additions, we have ionic and free radical reactions, or a combination of these two, which seem to be pH dependent. Then we have further complications when we involve ultraviolet radiation and photochemistry. We have the photochemical excited states and photosensitization. We can generate photosensitizing agents during the reaction to carbonyl compounds, and we can have interaction between the product and solvents, that is water, at each step of the reactions.

There are a few consoling features which appear in this apparent chemical nightmare. One is that the basic laws of chemistry are still authoritative. We have seen that nucleophilicity still plays a role; some of the products that have been isolated and identified are those that a chemist would predict, which is very nice to see.

We have also seen that we have the laws of aromatic substitution still operating, and free radical stability, and we are generating products of higher and higher oxidation state, finally going to water and carbon dioxide. With these general rules we can probably find our way toward the solution of the organics in drinking water problem.

Perhaps we can trust that all chemical knowledge is included in the periodic table of elements. The question then becomes one of how we are going to find it? We have a good foundation of free radical chemistry. We can lean back on organic reaction mechanisms, and to a lot of work done in photochemistry. The combined application of all these fields probably will be applied to the ultimate solution of the problem.

I didn't want to spoil the fun of the organic chemist and try to identify all the compounds that can be formed through ozonization, but probably we could split the field. There should be some people working towards the complete removal of organics because that should be our aim when it comes to drinking water, regardless of what the intermediates are. That may not provide a lot of publications for you, but it will solve the problem. You should aim for carbon dioxide, water, and perhaps chloride ion or hydrochloric acid, and with that the problem will be solved.

On the other hand, we have to give food for basic academic research and let other people work on and try to identify all the organic compounds. We have seen an extremely good example of a wide array of organic chemicals that can be formed during ozonization.

So to sum it up, the results presented here are very promising, and I can see that with this technology we may find solutions to the complete elimination of organics from drinking water.

Rip Rice: Thank you very much, Dr. Meresz. I didn't think a summation as good as that was possible.

Our next summationer is Dr. Philippe Hartemann from Nancy, France, who spoke to us on toxicology.

Dr. Hartemann is a medical doctor and has been active in the field of toxicology for many years.

Philippe Hartemann: I was also innocent this morning at 8:30 when Dr. Rice asked me to summarize this meeting from the point of view of toxicology. It is difficult enough for an American professor whom you can very easily understand; it is quite impossible for a French doctor with only a few poor experiments in this field. I think this meeting was very interesting, but we have only a few answers to the most important questions in the field of toxicology.

There are three major questions in this field. The first is the presence factor, the identification of by-products and end-products in ozonated water. It is very difficult to summarize in a few words what we learned during this meeting. Here in the States, you have a lot of experiments in the field of gas chromatography and mass spectrometry in wastewater after chlorination. But in Europe, except for the work from the laboratory of Professor Sontheimer in Germany, we have no work in this field, and we are not able to give you answers to these questions of the presence of organic products in water.

It appears that activated carbon is very effective for the removal of some products present in water after ozonation. It is probably the only answer to this question. It is evident that we need to have results in this field as soon as possible, and I hope that the program we have going on now in France to examine some waters after treatment by ozone or by prechlorination, ozonation and post-chlorination, will give us results quickly.

In my opinion, there is here a very large field for research workers interested in analytical methods, and I will be very happy to receive in France some graduate students able to work in this field, because in fact it is very difficult to find people able to work with mass spectrometry coupled with gas phase chromatography.

The second important question is the problem of epidemiological studies with the use of ozonated water. We have no answer to this question either.

Probably the reason that such information is quite impossible to obtain in France is that we use more wine, beer, brandy, or sometimes mineral water than ozonated drinking water. We are very intrigued by a survey done in Normandy, west of Paris, where we have the highest frequency of larynx and pharynx cancer. But this is caused by calvados, which is brandy made by distillation of apples. We have made no surveys with ozonated water as yet.

The third question is the problem of the results obtained with toxicity tests done with real drinking water after ozonization. There are many tests in this field. The first ones are long term tests done with animals. I know of results in Paris that Professor Truhaut has obtained with ozonated water. These results show that extracts are toxic, and sometimes more toxic after ozonization, after treatment, than is the parent water. But at the same time, we have no information on the same water after chlorination.

I have no information in the field of sewage treatment, except for a few tests done over a 19 day period. We heard some information on mutagenic assays, and some very nice work done at the Stanford Research Institute shows that it is possible to obtain mutagenic activity with some organic compounds after ozonation. But the problem with the Ames test is that it is necessary to use very high concentrations of products, and in my very poor experiments in this field with the Ames test and extracts of wastewater, we also find mutagenicity. But this was with wastewater ozonized in a laboratory treatment plant. At the same time we have performed experiments with the same water treated by chlorination, and there is no statistically significant difference in mutagenic activity between the two waters.

The latest tests used in this field are the short term toxicology tests, and I agree with Dr. Kühn when he said that ozonation gives more biodegradable products which are _normally_, I repeat, _normally_, less toxic products. But in this field I am not so confident. From my experiences, nearly all the time we find that ozonation of organic products gives compounds which are able to induce detoxification mechanisms in the liver. At the same time the same parent compounds and the same compounds after chlorination caused more significant liver necrosis. That means that the expression of the toxicity of these compounds is different.

But in the field of toxicology, in terms of toxicology, what does it mean? I know of some compounds which are able to give very good induction of cytochrome P-450 and other detoxifying enzymes, [benzo-(a)pyrene, and Arochlor], and they are quite toxic.

In conclusion, let me tell you a little story we tell in France in our labs, about a young assistant who got his training in the United States and comes to his old professor who was trained in France. And he said, "I am very happy, this week I found two carcinogenic compounds." And the professor shrugged and said, "What does that mean?" The assistant said, "You are not very happy? I found _two_ carcinogenic compounds." The professor answered, "I will be very happy if you tell me which compounds are _not_ carcinogenic," and I think this is our problem with extracts of treated waters.

This meeting was very interesting because it demonstrated how long our way is before we will know exactly what happens during drinking water treatment by ozonation, from the purview of toxicology. We have some results, a lot of methods, and a myriad of questions. But I think with the EPA research programs in these fields, and the growing interest of university research teams, there are now some graduate students working feverishly on ozone and toxicity. I don't remember the speaker who made this statement yesterday, but I hope we will see some very interesting results appear in just a couple of months. Thank you very much.

Rip Rice: Thank you very much Dr. Hartemann; an excellent summation of the toxicology situation. We will now move into the Round Table Workshop Discussion proper. With the prerogative of my being a Chairman, the last question that was asked of Dr. Symons had to do with instrumentation and automation of ozone systems, and Dr. Harvey Rosen of Union Carbide has asked to comment on that point. I would also like to put that same question to the people in the chlorine dioxide industry, because I am sure they have different techniques of measuring and controlling residuals.

Sidney Sussman, Olin Water Treatment Systems: We have been talking to several instrumentation people and here is one thought. There is a Swiss instrument available for colorimetric measurement and monitoring of the much higher chlorine dioxide concentrations in solutions that are used in the pulp and paper industry. We had hoped to be able to adapt that instrument to the low concentrations used in the water treatment industry, not so much for measurement but to demonstrate whether or not you have a chlorine-free chlorine dioxide.

But as it turned out, in looking further into the possibilities and discussing it with the manufacturers, they felt this could not be done. They thought they might be able to adapt a combination of a colorimetric device and an amperometric measuring device for this purpose if they could find an amperometric one that was sufficiently reliable as a monitor for continuous operation. I have not been able to help them on that one.

Rip Rice: Thank you. I assume that you are _not_ saying that when you use chlorine dioxide for water treatment there is no way to monitor the amount of ClO_2 that is present.

Sidney Sussman: No, we are not saying that. Yesterday when I presented the paper we ran through a quick summary of the testing procedures, largely colorimetric, that are useable.

Harvey Rosen, Union Carbide: Over the years there have been a great number of ozonation plants, primarily for drinking water, that have been installed all over the world, and various techniques have been used for measurement of dissolved ozone. These vary starting with an operator taking a sample of his ozonated water, putting in a variety of substances that are oxidized and change color and saying, "Ah ha! Obviously ozone is in the water, because there is some residual here and that's all I really require".

However, the whole area of analytical technology is becoming much more sophisticated. Systems have been designed with automatic control from the point of view of, for example in a wastewater situation, measuring and controlling the system by relating the flow to a constant dose of ozone.

This is one question that was asked this morning. For example, in wastewater where it is very difficult to measure an ozone residual because the demand for ozone is so high that it gets used up very quickly, we can control the ozone generator power and the flow of gas to the ozone generator and maintain a constant concentration of ozone in the gas stream, for example, at a 2 weight % level, and dose as a function of flow at a constant ozone dose level. This is quite easy to do with an analog system.

On the other hand, in potable water in the dose range from 1 to 4 mg/l that is nominally used and where the demand for ozone is not high, there is generally quite a high measurable ozone residual. There are a number of ways of automatically measuring that residual.

We heard on Wednesday about a membrane probe that Professor Johnson has developed at the University of North Carolina. That probe is available commercially and it is something which people are looking at and testing. There is also an automatic monitor that was developed at the National Bureau of Standards which is also being tested on ozone. It was developed for field testing chlorine at very low concentrations, for example, at very low residuals in the parts per billion range in rivers and estuaries, and to study the relationship of those concentrations to toxicity. It is based on a potassium iodide couple and can be used for measuring ozone as well.

There are ozone residual monitors. Many of the chlorine monitors are adaptable to ozone. Fisher & Porter, Analytical Instruments Development and others have ozone monitors which are mostly adaptations of their chlorine measuring systems. Once you have that signal, that's all you need in terms of a control, because the output of an ozone generator is controlled electrically, and with a signal from a residual monitor you can automatically tell the ozone generator to turn up or turn down and control ozone dosage as a function of maintenance of a residual after some fixed period of time that you design for.

So the answer to the question that was asked is

that it _is_ being done. It probably could be done better with respect to things like accuracy, speed of response etc., and there is a lot of money being spent by people in the business of developing new monitoring techniques for ozone. Remember also that we have some very, very sensitive techniques that are used for measuring ozone in a gas stream, as opposed to a liquid, and some of these are also adaptable to the control of ozonation processes in water and wastewater treatment. These include air pollution monitors, and in this country alone there are at least 15 manufacturers of air pollution monitors specific for ozone.

We can measure the ozone concentration coming out of our ozone generators to maintain a dose, and we can as easily measure ozone in the gas phase above an ozone contact chamber. Either or both monitors can be programmed into the ozone generator to signal it automatically to either increase or decrease the rate of ozone production. They can also be used to monitor ambient ozone in the plant for safety purposes.

I wanted to comment about statements that one of the "disadvantages" of ozone is that it is toxic. I have never heard of a good disinfectant that is _not_ toxic. In any well designed ozone system you won't know that it is an ozone system from the point of view of being able to smell any ozone, because if you can smell ozone, it is a poorly designed system. You can design tight systems so that you don't have those problems.

I did want to make one more point about the first EPA sponsored grant that was listed by Dr. Symons. As part of the survey of the state-of-the-art of ozone and chlorine dioxide for drinking water treatment by Public Technology Inc., one of the objectives of that study is to determine the status of process control, so that a survey of instrumentation and automation as currently practiced will be part of that report.

Both the IOI and the ASTM are reviewing methods of analysis, trying to improve them, testing them; automatic physical and chemical methods. Also, we are fully aware of the problems of separating ozone as an oxidant from other oxidizing materials that are in the water that might show an oxidant reaction and we are trying to separate those both from a specific research point of view as well as from that of practical engineering application.

Rip Rice: Thank you very much. I believe that answers that question. We now return to Organic Chemistry and Toxicology and the floor is open to questions.

George C. White, Consultant: I was asked by the French contingent to relate to you the climate that exists in Western Europe and in the British Isles in connection with their approach to treating drinking water, not simply disinfection. It's not an armed camp over there, with ozone people here, and chlorine people here, the chlorine dioxide people here, the permanganate people here and the activated carbon people here. Europeans will use anything and everything, and in almost any combination that they believe will do the job.

I think it would be appropriate if I explained to you one facet of how water treatment plants get built in Europe, and let's take Paris. I'm sure you have all heard of Cie Générale des Eaux. This is a large company that has four branches. One is the research and engineering group which is SETUDE, I don't remember what all those initials stand for, that's the research, development & design engineering group. They go out and study a project and find out what's best for treating that water.

Then they have the equipment group which is TRAILIGAZ, and SETUDE tells TRAILIGAZ, what kind of equipment to build to do the job. The third group of CGE builds the plant, and the fourth CGE division operates it. So you can see that they are not going to overlook anything that is necessary to do the job.

Now, one of the reasons they asked me to speak here is that I am fairly familiar with the water treatment plants around Paris, and I have seen everything that they use. For instance, they start off with breakpoint chlorination, followed with chlorine dioxide. They follow that with activated carbon; they follow that with ozonation and then they follow that with post-chlorination, when necessary, for distribution system cleanup. If you talk to the operators as I have, they will tell you that _every single step is necessary_ to give them the _quality_ and the taste of the water that the people like.

And don't kid yourself that the people in Paris don't drink the water, they do. I'm sorry that my friend of the other day got diarrhea but--- I drink the Paris water; my wife is a little bit nervous about it, but I've been drinking it for years with absolutely no problems.

So you see, the operators will tell you that if any _one_ of these processes fail, then they get an off-flavor. In other words the flavor is different, and it's _never_ better.

Dr. Kühn talked about ozone and ozone in combination with chlorine, and this is very common practice. As a matter of fact, Wallace & Tiernan in England builds ozonators. So nobody over there has any prejudices of one against the other, and they don't even _think_ in terms of "alternatives". This is the one _word_ that I've been hearing here, except for Jim Symons, the one word that I think is a mistake, that this or that compound is an "alternative". There _isn't_ any alternative. It's a combination of treatment methods.

Before I step down, I would like to mention something about chlorine dioxide. The sequence of water treatment in the United States is being changed because of what the EPA has found and is finding, I suppose that's where it began. They are substituting pre-chlorination with chlorine dioxide, and this has caused an upsurge of interest by manufacturers of chlorine dioxide equipment.

I discovered by accident the equipment that is made by a corporation by the name of CIFEC, in France. The principals are here and they asked me if I would do a translation of their brochure on chlorine dioxide which I did, and they said if you get it back to us right away we will have it all printed for the meeting at Cincinnati. Well it's done. I had to do it in long hand so there are a few typos in it, but they _do_ have some of these brochures and anybody that wants them, if you will leave your cards we will get them to you.

The point is that CIFEC makes an almost pure chlorine dioxide solution by their process. It is adaptable to any conventional chlorinator installation, and because they make an almost pure chlorine dioxide solution, you don't have to worry about the complicated analytical procedures that you need when you are only making partial chlorine dioxide solutions, as we have been doing in the United States for years. There just hasn't been the interest for any of the major manufacturers to try to design a piece of equipment like that because that's terribly expensive; there hasn't been the demand for it.

The French CIFEC method is available and it may be interesting to some of you, because you can automate it with conventional amperometric analyzers, you can calibrate it as you wish, and you don't have to go through and separate all the different fractional compounds.

We have quite a bit of information on wastewater that would be interesting to the people interested in wastewater. As I was telling the people from France, in the United States the interest in chlorine dioxide in wastewater is going to be very heavy on the Western side of the Rocky Mountains; on the Eastern side of the Mississippi it is going to be more in potable water. But they (CIFEC) have by a pilot plant study in actual operations found similar results to what Bernarde did with Professor Granstrom. I jotted down the following data: at a 2 mg/l chlorine dioxide dose on a secondary effluent with 10^8 15 minutes contact time, a total plate count of 10^8, now that's not coliforms, it's a total plate count of 10^8, was reduced after 15 minutes to a total of 100/100 ml, which was on the order of what Bernarde found. So here is a disinfectant that I think is great for tertiary effluents, and I am sure it has a good place in raunchy waters, a lot of which we have, so I am glad to see the rising interest in chlorine dioxide.

Now, there is one more thing that worries EPA, and as I mentioned the other day Harvey Rosen needs customers for ozone and the EPA needs things to worry about. The concern is for the formation of chlorite, which could be a problem. However, rumor has it that chlorine dioxide treatment followed by ozone converts the chlorite back to chlorine dioxide. Chlorite produces the same syndrome in babies as does high nitrates, but we don't know what the levels are.

Outside of that I have no information, except that 8 million people in Paris have been drinking water for 20 years that has been chlorinated and ozonated. In the last 10 to 12 years they have been drinking it with chlorine dioxide added. That's proof enough.

<u>Bill Ward</u>, Olin Water Services: I might add a post note to what Mr. White said. We have some experience in a round about way in treating wastewater for a municipal steam electric power plant on the West Coast that uses as its makeup a secondary treated, post-chlorinated sewage effluent. We have been treating this with chlorine dioxide for about 3 to 5 months now at the rate of about 1 mg/l dosage which results in, as nearly as we can measure <u>via</u> DPD testing, about 0.1 mg/l additional total combined chlorine in the effluent as blow-down.

The results have been a substantial drop in <u>E. coli</u> counts in the blow-down, a substantial drop in total plate counts in the recirculating cooling water, and a substantial drop in the corrosion rates in the metallurgy of the system. So, it would appear from our experience indirectly, in the steam electric power plant, that the use of chlorine dioxide does have some definite advantages for tertiary or secondary treated effluents.

<u>Joe Cotruvo</u>: One of the main benefits from a meeting of this type is that we have a lot of people with different backgrounds, from different countries. We have the opportunity to get those ideas to mix together and for the information to pass from one to the other. One thing I've noticed in the past few years is that, for whatever the reasons, there seem to be different philosophies that develop in one country versus another country, or in Europe versus the United States, and it is distressing to me that that's the case. Now I'm sure there are natural reasons for that; certainly there are unique problems in some areas and perhaps different economic considerations. But I have the feeling that one of the main reasons for the differences really is just the lack of under-

standing between the two groups, the lack of communication. And I hope if nothing else, at this time, we can get these different people together, and get those ideas moving back and forth.

One good benefit is the fact that we have represented here experience in the use of different materials, be they ozone, or chlorine dioxide, or granular activated carbon or combinations, whatever the permutations are. So there is certainly all that technical experience to be gained. Beyond that, there is the fact that people have been consuming waters for years that have been treated by those various methods in those areas. So, there certainly are opportunities for information to be collected, epidemiological types of information, and such, which is difficult enough to do within one country. But it gets even more difficult when you have to cross a border and get others to cooperate. So perhaps some of those opportunities can be identified and developed.

I also presume that in the light of that long experience that has existed in those countries, that there are some toxicological data that do exist somewhere, be they feeding studies in animals, on concentrates or individual compounds, that can be brought to light. I hope there is some mechanism for that to happen here. If anybody has some information along those lines, please speak up.

Sherman T. Mayne: I don't want to get fired from my job so this has nothing to do with the company that I work for. This is my own personal experience which happened while I was working with a state regulatory agency.

I happen to be a tremendous bass fisherman, I love to bass fish, and about ten years ago in a lake near my home I could consistently catch bass anywhere from five to ten pounds apiece. Then EPA came along, and needless to say in this one lake we have several major dischargers. While I was working for the state agency I know for a fact that they were using somewhere between 70 and 90 tons of chlorine a day for water treatment and waste treatment in this particular lake. And in the last seven years, ladies and gentlemen, I have not caught a bass over five pounds in that one lake.

In two lakes below that which contain no chlorinated discharges of any type, seven years ago they initiated a chlorination program. For the last five years there have been five to twenty thousand fish a week die in that lake. The state agency has been down there to investigate it, I was one of the investigators when I was working with the state. The incidence of leukemia in the people that drink water from this lake is eight times higher than that of any other county in North Carolina. The incidence of heart attacks is fourteen times higher than in any other county in North Carolina. This situation needs to be changed, either to eliminate chlorination or to eliminate the toxic effects of these chlorinated discharges.

Rip Rice: Earlier in the meeting we heard from chemists at Stanford Research Institute and the University of Colorado about the isolation and identification of specific organic compounds produced upon ozonation. Yesterday we heard workers at the Cleveland Regional Sewer District talk about improving the ability of activated carbon to remove organics from sewage effluents by preozonation and using the aerobic bacteria on carbon to remove oxidized organics from ozonized effluents.

This morning, Dr. Kühn discussed the identical concept applied to water supplies, preozonation followed by activated carbon, and using the aerobic bacteria on the carbon column to remove oxidized organics from drinking water.

Studies of the effects of preozonized carbon on organics removal, identification of specific organics at each stage, etc., would appear to be a fruitful research area in both water and wastewater treatment, and I wonder if EPA is planning some research studies in this area?

Gordon Robeck, EPA: This is, of course, the fundamental type of information that we are trying to obtain through inhouse work and extramural work. Al Stevens certainly related our efforts to do this with chlorine dioxide. As we see the future with other oxidants, as far as practicality and trying to determine just how much research money from our limited resources should go into certain oxidants, we certainly want to know what the by-products are wherever you put the oxidant in, be it at the head of the plant, at the head of the filters, or coagulation, or afterwards. So that I think the answer to your question is that we are aware of the dilemma, of the ignorance, and we will try to apply some of our limited resources in that direction. However, I don't anticipate that we will have quick answers for the accountability for all of the TOC and the by-products that come from many of the oxidants.

In relation to the matter of the challenge laid down to EPA a moment ago by Mr. Mayne, I'd like to say that the agency for many years has been sensitive as to the effects of chlorine and chloramines on aquatic life, and it has put a major effort into trying to find other ways of protecting the public health from the communicable disease standpoint, by experimenting with other oxidants and disinfectants such as ozone and chlorination/dechlorination. The Office of Water Programs recently came out with a policy to discontinue heavy chlorination of sewage plant effluents wherever there is no apparent downstream need for such protection. We are aware of the situation, and I sincerely hope that we don't use an overkill policy in drinking water any more than we do in sewage treatment.

For many years I have been going back and forth to Europe and trying to create conferences with AWWA joint sponsorship to try to do the very thing that George White talked about, namely, using some of these oxidants as supplements. I think the image that people have of U.S. drinking water is one of chlorinous taste, therefore somewhat objectionable, and it's high time that we do as well as some of the Europeans do in applying the kinds of technology that will produce the best quality drinking water under the circumstances.

Incidentally, EPA, HEW and other federal agencies have expressed their concern about the effects of aquatic pollutants, including halogenated ones, by conducting a special conference with the New York Academy of Sciences on Septemper 27-29, 1976. This conference, entitled "Aquatic Pollutants and

Biological Effects with Emphasis on Neoplasia," concerned the cause of tumors in fish living in polluted waters as well as the possible connection between human cancers and drinking water contaminants. Proceedings of that meeting will be available from the Academy in 1977.

<u>Allison Maggiolo</u>, Bennett College. I'm not going to make any bets today. I am prempted by the gentleman from EPA because I was going to go the defense of the people at EPA. They have been seriously concerned. If you will notice how slowly the government really moves, I don't think it is possible that EPA can move the Federal Government overnight. They have to work through political mechanisms. Now, let's be factual. If you think about what they are really doing for us, they are setting up for the first time guidelines that will become official, like OSHA has done. They are getting there very fast. And then the political climate will want to get on the bandwagon, and then the program will get moved, because of the large quantities of chlorinated organic compounds that we are now concerned about; we're getting down to what I call engineering numbers.

What does that mean? It means now that the dollars are going to be concerned with which treatment methods are the most practical and cheapest with varying types of wastes, in the textile industry, which I happen to be familiar with at the moment, and the oil industry, etc. What are we going to end up with? Not a political situation where some interest can politically have some clout. We will have a number that we're concerned about, and we will have to talk about engineering dollars. What I'm saying is that these numbers will move the program faster than even trying to fight the political system. And that's what EPA is doing.

What I am saying is that, basically, after twenty years on and off in ozone and related pollution problems, and seeing what is available for use, it's going to be a combination of technical approaches. But now, for the first time, it's not political. You are going to be fighting a number; how do you do it the cheapest way? I'm not going to say, because I'm biased. I'm going to fool you. We'll see what happens.

Joe Cotruvo: One of the comprehensive, very interesting papers that was presented at this Meeting was by Dr. Wolfgang Kühn this morning, in which he talked about the combinations of ozone and granular activated carbon and other adsorbents. I'd like to ask him to go into some more detail on that if possible, and particularly into the question of benefits versus disbenefits. At the bottom of one of his slides where he had the relationships of overtreatment versus undertreatment, one of the lines referred to overtreatment with granular activated carbon and that caused increased corrosiveness of the water. Among other things, I wish he would address that question in more detail.

George White: Dr. Cotruvo, could I ask him a specific question first?

Joe Cotruvo: By all means.

George White: Dr. Kühn, I was so impressed with what you had to say and what you've done in Germany, I wonder if you would just drive it by again, because I was taking notes and I'm sure others were, of how you achieved this high quality effluent that has this low chlorine demand. I think that's one of the most important facets of the whole paper. I'd like you to do that first and then answer Dr. Cotruvo's question.

Wolfgang Kühn: Our problem is, we also have a lot of problems, and one of the problems is the cooperation between chemists and toxicologists in Germany. There was such a lot of disagreement about, for example, the chloroform problem. In the United States they told us chloroform is carcinogenic. In Germany they told us it's not carcinogenic at all. They changed two weeks ago, and now they are telling us in Germany too, chloroform is carcinogenic. That means as a water chemist who has to provide protection for the people drinking the water, we cannot wait until the toxicologists develop the data. They needed quite a lot of time to be able to tell us that chloroform is carcinogenic, and it will take another year or two years until they can fix a standard of one microgram or five micrograms, or whatever it is. And then they will need another five years of work to determine what happens if two organics come together, therefore we have to work on the basis of what we call a "sure sign."

The toxicologists, for example, told me just four weeks ago in Berlin at a conference, "your co-parameter, total organic chlorine, that's nothing. You cannot do anything in toxicological aspects with this parameter." And I told them, "You are right. But if the TOCl is zero, it says everything about toxicology, there is just nothing present." This is the problem we have.

We have some good luck in the wastewater field. For example, when we do single compound analyses we are analyzing the lipophillic material, the biological material, and we are analyzing about two or three percent of the overall organics. And it is good luck for the chemist that these compounds are probably the most toxic ones.

The same thing is going on in water treatment. If you have, for example, activated carbon filtration, there is a high degree of breakthrough also in our filters, we know that about humic material, but we think humic material in water is a type of natural order. If you drink a cup of tea there is a lot of humic material in it. We cannot sell distilled water. Distilled water has corrosion problems and also toxicologists will tell you that distilled water is also toxic. The pressure in the living cell becomes too high. Nobody knows, what is "pure water." It is double distilled water or whatever. In Germany, we think "pure water" is a water which has some relation to the ground water which we have been using for two or three hundred years. We don't worry about a little more hardness or a little less hardness, or some humics or not, but, we should take out the industrially produced organic chemicals. And this we are trying to do in our treatment processes.

For example, forty years ago we started using ozonation for removal of manganese. There was a lot of manganese introduced during sand bank filtration. Sand bank filtration is probably old fashioned for you, but we very seldom take the water directly out of the river --- we make use of this sand bank filtration, and the removal of the TOC and COD is 80% in the sand bank. Sophisticated equipment used later on removes only 20% more.

But there is a high breakthrough of chlorinated hydrocarbons and pesticides in the sand bank because they are not biodegradable. There is a brand new water works in Germany costing around $40 million and they have all this purification equipment. They pump the water directly out of the Danube, and after all this purification they pump it down into the ground again and pump it out a hundred meters away for underground storage. This is what tells us that there should also be some biological activity in the water treatment. We can use the power of nature, as we call it. We use it in the wastewater field and we can use it in the drinking water field also.

In the beginning of treatment for manganese removal, ozone was used only to oxidize the manganese to make permanganate. Then they had to take out the permanganate, and this took activated carbon filters. People learned ten or fifteen years later that activated carbon filters are also good for taste and odor removal.

At that time activated carbon filters would run two or three years and there was no breakthrough of taste and odor.

The development of better analytical methods is the first step. A person first has to find an analytical procedure for chloroform, then he can find out if chloroform is a problem. When we found the analytical method for TOC we looked for breakthrough of the TOC, and we found that there is a breakthrough of TOC after seven months. So we concluded that we have to change our carbon filters after seven months. Nowadays we are looking for breakthrough of the TOCl, and we think now that we will have to regenerate the carbon after five weeks. And so a lot of drinking waterworks are just starting now to build a carbon reactivation plant beside the waterworks. On-site carbon reactivation is more economical under those circumstances.

One thing we are not worried about is the biological activity in the carbon filter. This is one of the most controversial subjects I have been involved in. When I came to the United States the first time they asked me quietly, "Are there some bacteria on your activated carbon?" I said yes, but that we are very glad to have these bacteria on the carbon. Not all bacteria are coliforms. We have many bacteria in our stomachs, and we can use them also in water treatment.

Now about corrosion, we have had problems with corrosion which was believed to be caused only by inorganic components of the water. But just a month ago data were obtained at the University of Karlsruhe to indicate that corrosion is also influenced by organic materials, and this depends on the precipitation of calcium carbonate, of discoloring in the pipes, and this is influenced by organics. The precipitation of calcium carbonate is much smoother if there are some organics present, and humic material especially influences this. After ozonation, for example, the precipitation of calcium carbonate and discoloring of pipes is much smoother and better than without anything. Therefore, if we would produce water with zero TOC there would be much more corrosion and we would have some problems probably with tin and with lead and with other heavy metals coming out of the pipe and into the water. Therefore, it is probably not necessary to attain a zero level, at least in TOC. If you eat an apple for example, there are also 200 individual organic chemical compounds in that and nobody worries about this because they are natural products. And so we think we do not have to worry about natural products in the water, because we eat them also.

Unknown: You made the point of considering that there is a difference in the degree of concern between natural and synthetic organic chemicals. As you well know, this was the basis in 1962 for setting out analyses in the drinking water standards for Carbon-Chloroform Extract rather than Total Organics, on the assumption that the extract is essentially synthetic. That has been adopted in a lot of drinking water standards everywhere except, oddly enough, in this country. Through what may very well be the acme of bad judgement, that has been dropped.

I'd like to ask you about something I know your center is working on, and that is the concept of carbon-14 analysis as distinguishing between natural organic material and industrial chemicals.

Wolfgang Kühn: We did a lot of work in C-14 measurement and determining what fraction of the organic compounds is produced by industry and is brought into the raw water by industry and which portions of organic materials are produced by nature. But I should perhaps say something about analytics. We have some parameters like TOC, UV measurements, and then there are the co-parameters, TOCl, TOBr, and the phenols, aromatics, etc., and these tend to compound the analytical procedures. But for the practical field, for the people working in the waterworks, we have to make very quick decisions, and the analytical method has to be quick and it should be cheap and easy to handle. And therefore, for the routine control of the waterworks, we use mostly co-parameters. If there is a change in the co-parameter --- if you measure a lot of co-parameters, everyone gets a feel for how to handle it, and how much chloroform also would be in if his total parameter would be this high, and so on.

Jim Symons: I might just add one thought. I think it's important to keep in mind when Wolfgang discusses the breakthrough of humic materials that German disinfection policies are somewhat different than ours. We have some experience using activated carbon columns which I didn't discuss during my talk because that wasn't the theme of the conference. Briefly, at first almost all the TOC is removed and chlorination or any disinfection can proceed without producing by-products of any kind that we can identify or see. And then some material begins to break through after three or four weeks and disinfection with chlorine begins to produce chloroform, a little bit at first, and then more and more.

We don't know what these materials are, and if they are, in fact, natural humics they wouldn't be important from a toxicological point of view themselves, but they may be a part of the reaction with the disinfectant then leading to something that might be of concern. So I don't disagree with what Wolfgang said. We all recognize that we live in an organic environment and most of them are beneficial to us in the way of foods, etc. But when you are mixing them with an oxidant, then the situation might be a little bit different.

Joe Cotruvo: Perhaps this is simplistic, but it appears that the point is, if you clean up the water well enough, then it doesn't make much difference what disinfectant you add at the end, because you won't have very much of any of the components, neither organics nor disinfectant.

Mike Kavanaugh, University of California, Berkeley: There were a couple of comments made here at the end which warmed my heart as an engineer. They referred to the problem of estimating the cost of achieving various objectives, particularly in light of the fact that we are now going to be receiving numbers on drinking water standards for a lot of compounds that we previously haven't had to meet. I certainly would like to make a plea, along with the comments related to the identification of the various compounds that are being formed by different oxidants. I would like to see as much kinetic data as possible which would permit me to make some kind of estimate as to the proper combination of techniques that ought to be organized to achieve a particular end.

Along the lines of that comment I'd like to ask a couple of questions of everybody. One regards the use of oxidants, such as ozone, or ozone in conjunction with ultraviolet, to achieve oxidation of various organic compounds to carbon dioxide. I've heard nothing but positive opinions here, or at least suggestions that it, in fact, can be done. I have talked to other people, and I'm a novice in ozone, I don't really know that much about it, and these people say that this is not a cost-effective way of doing it, particularly for example in removal of DOC from a secondary effluent.

The second question would be in regard to the formation of chloroform. Isn't the most cost-effective way the removal of the precursors? I guess this is in relation to what Dr. Symons presented. I didn't quite follow if you were removing those precursors before you were disinfecting. I guess you were. But is it possible to optimize that process before you, in fact, start using strong oxidants?

Jim Symons: You will remember Al Stevens' first slide yesterday when he suggested that precursors plus chlorine yield chloroform and other things.

We have discussed it, and we have frequently said in our publications and presentations that there are three ways to attack the problem. One is to change to an alternate disinfectant, one is to remove trihalomethanes after they are formed, and the third is to remove the precursor. I didn't go into great detail with that thrust, but we basically agree with you. From a cost-effective point of view, it seems to us that removal of precursor through adsorption seems to be the most cost-effective method of tackling the problem, with the additional benefit that you are removing other things as well during adsorption --- the trace organics that are there in the raw water. I was focusing on chloroform removal here because it seemed amenable to this conference, and I also talked precursor removal with oxidants, which was partially successful with chlorine dioxide and unsuccessful at economical doses with ozone. But it certainly is successful with adsorption.

Joe Cotruvo: Maybe we shouldn't separate the disinfectant and the treatment process. For example, Wolfgang Kühn also showed in his studies that ozone has an effect on the removability of various precursors, and by ozonating and then adsorption on GAC or post-coagulating that it was possible to achieve considerably enhanced reduction of the organic chemicals prior to a later disinfectant stage. In those studies ozone was part of the treatment process rather than the disinfectant.

Rip Rice: Mike Kavanaugh just said, "I have a need I would like to see filled, I would like to see certain kinetic data developed on certain reactions." I'd like everybody to think about what we need to do, what sorts of oxidant research programs we need to conduct.

Allison Maggiolo: On the question of UV and ozone, and kinetics and catalysis, from the little bit I have learned over the past 15 or 20 years and observing reactions actually done in a qualitative manner, I am going to prognosticate a little. You can use UV under certain circumstances to destroy ozone. But in solutions, if you don't have very much ozone and you don't have too high an organics concentration, I believe that in the presence of sufficient organic precursors that you will form a lot more free radicals with UV in the presence of

ozone. You don't need to use as much of either one.
These data have never been published and I believe
that area should be pursued, but again, pursued with
kinetic data, with engineering numbers. It is known,
qualitatively, that UV and ozone will make more free
radicals, and will destroy organics at a faster rate
kinetically.

Harvey Rosen: In answer to Mike Kavanaugh's
question, there is a piece missing between the
commercial installation of treatment schemes using
any or all of the alternatives that we have talked
about in a number of possible combinations.
This goes back to the problem of the raw water
source and, as you know, no two are the same.
Therefore there is a pilot stage, a treatability
study stage, something like that, that is going to
help you determine, in terms of what your water
quality objectives are, what are the possible
schemes that you might use in certain combinations
to achieve water quality. When you get there,
then you can start talking about the economics. I
don't think you can really start to do that before,
because it's a matter of cost-effectiveness. I
could give you a number for the cost of a pound of
ozone and it doesn't mean a thing. It really doesn't,
not in terms of what you are trying to achieve, what
other alternatives you have, and how you might use
that in combination with other treatment steps. I
think that's an important point bridging the basic
science that we have been talking about here, and
ultimately the installation and operation of a treatment plant.

Wolfgang Kühn: Probably I should have mentioned this.
I have heard such a lot about cost and money now. In
most fields I think that the United States is ahead of
the Europeans. There is no doubt of that. I saw a
most beautiful waterworks just last week at NASA, the
space lab waterworks, and so on. There are some
wastewater works in the United States that are much
more sophisticated than any works we have, using
reverse osmosis and everything. But why is there this
difference in the drinking water field? I think this
is only a small money problem.

Water is a big business, and we have just the
contrary consideration in Germany that you have in
the United States. We have to slow down the

people running the water plants. They come to us saying, "We want to buy a gas chromatograph, we want to buy a mass spec." And I tell them, that's a lot of money, why do you need a mass spec? He tells me, "That's not a lot of money, that's only five fire hydrants."

In calculating the cost of water in the United States, you separate treatment and distribution. I think that we have to keep in mind that the most expensive component of drinking water treatment is the distribution system. Only two, three, four, five percent is the cost of the water treatment, and fractions of one percent is the analytical cost. I think really, for ozonation or activated carbon, any cost calculation is really worthless because we are talking about fractions of a percent of the costs of the finished water. You have to take all those calculations into account with the distribution system.

Joe Cotruvo: I think we should start addressing the point that Rip made a few minutes ago. What ought we be discussing for the future, what sorts of additional work, what are the directions we ought to be going? Are there any comments on that?

Philippe Hartemann: From my point view as a toxicologist, I think that it is necessary to realize that there are now two approaches in the field of toxicology. Dr. Kühn has a problem with toxicologists. I am trying to be a toxicologist and now, in Europe, we have two approaches in toxicology. The first one, the old one, if you allow me to say the old one, says that it is necessary to have long term experiments conducted over two years or five years of feeding administration, and it is probably this type of toxicologist that Dr. Kühn met in Berlin.

The second approach is to say it is perhaps possible to assess the effects of some compounds with short term assays, with cell cultures and with subcutaneous tests. And if you look at work in the future I think there is no duality, because there is no identity between these two approaches. But it is necessary to use the two approaches at the same time. At present in Europe, we are using only the old approach.

We have problems now with organic compounds in the water, but we have been using ozonated water for sixty to seventy years, and we now have found some problems. It is impossible for toxicologists now to say that in two years we will be able to give you an answer to this problem. You needed sixty years to develop the chemistry; for us it is impossible in two years to give an answer. To do work in this field, then, it is necessary to have a complete overview of this toxicological problem. And don't forget to use some short term tests.

Harvey Rosen: Based on my earlier comment and the gap between the laboratory studies, the high concentrations, and what we might really see in a drinking water situation, I think it's time (and I think it's probably an EPA responsibility) to start to fill in that gap with respect to starting to add to this list of research laboratory type activities, the demonstration type situations that we have in wastewater, for trying new ways of doing things. So I would suggest that that might be thought about.

Joe Cotruvo: EPA has some activities like that underway, but I guess you are saying they should be expanded considerably.

Harvey Rosen: The problem is that there was an 80 city study that was expanded to 113 cities, and although there were some differences in treatment, predominately there was a lot of difference with respect to water source. There are a few plants which use carbon in different ways. There were two plants that have ozone, but again it's not really a practical situation if they don't do anything else to the water and I don't see that being really a practical alternative for most cases. So there is a gap there, as far as I'm concerned. In terms of reducing this to practice, when you start to set standards there will be a lot of people running around trying a lot of different things, I think maybe in a less organized way than they might get if they had the maximum in information.

Joe Cotruvo: Jim Symons did discuss some of the demonstration projects that are underway now in New Orleans, in Evansville and Miami, and a number of these kinds of treatments are being demonstrated at the pilot level.

Victor Kimm: The earlier comment on the demonstration funds prompts me to say that there is an authorization in the Safe Drinking Water Act which so far has been funded for only one specific project mandated by the Congress. That is the kind of a dilemna that I was trying to allude to a little earlier. That is, we don't have the luxury of going through the long chain research, development and demonstration process before we have to come up with some standards, at least as the statute is structured. I think that what we will find over the next few years is that there will be some standards. A lot of people are going to try a lot of different things, and hopefully out of that will come a much better understanding of an improved water works practice. I also think that within the framework of the resources that are available to the agency, we are working as diligently and in as many different areas as is humanly possible. And of course, one of the prime responsibilities we bear for any regulatory action is that we try, to the best of our ability, to assess the costs, and the economic impacts of those costs, which will be done in this case as well.

Otto Meresz, Ontario Ministry of the Environment: We have been talking a lot about haloform formation, and there's one area to which I would like to draw the toxicologists attention. The haloform problem not only consists of the haloforms we can measure directly, but also of the intermediates between the precursors and the haloforms. When we started our survey in Ontario we looked at the EPA method and found it very cumbersome; one could only do six or eight samples a day. So we developed a direct injection method which when we compared the two methods appeared to give values twice as high for haloforms. We found the difference to be in the intermediates which cannot be purged out.

For example, if you take chloral hydrate and purge it you don't get any haloforms. If you heat it first and then purge it, you can get good recovery. If you inject it directly into the water sample then you get the corresponding amount of chloroform.

Now if I may draw an analogy, most chemists are exposed to hydrochloric acid, and your body can defend itself from this in most cases, or else the toxic effects are immediately apparent. Now, if hydrochloric acid is delivered with an organic compound such as it is in Mustard Gas, which was used in chemical warfare, or as tear gas, then it can penetrate the cell and have a completely different toxic effect. Hydrochloric acid now is liberated within the cell, and is very toxic.

Now, if we are going to look at haloforms, then we should look at the haloform intermediates as well, because these may have completely different toxicological properties, because a haloform may be liberated later within the cells. We find that about 50% of the total haloforms are in such a latent state, and maybe these are the intermediates that some toxicological work should be directed to.

Joe Cotruvo: Yes, that's true. Our studies have also shown the same thing. Depending upon the method of analysis, you can get different measurements of those haloforms. I think that what you were referring to as the EPA method is the "first cut" at it that was taken about two years ago. But we are working in all of those areas, of course, looking at intermediates and at their conversion.

Mathilde Kland, Lawrence Berkeley Lab: No one here has mentioned the Chemical Industry Institute of Toxicology which has been organized recently and which is presently building a laboratory at Research Triangle Park, North Carolina in which they are going to do a lot of their own toxicological work eventually. Is there someone here who knows something about this Institute? No?

All I know about it is that it was originally housed in Research Triangle Park and that they are building a laboratory in the Raleigh, North Carolina area. They hope to do 70% of their toxicological work in-house eventually, but at present they are working through grants with institutions. The head of the CIIT was also head of the Department of Pharmacology of Michigan State University. I think that to some extent we are going to have to look to industry to do a lot more of the toxicological work that is involved in this area.

Rip Rice: If there are no more questions or comments, I would like to thank my Co-Chairman, Joe Cotruvo, all of the speakers, Myron Browning, Executive Director of the International Ozone Institute, our Co-Sponsor, the U. S. Environmental Protection Agency, and all attendees to this Workshop.